METHODS IN MOLECULAR BIOLOGY

Series Editor
John M. Walker
School of Life and Medical Sciences
University of Hertfordshire
Hatfield, Hertfordshire, UK

For further volumes:
http://www.springer.com/series/7651

For over 35 years, biological scientists have come to rely on the research protocols and methodologies in the critically acclaimed *Methods in Molecular Biology* series. The series was the first to introduce the step-by-step protocols approach that has become the standard in all biomedical protocol publishing. Each protocol is provided in readily-reproducible step-by-step fashion, opening with an introductory overview, a list of the materials and reagents needed to complete the experiment, and followed by a detailed procedure that is supported with a helpful notes section offering tips and tricks of the trade as well as troubleshooting advice. These hallmark features were introduced by series editor Dr. John Walker and constitute the key ingredient in each and every volume of the *Methods in Molecular Biology* series. Tested and trusted, comprehensive and reliable, all protocols from the series are indexed in PubMed.

Leptospira spp.

Methods and Protocols

Edited by

Nobuo Koizumi

Department of Bacteriology I, National Institute of Infectious Diseases, Tokyo, Japan

Mathieu Picardeau

Unité de Biologie des Spirochètes, Institut Pasteur, Paris, France

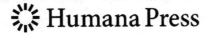

Editors
Nobuo Koizumi
Department of Bacteriology I
National Institute of Infectious
Diseases
Tokyo, Japan

Mathieu Picardeau
Unité de Biologie des Spirochètes
Institut Pasteur
Paris, France

ISSN 1064-3745   ISSN 1940-6029   (electronic)
Methods in Molecular Biology
ISBN 978-1-0716-0458-8   ISBN 978-1-0716-0459-5   (eBook)
https://doi.org/10.1007/978-1-0716-0459-5

This Humana imprint is published by the registered company Springer Science+Business Media, LLC, part of Springer Nature.
The registered company address is: 1 New York Plaza, New York, NY 10004, U.S.A.

Preface

This book is devoted to methods in molecular biology to study *Leptospira*, a fascinating group of bacteria, and their unique properties. *Leptospira* spp. are spirochetes that are comprised of pathogenic and saprophytic species. Saprophytes ubiquitously exist in the environment such as soil and water. On the other hand, pathogens, the causative agents of leptospirosis, a zoonotic disease with more than 1 million cases per year, colonize the proximal renal tubules of mammals and some other vertebrates. Pathogenic *Leptospira* spp. are excreted in the urine of reservoir hosts and contaminate environments, where they can survive for weeks. Humans and animals are infected with leptospires by exposure to water or soil contaminated with the urine of infected animals or direct exposure to the urine or tissues of infected animals.

The aim of this book is to provide tips for the manipulation of *Leptospira* spp. and the investigation of pathogenesis of leptospirosis based on cutting-edge experimental protocols. This book provides researchers detailed practical procedures and the Methods section describes comprehensive step-by-step procedures for each protocol. The Notes section is intended to give additional information and troubleshooting guides.

Leptospira spp. are slow-growing bacteria requiring rich media and several weeks of incubation for growth, and their isolation from environmental or biological samples is fastidious and challenging. Chapter 1 provides the procedure for cultivating and isolating leptospires from both clinical and environmental samples. The availability of whole genome sequences, as a result of improved methodology and reduced cost, has resulted in major improvements in the understanding of the spirochete biology, and whole genome sequencing (WGS) has emerged as an ultimate powerful tool for bacterial strain classification and epidemiological typing (Chapter 2). In the medical microbiology laboratory, Matrix-Assisted Laser Desorption/Ionization Time-of-Flight Mass Spectrometry is also an emerging tool for the identification of bacterial species, and this method can be applied for the identification of *Leptospira* species (Chapter 3).

Leptospira have unique morphological features. They exhibit thin, long, and helix-shaped cells. They also possess surface-exposed lipopolysaccharide (LPS) and periplasmic flagella (or endoflagella). Protocols for cell enumeration by flow cytometry (Chapter 4), RNA preparation (Chapter 5), and purification of LPS (Chapter 6) as well as a protocol for phage purification (Chapter 7) are included in this volume.

In comparison to other bacteria, our knowledge of the molecular basis of the pathogenesis of leptospirosis is limited. This is largely due to the fact that genetic manipulation of *Leptospira* was not possible or not efficient. Today major advances achieved in genetics and in all steps of functional studies in *Leptospira* spp. provide us with an opportunity to apply state-of-the-art approaches for the identification and characterization of virulence factors. Random transposon mutagenesis can be applied to both saprophytes and pathogens (Chapter 8). Transposon sequencing (Tn-seq) that combines transposon insertional mutagenesis with high-throughput sequencing of the transposon insertion sites is able to identify genes contributing to virulence in pathogens (Chapter 9). More recently, specific gene silencing using RNA-guided catalytically inactive Cas9 (dCas9) has been successfully used in *L. biflexa* (Chapter 10).

The study of host-pathogen interaction is pivotal to understand the pathogenesis of leptospirosis. Motility and chemotaxis of leptospires are essential for rapid dissemination in the host. An in vitro assay for the identification of chemoattractants is included in this volume (Chapter 11). Protocols for in situ structural analysis of periplasmic flagella in *Leptospira* spp. by electron cryotomography (Chapter 12) and measurement of cell-body rotation (Chapter 13) are provided. Bioluminescence imaging can be used to follow bacterial dissemination in a mouse in real time and in a noninvasive way (Chapter 14), and translocation of bacteria across cell monolayers can be determined (Chapter 15). After entering bloodstream, leptospires can resist host complement system, bind to host cells and the extracellular matrix, and evade host defense mechanism in macrophages. In this book, in vitro/ex vivo assays for the identification of leptospiral adhesion proteins (Chapter 16), leptospiral complement resistance (Chapter 17), and intracellular trafficking of leptospires in macrophages (Chapter 18) are included. Biofilm formation probably enhances the survival of *Leptospira* in the environment, and a protocol for quantification of biofilms is also included (Chapter 19). Assays for the measurement of viability of leptospires are also described (Chapter 20).

This book also offers protocols for in vivo analysis of transcriptional/proteomic changes of *L. interrogans* using dialysis membrane chamber implanted in the peritoneal cavity of rats (Chapter 21). Quantification of bacterial burden and host cytokine gene expression in the hamster model of acute leptospirosis is also described in Chapter 22. Hamsters are also used to evaluate vaccine candidates for leptospirosis (Chapter 23).

In addition to experimental protocols, this book includes review articles providing an overview of *Leptospira* and leptospirosis (Chapter 24) and laboratory diagnosis of leptospirosis (Chapter 25).

This book thus summarizes all aspects of manipulation of *Leptospira* spp., from strain isolation to cutting-edge approaches for studying the pathogenesis of leptospirosis. This book reflects the progress that has been made in the field in the last 10 years. All the contributors are leading researchers in the *Leptospira* and leptospirosis field, and we would like to thank them for providing their comprehensive protocols and techniques for this book. We also would like to thank Dr. John Walker, the Series Editor of *Methods in Molecular Biology*, for giving us the opportunity to edit this book and his continuous support.

We hope that this book will help new researchers join the field and study these fastidious but fascinating organisms.

Tokyo, Japan   *Nobuo Koizumi*
Paris, France   *Mathieu Picardeau*

Contents

Contributors

ANGELA SILVA BARBOSA • *Laboratory of Bacteriology, Butantan Institute, São Paulo, Brazil*

EMILIE BARSAC • *Leptospirosis Research and Expertise Unit, Institut Pasteur in New Caledonia, Institut Pasteur International Network, Noumea, New Caledonia*

NADIA BENAROUDJ • *Unité de Biologie des Spirochètes, Institut Pasteur, Paris, France*

DELPHINE BONHOMME • *Biology and Genetic of the Bacterial Cell Wall Unit, Innate Immunity and Leptospira Group, Institut Pasteur, Paris, France*

PASCALE BOURHY • *Unité de Biologie des Spirochètes, Institut Pasteur, Paris, France*

JULIE CAGLIERO • *Immunity and Inflammation Group, Institut Pasteur in New Caledonia, Institut Pasteur International Network, Noumea, New Caledonia*

MELISSA J. CAIMANO • *Departments of Medicine, Pediatrics, and Molecular Biology and Biophysics, University of Connecticut Health, Farmington, CT, USA*

KAREN V. EVANGELISTA • *Veterans Affairs Greater Los Angeles Healthcare System, Los Angeles, CA, USA; Department of Medicine, David Geffen School of Medicine, University of California Los Angeles, Los Angeles, CA, USA*

LUIS GUILHERME VIRGÍLIO FERNANDES • *Laboratorio de Desenvolvimento de Vacinas—Centro de Biotecnologia, Instituto Butantan, Sao Paulo, Brazil*

CÉLIA FONTANA • *Boehringer Ingelheim Animal Health, Saint-Priest, France*

DOMINIQUE GIRAULT • *Leptospirosis Research and Expertise Unit, Institut Pasteur in New Caledonia, Institut Pasteur International Network, Noumea, New Caledonia*

CYRILLE GOARANT • *Leptospirosis Research and Expertise Unit, Institut Pasteur in New Caledonia, Institut Pasteur International Network, Noumea, New Caledonia*

ANDRE ALEX GRASSMANN • *Department of Medicine, University of Connecticut Health, Farmington, CT, USA*

LINDA GRILLOVÁ • *Unité de Biologie des Spirochètes, Institut Pasteur, Paris, France*

KARL HUET • *Immunity and Inflammation Group, Institut Pasteur in New Caledonia, Institut Pasteur International Network, Noumea, New Caledonia*

LOURDES ISAAC • *Department of Immunology, Institute of Biomedical Sciences, University of São Paulo, São Paulo, Brazil*

MALIA KAINIU • *Leptospirosis Research and Expertise Unit, Institut Pasteur in New Caledonia, Institut Pasteur International Network, Noumea, New Caledonia*

AKIHIRO KAWAMOTO • *Graduate School of Frontier Biosciences, Osaka University, Osaka, Japan; Institute for Protein Research, Osaka University, Osaka, Japan*

NOBUO KOIZUMI • *Department of Bacteriology I, National Institute of Infectious Diseases, Tokyo, Japan*

AMBROISE LAMBERT • *ERRMECe, Equipe de Recherche sur les Relations Matrice Extracellulaire-Cellules, Institut des matériaux I-MAT, Université de Cergy-Pontoise, Maison Internationale de la Recherche, Neuville sur Oise, France*

KRISTEL LOURDAULT • *Department of Medicine, David Geffen School of Medicine at UCLA, Los Angeles, CA, USA*

MARIKO MATSUI • *Immunity and Inflammation Group, Institut Pasteur in New Caledonia, Institut Pasteur International Network, Noumea, New Caledonia*

JAMES MATSUNAGA • *Department of Medicine, David Geffen School of Medicine at UCLA, Los Angeles, CA, USA; Research Service, Veterans Affairs Greater Los Angeles Healthcare System, Los Angeles, CA, USA*

MD A. MOTALEB • *Department of Microbiology and Immunology, Brody School of Medicine, East Carolina University, Greenville, NC, USA*

CLÉMENCE MOUVILLE • *Unité de Biologie des Spirochètes, Institut Pasteur, Paris, France*

SHUICHI NAKAMURA • *Department of Applied Physics, Graduate School of Engineering, Tohoku University, Sendai, Miyagi, Japan*

ANA L. T. O. NASCIMENTO • *Laboratorio de Desenvolvimento de Vacinas—Centro de Biotecnologia, Instituto Butantan, Sao Paulo, Brazil*

CHRISTOPHER J. PAPPAS • *Department of Biology, Manhattanville College, Purchase, NY, USA*

MATHIEU PICARDEAU • *Unité de Biologie des Spirochètes, Institut Pasteur, Paris, France*

JEAN REYES • *Boehringer Ingelheim Animal Health, Saint-Priest, France*

OLIVIER SCHIETTEKATTE • *Unité de Biologie des Spirochètes, Institut Pasteur, Paris, France; Ecole doctorale BioSPC, Université Paris Diderot, Paris, France*

MARIE-ESTELLE SOUPÉ-GILBERT • *Leptospirosis Research and Expertise Unit, Institut Pasteur in New Caledonia, Institut Pasteur International Network, Noumea, New Caledonia*

TOSHIHIKO SUZUKI • *Department of Bacterial Pathogenesis, Infection and Host Response, Graduate School of Medical and Dental Sciences, Tokyo Medical and Dental University, Tokyo, Japan*

ALINE F. TEIXEIRA • *Laboratorio de Desenvolvimento de Vacinas—Centro de Biotecnologia, Instituto Butantan, Sao Paulo, Brazil*

ROMAN THIBEAUX • *Leptospirosis Research and Expertise Unit, Institut Pasteur in New Caledonia, Institut Pasteur International Network, Noumea, New Caledonia*

CLAUDIA TOMA • *Department of Bacteriology, Graduate School of Medicine, University of the Ryukyus, Okinawa, Japan*

FRÉDÉRIQUE VERNEL-PAUILLAC • *Biology and Genetic of the Bacterial Cell Wall Unit, Innate Immunity and Leptospira Group, Institut Pasteur, Paris, France*

CATHERINE WERTS • *Biology and Genetic of the Bacterial Cell Wall Unit, Innate Immunity and Leptospira Group, Institut Pasteur, Paris, France*

ELSIO A. WUNDER JR • *Department of Epidemiology of Microbial Diseases, Yale School of Public Health, New Haven, CT, USA*

HUI XU • *Department of Microbiology and Immunology, Brody School of Medicine, East Carolina University, Greenville, NC, USA*

CRISPIN ZAVALA-ALVARADO • *Unité de Biologie des Spirochètes, Institut Pasteur, Paris, France*

Chapter 1

Isolation and Culture of *Leptospira* from Clinical and Environmental Samples

Cyrille Goarant, Dominique Girault, Roman Thibeaux, and Marie-Estelle Soupé-Gilbert

Abstract

Leptospires, the etiological agents of leptospirosis, are fastidious slow-growing organisms. Here we describe the isolation and routine maintenance of leptospires from clinical (blood, urine, or tissue) and environmental (water or soil) samples. Using combinations of filtration, agar plating, and selective agents, leptospires can be isolated in pure cultures even from complex contaminated sources in standard EMJH culture medium.

Key words *Leptospira*, Isolation, Culture, Filtration, Selective agents, Soil, Water, Blood, Urine, Kidney

1 Introduction

Culture and isolation of *Leptospira* has long been a reference technique to biologically confirm acute leptospirosis in humans and animals as well as to evidence chronic shedding by animal reservoirs [1]. Successful culture of virulent leptospires was achieved in 1915 using complex culture media [2]. Only half a century later it was shown that leptospires could be cultured in a serum-free medium, leading to the development of the most widely used medium: the "EMJH" medium, named from Ellinghausen, McCullough, Johnson and Harris [3, 4]. Because leptospires are fastidious and slow-growing organisms, the chemical inhibition of contaminants was also considered at the same period in the 1960s [5, 6]. The advent of molecular techniques [7–9] has revolutionized the diagnosis turnaround time, leading to disregard *Leptospira* culture and isolation for diagnosis. In recent years however, the development of whole genome sequencing and genome-wide molecular studies has highlighted the usefulness of culture isolation [10], leading to a renewed interest in isolating leptospires from clinical, but also environmental samples [11]. Here, we summarize the

Nobuo Koizumi and Mathieu Picardeau (eds.), *Leptospira spp.: Methods and Protocols*, Methods in Molecular Biology, vol. 2134,
https://doi.org/10.1007/978-1-0716-0459-5_1, © Springer Science+Business Media, LLC, part of Springer Nature 2020

methods used to cultivate and isolate leptospires from clinical and environmental samples.

2   Materials

Prepare all solutions and media using ultrapure medical grade water, notably keeping in mind that leptospires might be present in any crude water source and able to pass through 0.2-μm pore size filters. Thoroughly rinse all glassware after cleaning and before autoclaving, also keeping in mind that leptospires are highly sensitive to detergents, which would kill them even in residual trace amounts (*see* **Note 1**).

1. EMJH supplement (Table 1): Dissolve 100 g of bovine serum albumin (BSA) in 600 mL of ultrapure sterile water in a 2-L beaker on a magnetic stirrer (heated at 37 °C). Stir gently to avoid foaming (*see* **Note 2**). Add 10 mL of 0.4% $ZnSO_4$, 10 mL of 1.5% $MgCl_2$, 10 mL of 1.5% $CaCl_2$, 10 mL of 10% sodium pyruvate, 20 mL of 20% (w/v) glycerol solution, 125 mL of 9.3% Tween 80 (11.58 mL of Tween 80 in 113.42 mL ultrapure water, *see* **Note 3**), two vials of vitamin B12 (cyanocobalamin vials at 1000 μg/2 mL), 1 mL of 0.3% $CuSO_4$, and 100 mL of 0.5% $FeSO_4$. Add 110-mL ultrapure sterile water making a total volume of 1000 mL of supplement. Adjust to pH = 7.4 using 10 N NaOH or 4 N HCl. Solutions of $ZnSO_4$, $MgCl_2$, $CaCl_2$, $CuSO_4$, sodium pyruvate, and glycerol are dissolved in ultrapure sterile water. These solutions can be autoclaved for 20 min at 110 °C and kept for 1 year at 4 °C. BSA and Tween 80 solutions must be prepared the day when the medium will be produced. $FeSO_4$ solution must be prepared not more than 30 min prior to be included in the EMJH supplement by dissolving 0.5 g of iron sulfate in 100-mL ultrapure sterile water in the dark. Tween must be added to water very slowly and progressively on a magnetic stirrer. This supplement can be aliquoted in 100-mL sterile vials, each usable to make 1 L of EMJH. When aliquoted it can be stored frozen at −18 °C for 1 year (*see* **Note 4**).

2. EMJH base: Dissolve 2.3 g of Difco Leptospira Medium Base EMJH (Becton Dickinson) in 900 mL of ultrapure sterile water. Adjust pH to 7.4 if needed.

3. EMJH base (2×): Dissolve 2.3 g of Difco Leptospira Medium Base EMJH in 400 mL of ultrapure sterile water (*see* **Note 5**).

4. EMJH medium: 900 mL of EMJH base, 100 mL of EMJH supplement. Using a disposable filtration device, filter at 0.22 μm before dispensing aseptically in sterile culture tubes (*see* **Note 6**). The amount of EMJH must be adapted to the tube, with a maximal depth of 5 cm and an air/EMJH ratio of

Table 1
Composition of EMJH

Zinc sulfate ($ZnSO_4$)	0.4 g in 100-mL ultrapure sterile water	Supplement
Magnesium chloride ($MgCl_2$)	1.5 g in 100-mL ultrapure sterile water	Supplement
Calcium chloride ($CaCl_2$)	1.5 g in 100-mL ultrapure sterile water	Supplement
Copper sulfate ($CuSO_4$)	0.3 g in 100-mL ultrapure sterile water	Supplement
Sodium pyruvate	10 g in 100-mL ultrapure sterile water	Supplement
Glycerol	3.18 mL in 16.82-mL ultrapure sterile water or 20 g in a total volume of 100 mL	Supplement
Tween 80	11.58-mL Tween 80 in 113.42-mL ultrapure water	Supplement
Bovine albumin	100-g bovine albumin in 600-mL ultrapure sterile water	Supplement
Iron (II) sulfate ($FeSO_4$)	0.5-g iron (II) sulfate in 100-mL ultrapure sterile water in the dark (*extemporaneous*)	Supplement
Leptospira medium Base	2.3 g per liter of $1\times$ EMJH (contains 100-mL supplement)	EMJH medium

4 to maintain aerobic conditions during culture. Incubate tubes at 37 °C in the dark for 48–72 h, and then visually inspect for possible bacterial growth to check for sterility. Tubes must then be stored in the dark at 4 °C and should be used within less than 2 months.

5. Two times concentrated EMJH medium: 400 mL of $2\times$ EMJH base, 100 mL of EMJH supplement. Similarly filter through a 0.22 μm filter.

6. Agar Noble solution: Mix 2.4 g of Agar Noble with 100 mL of ultrapure sterile water and autoclave at 121 °C for 15 min.

7. Semisolid EMJH medium: 100 mL of Agar Noble solution, 100 mL of $2\times$ concentrated EMJH medium. Two times concentrated EMJH medium is preheated and Agar Noble solution is kept at 45 °C. Dispense into Petri dishes; a volume of 20–25 mL is needed for each plate (to be used within 2–3 weeks). Petri dishes can be sealed with Parafilm to minimize dehydration during storage at 4 °C and incubation at 30 °C.

8. STAFF ($10\times$): 1 mL of 4 mg/mL sulfamethoxazole, 200 μL of 10 mg/mL trimethoprim, 5 μL of 100 mg/mL amphotericin B, 1 mL of 40 mg/mL fosfomycin, and 500 μL of 20 mg/mL solution of 5-fluorouracil (*see* **Note 7**). Add water to a final volume of 10 mL. Filter-sterilize through a 0.22-μm pore size filter. Aliquots can be kept frozen at −20 or −80 °C for months.

9. Disposable 1-L bottle-top vacuum filter with 0.22-μm pore size membrane.

10. Sterile 90-mm diameter Petri dishes.

11. Water bath (45–50 °C).

12. Disposable sterile 0.45 μm and 0.22-μm pore size syringe filters.

13. Disposable sterile 2-mL pipettes and pipetting device.

14. 5-mL single use sterile syringe.

15. Sterile filtration microtubes.

16. Incubator set at 30 °C (with optional shaker).

17. Dark field microscope with 10×, 20×, and 40× objectives.

18. Sterile phosphate buffer saline (10× PBS): 100 mM Na_2PO_4, 18 mM KH_2PO_4, 1.37 M NaCl, 27 mM KCl, pH 7.4.

19. Rabbit serum or fetal calf serum.

20. Glass culture tubes.

3   Methods

Some *Leptospira* are the etiological agents of leptospirosis, a potentially severe infectious disease. In addition, crude samples (either clinical or from the environment) may contain other infection hazards. It is strongly recommended that all culture and isolation procedures are conducted under a biosafety cabinet in BSL2 conditions. Leptospires are slow-growing, and cultures for isolation must be kept and checked for up to 4 months.

3.1   Routine Culture of Leptospira

Various *Leptospira* strains have different needs in culture. Routine culture in EMJH is usually performed in aerobic static conditions at 30 °C. The growth of some difficult strains may be improved by the addition of rabbit serum (*see* **Note 8**) and/or by shaking the culture on an orbital shaker in the incubator, because oxygen is frequently a factor for growth. Routine subculture in liquid medium is usually performed weekly. A volume of 500 μL to 1 mL of the old culture is used to inoculate a new 10-mL EMJH culture tube.

3.2   Elimination of Contaminants

Contaminants easily overgrow *Leptospira* in culture. Contaminants can be eliminated chemically by treating cultures with selective agents. Adding 0.1 volume of 10× STAFF will prove successful in killing most contaminants. Contaminants can also be eliminated by filtration, taking benefit of the small size of leptospires, which will pass through 0.22-μm pore size filters. First filter through a 0.45-μm pore size filter and subculture. If contaminants remain, filter through a 0.22-μm pore size filter (*see* **Note 9**).

Another strategy is to inoculate EMJH agar plates, where *Leptospira* colonies might be larger than contaminants' allowing to separate leptospires from contaminants after prolonged incubation.

3.3 Culture from Blood or Cerebrospinal Fluid

Aseptically add 0.5–1 mL of whole fresh (*see* **Note 10**) blood collected on a heparin tube in a 10-mL EMJH culture tube named R1, and gently homogenize. Then take 1 mL to add in a new 10-mL EMJH culture tube named R2. Similarly do two additional serial dilutions leading to R3 and R4 culture tubes (blood can be inhibitory and dilutions are important). All four tubes must then be incubated at 30 °C aerobically (cap not tightened) in the dark for up to 18 weeks (*see* **Note 11**). Check daily during the first 2 weeks, then weekly for *Leptospira* growth by placing a drop of culture medium on a microscope glass slide with the dark-field microscope at 100× (direct observation in the drop), and 200× and/or 400× (with a glass coverslip) magnification. Alternatively, using fresh serum or plasma, up to ten drops can be seeded in a 10-mL EMJH tube. After gentle homogenization, 1 mL is transferred to a new 10-mL EMJH tube. Both tubes are also incubated at 30 °C aerobically. Cerebrospinal fluid can be treated identically.

3.4 Culture from Urine

Fresh urine can be used directly to inoculate EMJH culture tubes. A volume of 500 µL of urine is usually used for this purpose (*see* **Note 12**). If not cultured immediately, urine pH can be buffered by adding 0.1 volume of 10× sterile PBS. Voided urine is prone to contamination and should be refrigerated if not cultured immediately. Cultures must be checked early and frequently. The addition of 0.1 volume of 10× STAFF in the culture might prove useful to control contaminants to overgrow leptospires. If collected aseptically from the bladder, urine culture is less prone to contamination.

3.5 Culture from Organs

In the acute stage of the infection, leptospires might be cultured from any organ. However, most frequently the liver or kidneys will be used as material to initiate *Leptospira* culture. In animals, fetus organs, placenta, or abortion products can also be used. In kidneys, because leptospires are located in the glomeruli and the proximal renal tubule, a tiny amount (e.g., a thin slice cut with a sterile scalpel) of cortical kidney tissue can be placed directly in a 10-mL EMJH tube. Alternatively, up to 1 g of tissue can be macerated or ground using a sterile pellet pestle and then used to inoculate serial dilutions of EMJH. After a few days, 1 mL of the initial tube can be transferred into a second EMJH culture tube, because tissue degradation products can be inhibitory. When organs are collected postmortem, the organ capsule can be disinfected with 70% ethanol (for 30–60 s before rinsing with sterile water for another 30–60 s) before a small sample is taken. Diluted subcultures are also recommended.

3.6 Culture from Water

Crude water samples will contain a variety of microorganisms prone to overgrow *Leptospira* in culture. It is recommended to systematically use STAFF to favor selective growth of leptospires. Alternatively, water can first be filtered through a 0.45-μm or a 0.22-μm pore size filter before being used for culture (*see* **Note 13**). Early and frequent observation of cultures is needed, and additional filtration steps may be needed to separate leptospires from contaminants.

3.7 Culture from Soil Samples

Direct culture from soil relies on extracting leptospires from soil particles in water and then using water for culture. Put ca. 2–5 g of soil in a 15-mL conical tube. Add 5 mL of sterile 1× PBS. Shake vigorously. Allow soil particles to settle for 5–15 min. Filter supernatant using a 0.45-μm pore size syringe filter, and take 500 μL of filtered supernatant to inoculate a 10-mL EMJH tube with STAFF (*see* **Note 14**). Direct inoculation of unfiltered suernatant can also be used. This procedure is summarized in Fig. 1.

3.8 Isolation of a Single Colony

To obtain a pure monoclonal isolate, positive liquid cultures must be plated on EMJH agar. Two hundred microliters of serial tenfold dilutions are plated on EMJH agar plates. The plates are incubated at 30 °C until leptospiral colonies are visible. To subculture a clonal isolate, select an isolated individual colony, take a small piece of the colony using a sterile micropipette tip, and transfer it in an EMJH culture tube (*see* **Note 15**).

4 Notes

1. Because of the very high sensitivity of leptospires to detergents, some laboratories even use only new glass tubes for EMJH. If tubes are recycled and washed, it is recommended that they are rinsed three times with distilled water.

2. Not all bovine serum albumin brands and batches perform equally for EMJH. Even in the same brand, some batches are not as good as others. This most probably depends on the amount of residual lipids in the albumin: the ability of lipids to form peroxides can lead to toxicity to leptospires.

3. Tween 80 is also named Polysorbate 80. It spontaneously undergoes autoxidation over time leading to the formation of peroxides that are toxic to leptospires, so prefer not to stock for prolonged periods and to store in cool conditions.

4. To freeze bottles of supplement, do not close the cap but cover with Parafilm until frozen, and then tighten the screw cap.

5. Preparing concentrated 2× EMJH is an interesting option allowing to try isolation from a larger volume of water without dilution of the culture medium. In addition, 2× EMJH can be

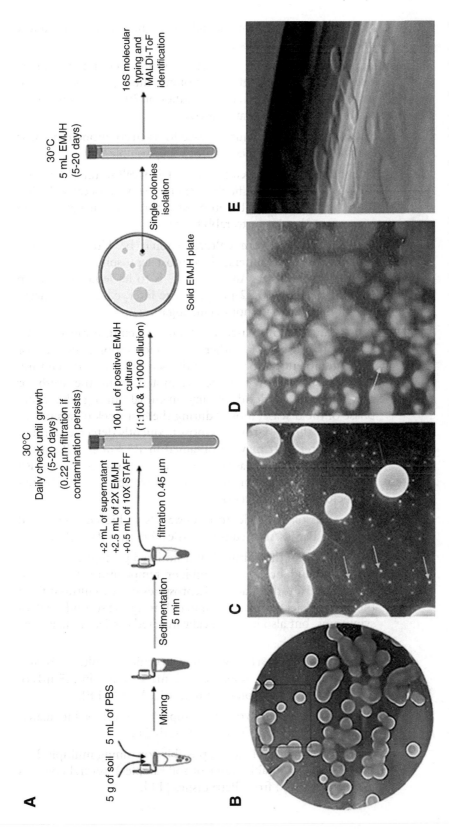

Fig. 1 *Leptospira* isolation on EMJH agar plates. (**a**) Global isolation strategy. (**b**) Macroscopic aspect of colonies. (**c**) Closer view from (**a**) showing colonies of different sizes. (**d**) Opaque (yellow arrow) and transparent (red arrow) colonies. (**e**) Translucent colonies. (Original figure published under Creative Commons Attribution License in Thibeaux R, Girault D, Bierque E, Soupé-Gilbert ME, Rettinger A, Douyère A, Meyer M, Iraola G, Picardeau M, Goarant C. Biodiversity of environmental Leptospira: improving identification and revisiting the diagnosis. Front Microbiol. 2018 May 1;9:816. doi:https:/doi.org/10.3389/fmicb.2018.00816)

stored for up to 2 months and used when needed to rapidly prepare EMJH agar plates.

6. Because EMJH is an expensive culture medium, it can be relevant to prepare both small volume tubes (3- or 4-mL) and larger volumes (10-mL culture tubes or 20-mL culture flasks) to adjust to the needs at best cost.

7. Appropriate solvents must be used for each component because many of these molecules are not water soluble.

8. In our experience, cell culture tested fetal calf serum (FCS) also improves growth of difficult *Leptospira* strains. Because FCS is more frequently available in most laboratories, it provides an interesting alternative to rabbit serum.

9. Only dense *Leptospira* cultures should be filtered through 0.22-μm pore size filters, since most leptospires will also be retained in the filter. An alternative is to let leptospires swim actively through a 0.22-μm pore size filter, e.g., using micro-filtration columns without centrifugation.

10. Blood samples to be used for *Leptospira* culture must not be refrigerated. It must be either inoculated as soon as possible or kept at ambient temperature until used for *Leptospira* culture. *Leptospira* isolation may be attempted from the fresh or non-refrigerated blood of any mammal during the acute phase of the disease, mostly during the first week of symptoms in an acute illness. Growth is most frequently detected in tubes R3 and/or R4. It may be obtained in R3 but not R4 if the initial concentration was very low, but also in some cases in R4 but not R3, for example if patient was treated with antibiotics just before blood collection.

11. Successful isolation of leptospires was also achieved from blood seeded in BacT/Alert culture bottles for a few days [12].

12. As for blood, if the culture is not initiated promptly, the urine sample should be kept at ambient temperature if collected aseptically from the bladder. Leptospires can be cultured from the urine of sick humans or animals during the second week of symptoms, but also in chronically infected carrier animals (reservoir animals).

13. The use of 2× EMJH will allow to initiate culture from a significant volume of water (e.g., 2 mL of water in 2.5 mL of 2× EMJH and addition of 500 μL of 10× STAFF).

14. Alternatively, a larger volume of supernatant can be inoculated in 2× EMJH as done in a recent study [11].

15. When working from material possibly containing multiple *Leptospira* strains (notably water or soil samples), several colonies might be isolated from Petri dishes [11].

References

1. Turner LH (1967) Leptospirosis. I. Trans R Soc Trop Med Hyg 61(6):842–855

2. Inada R, Ido Y, Hoki R, Kaneko R, Ito H (1916) The etiology, mode of infection and specific therapy of Weil's disease (Spirochaetosis Icterohaemorrhagica). J Exp Med 23 (3):377–402

3. Ellinghausen HC Jr, McCullough WG (1965) Nutrition of Leptospira pomona and growth of 13 other serotypes: fractionation of oleic albumin complex and a medium of bovine albumin and polysorbate 80. Am J Vet Res 26:45–51

4. Johnson RC, Harris VG (1967) Differentiation of pathogenic and saprophytic Leptospires. I. Growth at low temperatures. J Bacteriol 94(1):27–31

5. Johnson RC, Rogers P (1964) 5-Fluorouracil as a selective agent for growth of Leptospirae. J Bacteriol 87:422–426

6. Cousineau JG, McKiel JA (1961) In vitro sensitivity of Leptospira to various antimicrobial agents. Can J Microbiol 7:751–758

7. Merien F, Amouriaux P, Perolat P, Baranton G, Saint Girons I (1992) Polymerase chain reaction for detection of Leptospira spp. in clinical samples. J Clin Microbiol 30(9):2219–2224

8. Gravekamp C, Van de Kemp H, Franzen M, Carrington D, Schoone GJ, Van Eys GJ, Everard CO, Hartskeerl RA, Terpstra WJ (1993) Detection of seven species of pathogenic leptospires by PCR using two sets of primers. J Gen Microbiol 139(8):1691–1700

9. Guernier V, Allan KJ, Goarant C (2018) Advances and challenges in barcoding pathogenic and environmental *Leptospira*. Parasitology 145(SI5):595–607. https://doi.org/10.1017/S0031182017001147

10. Thibeaux R, Iraola G, Ferrés I, Bierque E, Girault D, Soupé-Gilbert ME, Picardeau M, Goarant C (2018) Deciphering the unexplored Leptospira diversity from soils uncovers genomic evolution to virulence. Microb Genom 4 (1):000144. https://doi.org/10.1099/mgen.0.000144

11. Thibeaux R, Girault D, Bierque E, Soupé-Gilbert ME, Rettinger A, Douyere A, Meyer M, Iraola G, Picardeau M, Goarant C (2018) Biodiversity of environmental Leptospira: improving identification and revisiting the diagnosis. Front Microbiol 9:816. https://doi.org/10.3389/fmicb.2018.00816

12. Girault D, Soupé-Gilbert ME, Geroult S, Colot J, Goarant C (2017) Isolation of *Leptospira* from blood culture bottles. Diagn Microbiol Infect Dis 88:17–19. https://doi.org/10.1016/j.diagmicrobio.2017.01.014

Chapter 2

Core Genome Multi-locus Sequence Typing Analyses of *Leptospira* spp. Using the Bacterial Isolate Genome Sequence Database

Linda Grillová and Mathieu Picardeau

Abstract

With the advent of whole-genome sequencing (WGS), comparative analysis has led to the use of core genome MLST (cgMLST) schemes for the high-resolution reproducible typing of bacterial isolates. In cgMLST, hundreds of loci are used for gene-by-gene comparisons of assembled genomes for studying the genetic diversity of clinically important pathogens. Combination of the cgMLST data and metadata of the isolates is useful for epidemiological investigations.

Here we present a cgMLST scheme for the high-resolution typing of isolates from the whole *Leptospira* genus, enabling identification at the level of species, clades, clonal groups, and sequence types. We show several examples how the cgMLST *Leptospira* database, which is a publicly available web-based database, can be used for the analyses of WGS data of *Leptospira* isolates. This effort was undertaken in order to facilitate international collaborations and support the global surveillance of leptospirosis.

Key words Genome, Core genome multi-locus sequence typing, Bacterial Isolate Genome Sequence Database (BIGSdb)

1 Introduction

Molecular typing of bacterial isolates is a powerful tool for surveillance and epidemiology of diseases. Discrimination of genetic variants and characterization of the predominant *Leptospira* strains in the environment, patients, or animal populations are essential for identifying sources of infection and developing evidence-based infection control and prevention strategies. Different molecular typing schemes are currently available for pathogenic *Leptospira* [1–4], but a harmonized typing tool needs to be established not only for pathogenic species but also for the whole genus. A core genome multi-locus sequence typing (cgMLST; based on 545 core genes) recently designed based on high-quality genome sequences representing all known *Leptospira* species [5, 6]. This scheme can significantly increase our understanding of the

Nobuo Koizumi and Mathieu Picardeau (eds.), *Leptospira spp.: Methods and Protocols*, Methods in Molecular Biology, vol. 2134,
https://doi.org/10.1007/978-1-0716-0459-5_2, © Springer Science+Business Media, LLC, part of Springer Nature 2020

epidemiology and general biology of *Leptospira* spp. First, cgMLST can be applied to pathogenic (sub-clade P1), intermediate (sub-clade P2), and saprophytic isolates (sub-clades S1 and S2) and, thus, has a potential to elucidate the role of intermediates, which is a group of strains of unclear pathogenicity from the clinical perspective phylogenetically related to pathogens. cgMLST has a high discrimination power resulting in the identification of species, clades, clonal groups (CGs), sequencing types (STs), and most probable serogroups. This would allow tracking of *Leptospira* strains and could help, for example, the detection of new genotypes and length of time for which a given genotype persists. The widespread use of this cgMLST scheme should enable the identification of such relationships at the global level and over time. The *Leptospira* cgMLST database (https://bigsdb.pasteur.fr/leptospira/) is a publicly available web-based database hosted at the Institut Pasteur. At present, the database contains data from 1007 *Leptospira* strains (08/28/2019).

The Bacterial Isolate Genome Sequence Database (BIGSdb) platform was initially developed by Jolley and Maiden [7] for automatic ST and CG assignments, for determination of new alleles, for storage of sample metadata, for identification of new associations between genotypes and metadata using various tools, and for user-friendly visualization of molecular typing data using breakdown options and external plug-ins such as Interactive Tree of Life (iTOL) [8] and GrapeTree [9]. This chapter presents some examples of how the WGS data can be used in BIGSdb for user-friendly visualization of phylogenic relationships among STs of *Leptospira*.

2   Materials

2.1   Genome Requirements

For submission, the treated WGS data are needed (i.e., low Phred score base, trimming, exogenous oligonucleotide clipping, sequencing error correction, and read coverage homogenization) (*see* **Note 1**). Draft genomes with $50\times$ minimum coverage and a minimum N50 of 10,000 nt are required (*see* **Notes 2–4**).

2.2   Information on Isolates

Relevant information of isolates such as isolate identification name/number, country of origin, biological source of sample, year of isolation, serogroup and serovar, etc. are required.

2.3   Hardware/ Software Requirements

The BIGSdb is an online database. As such, users do not need to install a particular software.

3   Methods

3.1   Data Submission

New users can contact the curators by e-mail (leptospiraMLST@-pasteur.fr). Subsequently, the curators will create an account and provide the log-in details to new users, who are then able to submit their data (*see* **Note 5**). Whole-genome sequence data in FASTA format can be submitted to BIGSdb as (1) a single contig of a closed chromosome, (2) a multi-FASTA file with closed chromosome and plasmids from the same isolate, (3) multi-contig files (whole-genome shotgun), or (4) scaffold files. There are a number of fields that must be filled in so the curators know how the data were obtained, e.g., the sequencing platform used, read length, coverage, and assembly (de novo or mapped). Make sure the "e-mail submission updates" box is checked if you wish to receive e-mail notification of the result of your submission. Subsequently, curators will check the quality of the data, and the sequences will be automatically scanned for cgSTs and cgCGs assignments. cgSTs represent profiles that differ by no allele other than for missing data. cgSTs that share all but one or few alleles are considered to be strongly related even if the differing alleles contain multiple single-nucleotide variants (SNVs) due to recombination. cgCGs are defined by a single-linkage clustering threshold of 40 allelic mismatches; i.e., CG is defined as a group of cgMLST allelic profiles differing by no more than 40 allelic mismatches, out of 545 gene loci, from at least one other member of the group.

Every sequence entry should be accompanied by metadata of the sample. The researchers are encouraged to upload as much information about patients and isolates as available. The template for *Leptospira* isolate metadata can be downloaded at the Institut Pasteur MLST webpage (https://bigsdb.pasteur.fr/leptospira/). Some fields are mandatory and cannot be left blank. Check the "Description of database fields" link on the database contents page to see a description of the fields and allowed values where these have been defined.

3.2   Data Export

Different data can be exported from BIGSdb. You can export the isolate recordsets by clicking the "Export dataset" link in the Export section of the main contents page, or you can export recordsets of isolates returned from a database query by clicking the "Dataset" button in the Export list at the bottom of the results table. You can then download the data in tab-delimited text or Excel formats. In the advanced options, choose the cgMLST scheme in order to export cgSTs and "Test-40" under the "Classification scheme" to obtain the cgCGs.

Similarly, the original submitted data as well as the sequences of core genes extracted from the original data can be exported. By default, the data will be extracted unaligned, but you can also

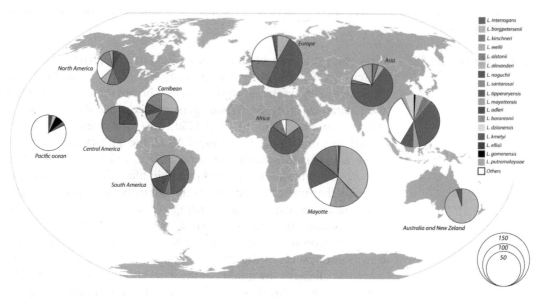

Fig. 1 The prevalence of *Leptospira* species around the globe. The data were generated using the "Two field breakdown" option in the BIGSdb contents page (first of July, 2019)

choose to align the sequences by checking the "Align sequences" checkbox (use the MAFFT as the aligner). The export of aligned data is more time-consuming than the export of unaligned data and is restricted for exporting 200 isolates only.

3.3 Data Analyses

3.3.1 General Overview

An easy way to get a general overview of all or selected data present in BIGSdb is to use the breakdown option plug-ins. For example, when you want to know the prevalence of different *Leptospira* species in different geographical areas, you can click the "Two field breakdown" link on the main contents page. This plug-in exports a table breaking down one field against another (the breakdown of "species" by "country") (Fig. 1) (*see* **Note 6**).

3.3.2 Interactive Tree of Life (iTOL)

The iTOL [8] plug-in incorporated to BIGSdb enables generation and visualization of phylogenetic trees calculated from concatenated sequence alignments of core genes ($n = 545$) using the neighbor-joining clustering method. It can be assessed from the contents page or following the query by clicking the "iTOL" link. Since this analysis requires the preassembly of the data, it is possible to export only 200 isolates or less. The simple neighbor-joining method produces unrooted trees, but it does not assume a constant rate of evolution across lineages [10]. In contrast, the maximum likelihood method uses a more complex evolution model and is known to be stronger than the neighbor-joining method for reconstructing sequence histories [11]. In general, iTOL plug-ins which generate the phylogenetic tree based on neighbor-joining method are good enough for an initial overview of the phylogeny; however, for publication purposes, we recommend employing the more precise and time-consuming maximum likelihood method, which is not a part of the BIGSdb function. Additional fields can be selected to be included as metadata for use in coloring nodes— select any fields you wish to include in the "iTOL datasets" list. For detailed explanation of the iTOL function, *see* the following link: https://itol.embl.de/help.cgi.

3.3.3 Species Identification

If you are not sure which species of *Leptospira* you are working with, you can check using iTOL plug-in by generating the phylogenetic tree based on the concatenated core gene sequences of your unknown sample(s) together with the reference sequences of *Leptospira* species ($n = 64$) [6] which are present in BIGSdb (Table 1 and Fig. 2). Selecting the BIGSdb IDs of reference strains and your unknown sample(s) in the iTOL plugin will generate a phylogenetic tree which will cluster your isolates together with one of the reference strains. If the clustering is not clear, the average nucleotide identity (ANI) of draft genomes should be performed, for example, using the ANI calculator [14] at the following link: https://www.ezbiocloud.net/tools/ani.

3.3.4 Prediction of Possible Serogroup

Strains belonging to the same serogroups are usually subdivided into several cgCGs; however, when strains are part of the same clonal group, they should belong to the same serogoup (based on the all available isolates at the time of writing, $n = 1007$); i.e., the branching based on the concatenated cgMLST sequences could be useful for determination of the potential serogroup.

3.3.5 GrapeTree

GrapeTree allows for exploration of the fine-grained population structure and phenotypic properties of large number of genomes (more than 200) in a web browser. It generates and displays the minimum spanning tree (MSTree) based on cgSTs [9]. The GrapeTree algorithm is able to export large datasets and is compatible to handle larger amount of missing sequences and thus is perfect for handling cgMLST data. The datasets can include metadata, which allows nodes in the result tree to be colored interactively. It can be accessed from the contents page or following the query by clicking the "GrapeTree" link. In the *Leptospira* setting, it could be very useful in an easy identification of the potential source of infection by determination of cgCGs which are shared among human and animal isolates (Fig. 3). In Fig. 3, it is evident that several cgCGs are unique to particular hosts (e.g., cgCG176 was found only in dogs and cgCG81 was found only in patients), while other cgCGs were shared among multiple hosts (e.g., cgCG6 was shared among humans, dogs, and rats and cgCG5 was found in humans, cows, hedgehogs, dogs, and other mammals). Another example of how GrapeTree can be used in *Leptospira* molecular epidemiology is the tracking of the different distributions of CGs over a specific time period.

Table 1
Reference strain of *Leptospira* species

Species	Isolate	BIGSdb id	Phylogenetic group	Country	Source
L. alexanderi	L 60	20	P1	China	Human
L. alstonii	79601	21	P1	China	Amphibian
L. interrogans	56601	27	P1	China	Human
L. kmetyi	Bejo-Iso9	52	P1	Malaysia	Environment
L. noguchii	CZ 214T	94	P1	Panama	Other mammal
L. santarosai	LT821	96	P1	Panama	Rat
L. kirschneri	200702274	110	P1	France	Human
L. mayottensis	200901116	149	P1	Mayotte	Human
L. borgpetersenii	L550	225	P1	Australia	Human
L. weilii	LT2116	232	P1	Australia	Human
L. barantonii	201602184	368	P1	New Caledonia	Environment
L. dzianensis[a]	201601115	378	P1	Mayotte	Environment
L. adleri	201602187	434	P1	New Caledonia	Environment
L. ellisii	SSW8	529	P1	Malaysia	Environment
L. putramalaysiae[a]	SCW20	535	P1	Malaysia	Environment
L. tipperaryensis[b]	GWTS	551	P1	Ireland	Other mammal
L. gomenensis	201800299	677	P1	New Caledonia	Environment
L. wolffii	Khorat-H2	83	P2	Thailand	Human
L. broomii	5399	102	P2	Denmark	Human
L. fainei	BUT 6	226	P2	Australia	Pig
L. licerasiae	VAR010	279	P2	Peru	Human
L. inadai	10	298	P2	USA	Human
L. dzoumogneensis	201601113	416	P2	Mayotte	Environment
L. fletcheri	SCW15	532	P2	Malaysia	Environment
L. selangorensis	SCW17	533	P2	Malaysia	Environment
L. langatensis	SCW18	534	P2	Malaysia	Environment
L. fluminis	SCS5	540	P2	Malaysia	Environment

(continued)

Table 1
(continued)

Species	Isolate	BIGSdb id	Phylogenetic group	Country	Source
L. semungkisensis	SSS9	542	P2	Malaysia	Environment
L. saintgironsiae	SCS5	549	P2	Malaysia	Environment
L. johnsonii	E8	563	P2	Japan	Environment
L. sarikeiensis	201702455	575	P2	Malaysia	Environment
L. venezuelensis	CLM-U50	635	P2	Venezuela	Human
L. koniamboensis	201800265	644	P2	New Caledonia	Environment
L. andrefontaineae	201800301	679	P2	New Caledonia	Environment
L. perolatii	FH1-B-B1	812	P2	New Caledonia	Soil
L. neocaledonica	ES4-C-A1	813	P2	New Caledonia	Soil
L. haakeii	ATI7-C-A2	814	P2	New Caledonia	Soil
L. hartskeerlii	MCA1-C-A1	815	P2	New Caledonia	Soil
L. biflexa	Patoc 1 (Paris)	128	S1	Italy	Environment
L. bandrabouensis	201601111	185	S1	Mayotte	Environment
L. bouyouniensis	201601297	196	S1	Mayotte	Environment
L. mtsangambouensis	201601298	197	S1	Mayotte	Environment
L. meyeri	Went 5	204	S1	Canada	Unknown
L. wolbachii	CDC	224	S1	USA	Environment
L. yanagawae	Sao Paulo	260	S1	Brazil	Environment
L. brenneri	201602177	365	S1	New Caledonia	Environment
L. levettii	201602181	367	S1	New Caledonia	Environment
L. harrisiae	201602189	379	S1	New Caledonia	Environment
L. vanthielii	Waz Holland	419	S1	Netherlands	Environment
L. jelokensis	201702419	566	S1	Malaysia	Environment
L. congkakensis	201702421	568	S1	Malaysia	Environment
L. kemamanensis	201702454	573	S1	Malaysia	Environment

(continued)

Table 1
(continued)

Species	Isolate	BIGSdb id	Phylogenetic group	Country	Source
L. perdikensis	201702692	585	S1	Malaysia	Environment
L. ellinghausenii	201800220	632	S1	Japan	Environment
L. terpstrae	ATCC 700639	634	S1	China	Environment
L. montravelensis	201800279	658	S1	New Caledonia	Environment
L. bourretii	201800280	659	S1	New Caledonia	Environment
L. noumeaensis	201800287	666	S1	New Caledonia	Environment
L. kanakyensis	201800292	670	S1	New Caledonia	Environment
L. ilyithenensis	201400974	2	S2	Algeria	Environment
L. ognonensis	201702476	580	S2	France	Environment
L. kobayashii	E30	587	S2	Japan	Environment
L. ryugenii	YH101	588	S2	Japan	Environment
L. idonii	201300427	630	S2	Japan	Environment

[a]*L. yasudae* and *L. stimsonii* [12] are presented here as *L. dzianensis* and *L. putramalaysiae* [6]

[b]*L. tipperaryensis* was originatelly described as *L. alstonii* [13] and later on reclassified as a new species and named *L. tipperaryensis* [6]

4   Notes

1. Platforms usually provide some quality control measures and follow protocols for filtering sequencing artifacts.

2. *Leptospira* strains have genomes that are 3.8–4.6 Mb in size with 35–45% GC content.

3. It is important to perform whole-genome sequencing from a clonal culture to avoid comparing mixed genomes. Isolation from an individual colony on an agar plate is, therefore, recommended to recover clonal *Leptospira* cultures [15].

4. A simulation of the effect of missing data (uncalled cgMLST alleles) on the clustering results showed that cluster assignment is robust even with high amounts of missing data (affecting up to 400 loci out of 545) [5], indicating that even incomplete genomes should be typeable by cgMLST.

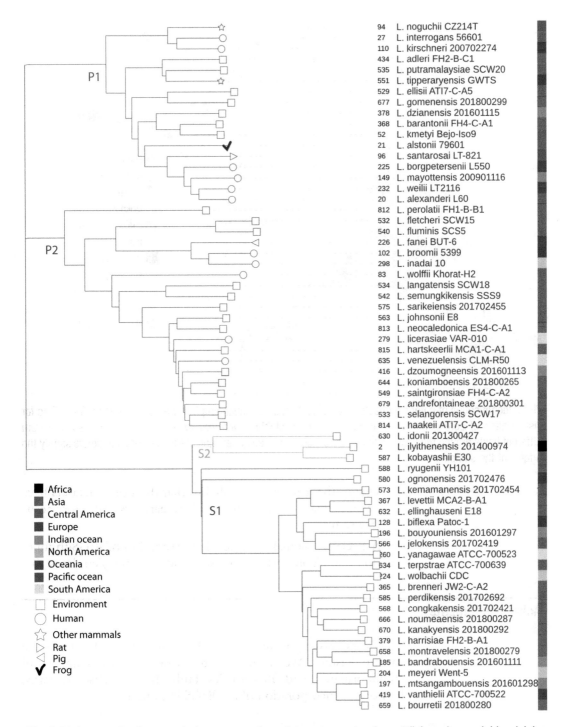

94	L. noguchii CZ214T
27	L. interrogans 56601
110	L. kirschneri 200702274
434	L. adleri FH2-B-C1
535	L. putramalaysiae SCW20
551	L. tipperaryensis GWTS
529	L. ellisii ATI7-C-A5
677	L. gomenensis 201800299
378	L. dzianensis 201601115
368	L. barantonii FH4-C-A1
52	L. kmetyi Bejo-Iso9
21	L. alstonii 79601
96	L. santarosai LT-821
225	L. borgpetersenii L550
149	L. mayottensis 200901116
232	L. weilii LT2116
20	L. alexanderi L60
812	L. perolatii FH1-B-B1
532	L. fletcheri SCW15
540	L. fluminis SCS5
226	L. fanei BUT-6
102	L. broomii 5399
298	L. inadai 10
83	L. wolffii Khorat-H2
534	L. langatensis SCW18
542	L. semungkikensis SSS9
575	L. sarikeiensis 201702455
563	L. johnsonii E8
813	L. neocaledonica ES4-C-A1
279	L. licerasiae VAR-010
815	L. hartskeerlii MCA1-C-A1
635	L. venezuelensis CLM-R50
416	L. dzoumogneensis 201601113
644	L. koniamboensis 201800265
549	L. saintgironsiae FH4-C-A2
679	L. andrefontaineae 201800301
533	L. selangorensis SCW17
814	L. haakeii ATI7-C-A2
630	L. idonii 201300427
2	L. ilyithenensis 201400974
587	L. kobayashii E30
588	L. ryugenii YH101
580	L. ognonensis 201702476
573	L. kemamanensis 201702454
367	L. levettii MCA2-B-A1
632	L. ellinghauseni E18
128	L. biflexa Patoc-1
196	L. bouyouniensis 201601297
566	L. jelokensis 201702419
260	L. yanagawae ATCC-700523
634	L. terpstrae ATCC-700639
224	L. wolbachii CDC
365	L. brenneri JW2-C-A2
585	L. perdikensis 201702692
568	L. congkakensis 201702421
666	L. noumeaensis 201800287
670	L. kanakyensis 201800292
379	L. harrisiae FH2-B-A1
658	L. montravelensis 201800279
185	L. bandrabouensis 201601111
204	L. meyeri Went-5
197	L. mtsangambouensis 201601298
419	L. vanthielii ATCC-700522
659	L. bourretii 201800280

Legend:
- ■ Africa
- ■ Asia
- ■ Central America
- ■ Europe
- ■ Indian ocean
- ■ North America
- ■ Oceania
- ■ Pacific ocean
- ■ South America
- ☐ Environment
- ○ Human
- ☆ Other mammals
- ▷ Rat
- ◁ Pig
- ✔ Frog

Fig. 2 Phylogeny of reference strains representing all known species (*n* = 64) based on neighbor-joining clustering method extracted from BIGSdb iTOL plug-in

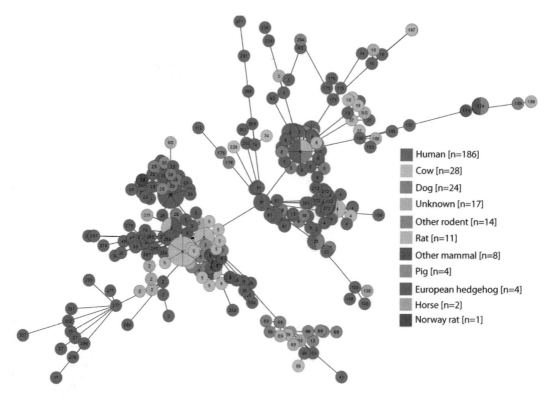

Fig. 3 Identification of potential infection sources. A minimum spanning tree was created using GrapeTree for visualization of core genomic relationships of all available *L. interrogans* strains ($n = 299$) isolated from different hosts around the globe. The numbers inside the tree nodes indicate the cgCGs, and colors signify the origins of the samples

5. One of the features of BIGSdb is that the stored datasets have been manually curated to provide researchers with more accurate results.

6. A BIGSdb manual for a detailed description of BIGSdb function is available (https://bigsdb.readthedocs.io/en/latest/).

Acknowledgments

This work was supported by a PTR grant (PTR30-17) from the Institut Pasteur. We would like to thank to Julien Guglielmini (Bioinformatics and Biostatistics Hub, Institut Pasteur, Paris, France) for incorporation of the BIGSdb plug-ins.

References

1. Herrmann JL, Bellenger E, Perolat P, Baranton G, Saint Girons I (1992) Pulsed-field gel electrophoresis of NotI digests of leptospiral DNA: a new rapid method of serovar identification. J Clin Microbiol 30 (7):1696–1702

2. Majed Z, Bellenger E, Postic D, Pourcel C, Baranton G, Picardeau M (2005) Identification

of variable-number tandem-repeat loci in *Leptospira interrogans* sensu stricto. J Clin Microbiol 43(2):539–545

3. Slack AT, Dohnt MF, Symonds ML, Smythe LD (2005) Development of a multiple-locus variable number of tandem repeat analysis (MLVA) for *Leptospira interrogans* and its application to *Leptospira interrogans* serovar Australis isolates from Far North Queensland, Australia. Ann Clin Microbiol Antimicrob 4:10

4. Ahmed N, Devi SM, Valverde M d l A, Vijayachari P, Machang'u RS, Ellis WA et al (2006) Multilocus sequence typing method for identification and genotypic classification of pathogenic *Leptospira* species. Ann Clin Microbiol Antimicrob 5:28

5. Guglielmini J, Bourhy P, Schiettekatte O, Zinini F, Brisse S, Picardeau M (2019) Genus-wide *Leptospira* core genome multilocus sequence typing for strain taxonomy and global surveillance. PLoS Negl Trop Dis 13(4): e0007374

6. Vincent AT, Schiettekatte O, Goarant C, Neela VK, Bernet E, Thibeaux R et al (2019) Revisiting the taxonomy and evolution of pathogenicity of the genus *Leptospira* through the prism of genomics. PLoS Negl Trop Dis 13(5): e0007270

7. Jolley KA, Maiden MC (2010) BIGSdb: scalable analysis of bacterial genome variation at the population level. BMC Bioinformatics 11 (1):595

8. Letunic I, Bork P (2016) Interactive tree of life (iTOL) v3: an online tool for the display and annotation of phylogenetic and other trees. Nucleic Acids Res 44(W1):W242–W245

9. Zhou Z, Alikhan NF, Sergeant MJ, Luhmann N, Vaz C, Francisco AP et al (2018) GrapeTree: visualization of core genomic relationships among 100,000 bacterial pathogens. Genome Res 28(9):1395–1404

10. Saitou N, Nei M (1987) The neighbor-joining method: a new method for reconstructing phylogenetic trees. Mol Biol Evol 4(4):406–425

11. Fukami K, Tateno Y (1989) On the maximum likelihood method for estimating molecular trees: uniqueness of the likelihood point. J Mol Evol 28(5):460–464

12. Casanovas-Massana A, Hamond C, Santos LA, de Oliveira D, Hacker KP, Balassiano I et al (2019) *Leptospira yasudae* sp. nov. and *Leptospira stimsonii* sp. nov., two new species of the pathogenic group isolated from environmental sources. Int J Syst Evol Microbiol. https://doi.org/10.1099/ijsem.0.003480

13. Nally JE, Bayles DO, Hurley D, Fanning S, McMahon BJ, Arent Z (2016) Complete genome sequence of *Leptospira alstonii* serovar Room22 strain GWTS #1. Genome Announc 4 (6):e01230-16

14. Yoon SH, Ha SM, Lim J, Kwon S, Chun J (2017) A large-scale evaluation of algorithms to calculate average nucleotide identity. Antonie Van Leeuwenhoek 110(10):1281–1286

15. Thibeaux R, Girault D, Bierque E, Soupé-Gilbert M-E, Rettinger A, Douyère A et al (2018) Biodiversity of environmental *Leptospira*: improving identification and revisiting the diagnosis. Front Microbiol 9:816

Chapter 3

Use of MALDI-ToF Mass Spectrometry for Identification of *Leptospira*

Dominique Girault, Malia Kainiu, Emilie Barsac, Roman Thibeaux, and Cyrille Goarant

Abstract

Medical microbiology has used phenotypical and metabolic criteria to identify bacterial pathogens for decades. However, no such criteria have been applied to identify leptospires at the species level. In the recent years, matrix-assisted laser desorption/ionization time-of-flight (MALDI-ToF) mass spectrometry (MS) has emerged as new tool for the identification of bacterial species in the medical microbiology laboratory. This technology has rapidly gained more and more popularity. Actually, this technique is sensitive and economic, saving both labor and bench costs, but also rapid, significantly reducing turnaround time from isolation to identification. MALDI-ToF MS provides an unprecedented tool for the rapid identification of *Leptospira* at the species level.

Key words *Leptospira*, Isolate, Identification, MALDI-ToF MS, Spectra database

1 Introduction

Bacteria in the genus *Leptospira* have long been classified into only two species, namely, "*L. interrogans sensu lato*" for all pathogenic strains and "*L. biflexa sensu lato*" for all saprophytic strains. The ability to grow at 13 °C and in the presence of the purine analogue 8-azaguanine were considered unique features of *L. biflexa* and were used as metabolic criteria for identification [1, 2]. However, *Leptospira* taxonomy has been revolutionized since the advent of molecular approaches [3, 4] and the discovery of a new "intermediate" cluster in the genus [5]. Additional novel species were recently described, mostly from the environment, now bringing the total number of species to 64 [6–8]. No phenotypical identification key is available to identify such a large number of species. Therefore, DNA sequencing is most frequently used to identify *Leptospira* isolates at the species level. Here we summarize the steps to identify *Leptospira* isolates at the species level using MALDI-ToF MS in the medical microbiology laboratory [7, 9–11].

Nobuo Koizumi and Mathieu Picardeau (eds.), *Leptospira spp.: Methods and Protocols*, Methods in Molecular Biology, vol. 2134, https://doi.org/10.1007/978-1-0716-0459-5_3, © Springer Science+Business Media, LLC, part of Springer Nature 2020

2   Materials

2.1   Mass Spectrometer and Software

All procedures described use the Microflex mass spectrometer (Bruker Daltonics) system and attached software. *Leptospira* isolates (*see* Chapter 1) are used as the starting biological material.

1. MALDI-ToF Microflex (Bruker Daltonics).
2. Maldi Biotyper 3.0 (Bruker Daltonics).

2.2   Leptospira Culture

For MALDI-ToF MS analysis, fresh cultures at the late exponential phase are used (*see* **Note 1**).

1. *Leptospira* spp.-type strains: These strains are used for the creation of a MS database.
2. *Leptospira* isolates: *Leptospira* are isolated as described and routinely subcultured in EMJH (Chapter 1).
3. Ellinghausen-McCullough-Johnson-Harris (EMJH) culture medium [2, 12]: Dissolve 2.3 g of Difco™ Leptospira Medium Base EMJH (Becton Dickinson) in 900 mL of purified water. Add 100 mL of Difco™ Leptospira Enrichment EMJH (Becton Dickinson).

2.3   Media and Reagents

1. Phosphate buffered saline (PBS): 10 mM Na_2PO_4, 1.8 mM KH_2PO_4, 137 mM NaCl, 2.7 mM KCl, pH 7.4.
2. Sterile ultrapure water.
3. Absolute ethanol.
4. Formic acid: 70% solution in water.
5. Acetonitrile.
6. HCCA matrix solution: 10 mg/mL of α-cyano-4-hydroxy-cinnamic acid in 50% acetonitrile/2.5% trifluoroacetic acid (*see* **Note 2**).

2.4   Target Plate, Other Material, Plasticware, and Consumables

1. Ground steel 96-spot target plate.
2. Benchtop centrifuge for microtubes.
3. Micropipettes (1–10; 10–100; 100–1000).
4. Pipette tips.
5. Biosafety cabinet.
6. Microtubes (1.5 mL).
7. Nitrile gloves.
8. Safety glasses.

3 Methods

3.1 Preparation of the Cell Pellet

1. Pipette 1 mL of *Leptospira* culture in a 1.5 mL microtube.

2. Centrifuge for 10 min at 16,000–18,000 × *g* at ambient temperature.

3. Discard EMJH supernatant and resuspend leptospires in 200 µL PBS by pipetting.

4. Centrifuge for 10 min at 16,000–18,000 × *g* at ambient temperature.

5. Discard supernatant and resuspend pellet with 300 µL ultrapure water by pipetting and vortexing if needed. A homogenous cloudy suspension must be obtained.

6. Add 900 µL absolute ethanol and homogenize by pipetting and/or vortexing (*see* **Note 3**).

7. Centrifuge for 10 min at 16,000–18,000 × *g* at ambient temperature.

8. Discard the supernatant and allow the pellet to dry completely by opening the tube lid under the biosafety cabinet (*see* **Note 4**).

9. Add 10–20-µL formic acid depending on the size of the pellet (*see* **Note 5**). Vigorously homogenize by pipetting up and down multiple times.

10. Add the same volume acetonitrile and homogenize by pipetting up and down.

11. Centrifuge for 2 min at 15,000 × *g* at ambient temperature.

12. Spot 1 µL of this preparation on a spot of the MALDI-ToF target plate (*see* **Note 6**).

13. Dry at ambient temperature.

14. Within 10 min after complete drying, add 1.2 µL HCCA matrix and dry at ambient temperature.

15. Place the target plate into the MALDI-ToF.

3.2 Creation of a Reference Spectrum (Main Spectral Projection, MSP)

To create a reference spectrum, a minimum of eight spots of the strain to be analyzed must be prepared. Each of these eight spots must then be analyzed three times, leading to a total of 24 individual mass spectra, which will be used to create the reference spectrum of the strain, named "MSP" for main spectral projection (*see* **Note 7**). Individual spectra are normalized and compared visually. Atypical or poor-quality spectra can be excluded from the complete set, but at least 20 individual spectral acquisitions must be kept to create the reference MSP. Figure 1 shows the MSP-derived phylogeny of 35 *Leptospira* species.

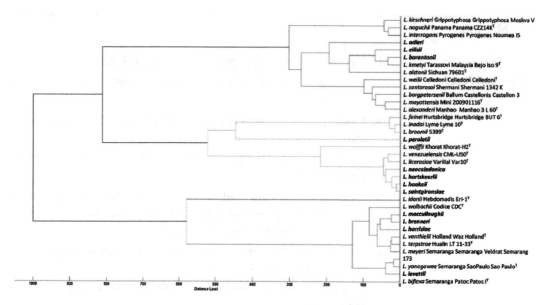

Fig. 1 MSP-derived phylogeny of 35 species of the genus *Leptospira*. (Original figure published under Creative Commons Attribution License in Thibeaux R, Girault D, Bierque E, Soupé-Gilbert ME, Rettinger A, Douyère A, Meyer M, Iraola G, Picardeau M, Goarant C. Biodiversity of environmental leptospira: improving identification and revisiting the diagnosis. Front Microbiol. 2018 May 1;9:816. doi: https://doi.org/10.3389/fmicb.2018.00816)

3.3 Creation of a MS Database

Using the features of the software, create a MS database with well-identified clearly defined strains of each species (e.g., type strains). This database will then be used for pairwise comparison of mass spectra of unknown isolates to allow identification (*see* **Note 8**). Figure 2 shows normalized MSP of a representative subset of *Leptospira* reference strains and isolates.

3.4 Identification of an Unknown Isolate

Unknown isolates are prepared as described in Subheading 3.2. A single spot is usually sufficient to obtain an identification but replicates might be preferred.

The MSP database is loaded into MALDI Biotyper 3.0 software and used to compare the MS of the isolate with the MS database. Using pairwise comparisons, a similarity score is calculated considering the proportion of matching peaks between the MS of the unknown isolate and all MS in the reference database.

The similarity score ranging from 0 to 3 is interpreted as per Bruker recommendations: a score ≥2.3 is considered as a valid identification of the species, and a score in the range 2.0–2.3 is considered as inconclusive for the species. Scores below 2.0 indicate no reliable match to any of the MSP in the database (*see* **Note 9**). Table 1 summarizes the recommendations for reliable identification using identification scores.

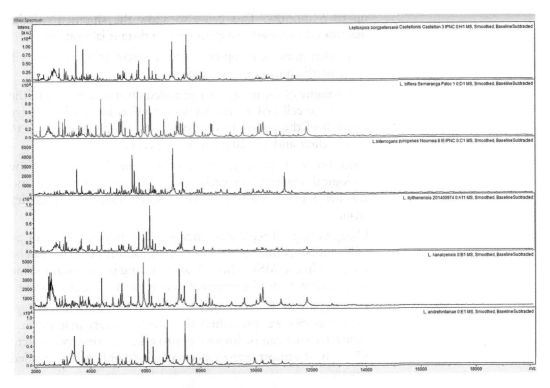

Fig. 2 Normalized MSP of a representative subset of *Leptospira* isolates

Table 1
Correspondence of scores as instructed by Bruker

Value	Description	Symbol	Color
2.300 ... 3.000	High probability of species identification	(+++)	Green
2.000 ... 2.299	Reliable identification of genus, species identification probable	(++)	Green
1.700 ... 1.999	Genus identification probable	(+)	Yellow
0.000 ... 1.699	Insufficient confidence for identification	(-)	Red

4   Notes

1. This particular matrix is best suited for the analysis of proteins by MALDI-ToF. Consequently, the mass spectra observed mostly consist of proteins.

2. Late exponential phase cultures of *Leptospira* are usually in the range $1–3 \times 10^8$ cells/mL. This corresponds to 5–7-day-old culture for most leptospires.

3. At this step, the procedure can be stopped for delayed analysis. The *Leptospira* suspension can be kept at $-18\,°C$ and stored for

up to 6 months. This can also allow the shipment of the inactivated *Leptospira* suspension to a distant laboratory.

4. The pellet must be completely dry. Any remaining ethanol will decrease the quality of the reads.

5. The volume of formic acid to be added must be adjusted to the size of the cell pellet, still paying attention to avoid excessive dilution of the pellet. In case of doubt, simply add 10 μL to small pellets and 15–20 μL to large pellets.

6. Once the spots are dry, you can check for the density of the biological material on each spot. If needed, you can add another 1 μL to spots with low material density and let dry again.

7. Using biological replicates (independent preparation from an independent culture of the same strain), additional reads can be included in the MSP. This will allow taking into account some peaks that may not be systematically present for a given strain.

8. Of note, the MSP is portable. A database with MSP for 35 *Leptospira* species was published as a supplementary material as a btmsp file that can be imported into a Bruker Biotyper system [7]. It is of utmost importance to include only well-identified strains.

9. Scores below 2.0 or even below 1.5 do not preclude the identification of the genus *Leptospira* as observed with novel species in a former study [7]. Scores of 2.300 or higher might be observed for closely related species, notably species with 100% nucleotide identity of their respective 16S rRNA genes, as also observed between *Escherichia coli* and *Shigella*.

References

1. Johnson RC, Rogers P (1964) Differentiation of pathogenic and saprophytic leptospires with 8-azaguanine. J Bacteriol 88:1618–1623

2. Johnson RC, Harris VG (1967) Differentiation of pathogenic and saprophytic Leptospires. I. Growth at low temperatures. J Bacteriol 94(1):27–31

3. Yasuda PH, Steigerwalt AG, Sulzer KR, Kaufmann AF, Rogers F, Brenner DJ (1987) Deoxyribonucleic acid relatedness between serogroups and serovars in the family *Leptospiraceae* with proposals for seven new *Leptospira* species. Int J Syst Bacteriol 37(4):407–415. https://doi.org/10.1099/00207713-49-2-839

4. Ramadass P, Jarvis BD, Corner RJ, Penny D, Marshall RB (1992) Genetic characterization of pathogenic Leptospira species by DNA hybridization. Int J Syst Bacteriol 42 (2):215–219. https://doi.org/10.1099/00207713-42-2-215

5. Perolat P, Chappel RJ, Adler B, Baranton G, Bulach DM, Billinghurst ML, Letocart M, Merien F, Serrano MS (1998) *Leptospira fainei* sp. nov., isolated from pigs in Australia. Int J Syst Bacteriol 48:851–858. https://doi.org/10.1099/00207713-48-3-851

6. Thibeaux R, Iraola G, Ferrés I, Bierque E, Girault D, Soupé-Gilbert ME, Picardeau M, Goarant C (2018) Deciphering the unexplored Leptospira diversity from soils uncovers genomic evolution to virulence. Microb Genom 4 (1):000144. https://doi.org/10.1099/mgen.0.000144

7. Thibeaux R, Girault D, Bierque E, Soupé-Gilbert ME, Rettinger A, Douyere A, Meyer M, Iraola G, Picardeau M, Goarant C (2018) Biodiversity of environmental

Leptospira: improving identification and revisiting the diagnosis. Front Microbiol 9:816. https://doi.org/10.3389/fmicb.2018.00816

8. Vincent AT et al (2019) Revisiting the taxonomy and evolution of pathogenicity of the genus Leptospira through the prism of genomics. PLoS Negl Trop Dis 13(5):e0007270

9. Djelouadji Z, Roux V, Raoult D, Kodjo A, Drancourt M (2012) Rapid MALDI-TOF mass spectrometry identification of Leptospira organisms. Vet Microbiol 158:142–146. https://doi.org/10.1016/j.vetmic.2012.01.028

10. Rettinger A, Krupka I, Grunwald K, Dyachenko V, Fingerle V, Konrad R, Raschel H, Busch U, Sing A, Straubinger RK, Huber I (2012) Leptospira spp. strain identification by MALDI TOF MS is an equivalent tool to 16S rRNA gene sequencing and multi locus sequence typing (MLST). BMC Microbiol 12(1):185. https://doi.org/10.1186/1471-2180-12-185

11. Calderaro A, Piccolo G, Gorrini C, Montecchini S, Buttrini M, Rossi S, Piergianni M, De Conto F, Arcangeletti MC, Chezzi C, Medici MC (2014) Leptospira species and serovars identified by MALDI-TOF mass spectrometry after database implementation. BMC Res Notes 7(1):330. https://doi.org/10.1186/1756-0500-7-330

12. Ellinghausen HC Jr, McCullough WG (1965) Nutrition of Leptospira pomona and growth of 13 other serotypes: fractionation of oleic albumin complex and a medium of bovine albumin and polysorbate 80. Am J Vet Res 26:45–51

Chapter 4

Cell Enumeration of *Leptospira* by Flow Cytometry

Célia Fontana and Jean Reyes

Abstract

Rapid and reliable enumeration of *Leptospira* spp., the causative agent of leptospirosis, represents a technical challenge because leptospires are thin, highly motile, and slow-growing bacteria. The current gold standard for cell enumeration is the use of a Petroff-Hausser counting chamber and a dark-field microscope, but this method remains time-consuming and lacks reproducibility. New alternative techniques are then of great interest. Here we describe the protocol for counting leptospires by flow cytometry. This method is rapid, reproducible, sensitive, and hence suitable to become a new standard to enumerate *Leptospira* spp.

Key words *Leptospira*, Enumeration, Flow cytometry, Fluorescent dyes, Cell viability

1 Introduction

Tools to manipulate *Leptospira* spp. are still in their infancy. Even today, rapid and reliable enumeration of these atypical bacteria represents a technical challenge. Classical methods like colony-forming unit (CFU) counts are not suitable for routine testing considering the slow growth rate of leptospires—more than 4 weeks are necessary to obtain colonies of pathogenic strains on agar plates [1]. The current standard method for enumerating *Leptospira* spp. is the count of motile bacteria on a Petroff-Hausser chamber under dark-field microscopy. Due to high motility of *Leptospira* spp. within or between microscope fields, this technique remains time-consuming, requires a well-trained operator, and lacks reproducibility [2]. Other approaches like bioluminescence have been described using an ATP assay, but the protocol was labor-intensive and valid only in the limited range of 4×10^8–8×10^9 leptospires/mL [3]. More recently, the modification of *L. interrogans* and *L. biflexa* strains by inserting the *luxCDABE* cassette in their genomes resulted in an efficient counting method of bioluminescent strains but restricted to these genetically modified strains [2]. Alternatively, a viability PCR using propidium monoazide was implemented for efficient quantification of living *Leptospira*

Nobuo Koizumi and Mathieu Picardeau (eds.), *Leptospira spp.: Methods and Protocols*, Methods in Molecular Biology, vol. 2134, https://doi.org/10.1007/978-1-0716-0459-5_4, © Springer Science+Business Media, LLC, part of Springer Nature 2020

cells in water and soil with a high sensitivity, but this technique requires time-consuming DNA extraction and qPCR [4]. Other enumeration tools have been developed such as determination of optical density or measurement of electric resistance using a Coulter counter [5], but none of them quantify viability.

Flow cytometry (FCM) has been extensively used over the past years for cellular biology applications, in particular in the immunology field. Lately, constructors commercialized new generation devices capable of detecting smaller events such as bacteria and viruses of 70 nm [6]. For this reason, FCM has been increasingly used in diverse fields of microbiology like medicine, environmental studies, food and pharmaceutical industries. In 2007 and 2010, FCM was used to efficiently enumerate the spirochetes *Borrelia burgdorferi* [7] and *Treponema denticola* [8]. Recently, we demonstrated that enumeration of *Leptospira* spp. by this technology is also effective [9]. We showed that this enumeration technique that quantifies viability is instantaneous, with only 5 min necessary for a complete analysis, highly reproducible, sensitive, and statistically equivalent to the standard Petroff-Hausser method for different strains and at different physiologic states.

This procedure describes the protocol for enumerating *Leptospira* spp. by FCM.

2   Materials

2.1   Culture of Leptospira

1. Ellinghausen-McCullough-Johnson-Harris (EMJH) medium: Difco Leptospira Medium Base EMJH (Becton Dickinson), Difco Leptospira Enrichment EMJH (Becton Dickinson). For 1 L of EMJH liquid medium, dissolve 2.3 g of Leptospira Medium Base EMJH into 900 mL distilled water and sterilize by autoclaving. Add 100 mL of Leptospira Enrichment EMJH.

2. *L. interrogans* or *L. biflexa* strain.

2.2   Dilution of Leptospira Prior to Flow Cytometry Analysis

1. Spectrophotometer for $OD_{450\ nm}$ measurement.

2. Dark-field microscope.

3. 12×75 mm or microcentrifuge tubes.

4. Saline solution: 0.9% NaCl in distilled water, pH 7.5 (*see* **Note 1**).

2.3   Staining of Leptospira

1. SYTO® 9 and propidium iodide (PI) fluorescent dyes: SYTO® 9 stains DNA of all cells, and PI only stains DNA of altered membrane cells corresponding to dying or dead cells. Maximal wavelengths of excitation and emission are, respectively, 485 and 498 nm for SYTO® 9 and 535 and 617 nm for PI (Fig. 1). Both dyes are supplied in the LIVE/DEAD® BacLight™ Bacterial Viability kit (Life Technologies) (*see* **Note 2**).

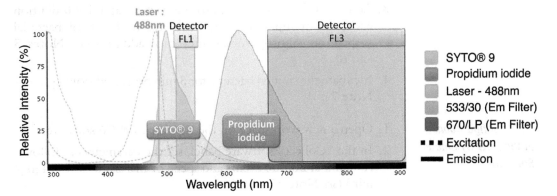

Fig. 1 Excitation and emission spectra of SYTO® 9 and PI dyes superimposed with BD Accuri™ C6 laser emission and detector wavelengths. (Adapted from Fluorescence SpectraViewer (https:/www.thermofisher. com/de/de/home/life-science/cell-analysis/labeling-chemistry/fluorescence-spectraviewer.html))

2.4 Flow Cytometry Analysis

1. BD Accuri™ C6 flow cytometer combined with BD Accuri™ C6 software (BD Biosciences): This flow cytometer is composed of a blue laser emitting at 488 nm, a forward scatter detector (FSC) at 0°, a side scatter detector (SSC) at 90°, and fluorescence detectors in particular a green fluorescence one, FL1, detecting at 533 ± 15 nm and a red one, FL3, detecting at >670 nm (Fig. 1). An automatic calculation allows the determination of the absolute count of events per microliter (*see* **Note 3**).

2. 0.22 μm filtered, deionized water.

3. Validation beads supplied by flow cytometer manufacturer.

3 Methods

3.1 Flow Cytometer Start-Up

1. Launch the BD Accuri™ C6 flow cytometer according to manufacturer's protocol. The start-up process including running manufacturer's validation beads to monitor flow cytometer performance takes approximately 20 min.

3.2 Preparation of Leptospira Prior to Flow Cytometry Analysis

1. Measure culture $OD_{450 \, nm}$ to evaluate its density. If no spectrophotometer is available, evaluate the density by observing the bacteria under dark-field microscopy.

2. For non-detectable $OD_{450 \, nm}$, usually corresponding to culture less concentrated than 10^7–10^8 leptospires/mL, dilute the bacteria at 1:10 in saline solution. For $0 < OD_{450 \, nm} \leq 0.3$, dilute the bacteria at 1:100 in saline solution (two serial dilutions at 1:10). For $OD_{450 \, nm} > 0.3$, dilute the bacteria at 1:1000 in saline solution (three serial dilutions at 1:10) (*see* **Note 4**).

3. Add SYTO® 9 and PI dyes to the bacteria at 1:1000 dilution each, for example, 1 μL of each dye to 1 mL of bacterial suspension. Shortly vortex after each addition (*see* **Notes 5** and **6**).

4. Incubate the stained bacteria for 3 min protected from light (*see* **Note 7**).

3.3 Configuration of the Flow Cytometer Software

1. Open a new workspace on the BD Accuri™ C6 software.

2. In the "Collect" software panel, set the parameters as follows: run limit at 20,000 events and speed of analysis at slow (14 μL/min) (*see* **Note 8**).

3. In the "Data display" panel, add the following dot plots: SSC-A versus FSC-A to visualize the shape of the population and FL3-A (red fluorescence) versus FL1-A (green fluorescence) to visualize fluorescent leptospires and enumerate live bacteria. Set the plot axes in logarithmic scales.

3.4 Controls to Perform Prior to First Leptospira Enumerations

Perform the steps described below before the first *Leptospira* enumerations to understand and fix proper flow cytometer parameters. For routine analyses, go directly to Subheading 3.5.

1. Prepare 12 × 75 mm or microcentrifuge tubes of at least 100 μL of (1) unstained saline solution, (2) unstained EMJH medium, (3) stained saline solution, and (4) stained EMJH medium.

2. Shortly vortex the tubes prior to analysis.

3. Place the first tube to analyze under the sample injection port (SIP).

4. In the "Collect" panel of BD Accuri™ C6 software, fix a permissive threshold at 1000 on FSC.

5. Click on the "Run" button. Events are automatically recorded on FSC/SSC and FL1/FL3 dot plots. Zoom on the recorded dots using the magnifying glass button to correctly visualize the obtained profiles.

6. Analyze the other tubes following the same procedure. Events of these four analyses all correspond to background noise and should be removed for further analyses (Fig. 2a, b). This explains the necessity of applying a higher threshold.

7. Fix a threshold at 1000 on FL1 and repeat analyses of the tubes from step 1 (*see* **Note 9**). A minor quantity of events should be recorded, meaning that background noise is correctly removed from analysis (Fig. 2c).

8. Repeat **steps 1–7** with tubes of (1) unstained fresh *Leptospira* culture and (2) stained fresh *Leptospira* culture, diluted as described in Subheading 3.2. Unstained *Leptospira* culture is

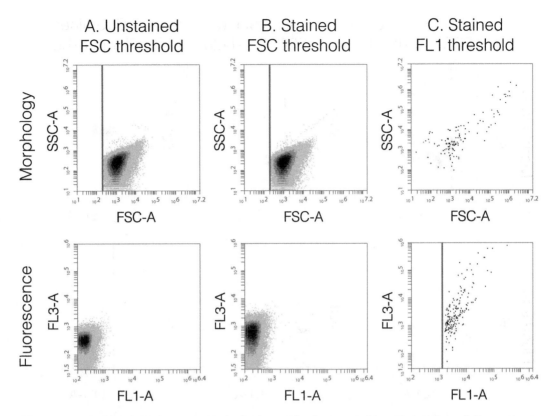

Fig. 2 Determination of proper thresholds for background noise removal based on saline solution analyses. FCM morphology (FSC/SSC) and fluorescence (FL1/FL3) profiles of (**a**) unstained saline with a threshold at 1000 on FSC, (**b**) stained saline with a threshold at 1000 on FSC, and (**c**) stained saline with a threshold at 1000 on FL1. Analysis of EMJH gives identical results (not shown). The red lines represent the thresholds

undifferentiated from background noise, contrarily to stained ones (Fig. 3a, b). In addition, application of a fluorescence threshold leads to significant enhancement of bacterial population detection (Fig. 3c) (*see* **Note 10**).

9. Repeat **steps 1–7** with tubes of stained (1) fresh viable *Leptospira* culture, (2) the same culture heat-killed at 70 °C for 15 min, and (3) a mix of 50% fresh and 50% heat-killed cultures. Live bacteria should emit high green fluorescence and low red fluorescence on FL1/FL3 dot plot, contrarily to dead cells, emitting high red fluorescence and low green fluorescence (Fig. 4). Based on these analyses, draw a polygonal gate on FL1/FL3 dot plot containing only live leptospires. Name this gate "Live." Dead bacteria should not appear in the drawn "Live" gate but be in the upper and/or left part of the FL1/FL3, as shown in Fig. 4. Reuse this gate for all future analyses.

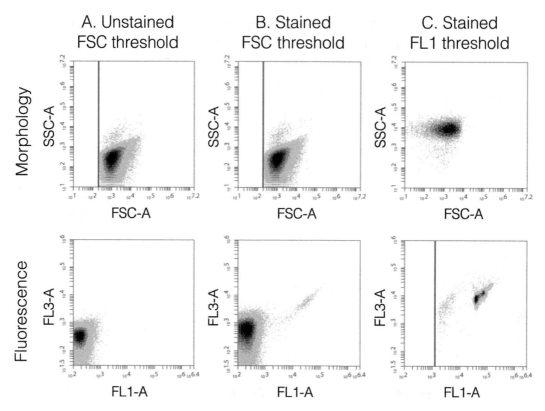

Fig. 3 Enhancement of *Leptospira* detection by staining bacteria and thresholding on fluorescence. FCM morphology (FSC/SSC) and fluorescence (FL1/FL3) profiles of (**a**) unstained bacteria with a threshold at 1000 on FSC, (**b**) stained bacteria with a threshold at 1000 on FSC, and (**c**) stained bacteria with a threshold at 1000 on FL1. The red lines represent the thresholds

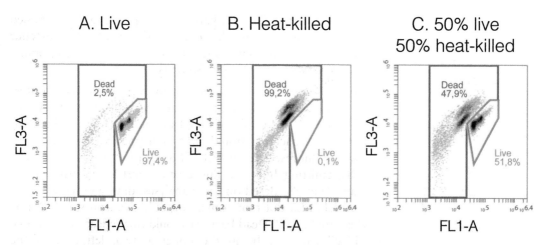

Fig. 4 Discrimination of live and dead bacteria based on fluorescence profiles. FCM fluorescence (FL1/FL3) profiles of a stained (**a**) fresh viable *Leptospira* culture, (**b**) heat-killed *Leptospira* culture, and (**c**) mix of 50% fresh and 50% heat-killed *Leptospira* cultures. The drawn green "Live" gate contains viable bacteria whereas the red "Dead" gate contains dying and dead ones. Percentages represent population proportion in each gate

3.5 Enumeration of Leptospira by Flow Cytometry

1. As determined in Subheading 3.3, in the "Collect" panel of BD Accuri™ C6 software, set the threshold on FL1-H (green fluorescence) at 1000 to remove background noise.

2. Shortly vortex the tube containing the stained bacteria prior to analysis.

3. Place the tube under the SIP.

4. Click on the "Run" button to launch analysis. The recorded events will automatically be displayed on FSC/SSC and FL1/FL3 dot plots (Fig. 5a) (*see* **Notes 11** and **12**).

5. To assess culture viability, look at the calculated percentage of events contained in the "Live" gate (*see* **Note 13**).

6. To obtain sample enumeration titer, go to the "Statistics" software panel, and choose to display "Events/μL" contained in the "Live" gate (Fig. 5b). Calculate the enumeration titer as follows: number of leptospires/mL = number of events in the "Live" gate per microliter × dilution factor × 1000 (Fig. 5c) (*see* **Note 14**).

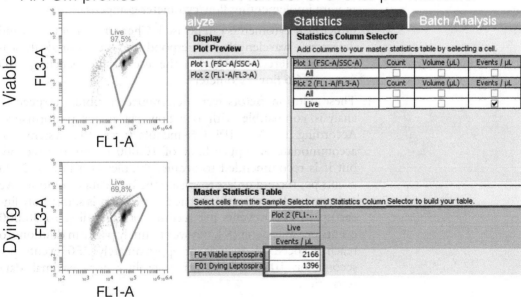

A. FCM profiles

B. Number of events per microliter

C. Enumeration titer

Viable *Leptospira* : 2166 × 1000 × 1000 = 2.166 × 10^9 leptospires/ml
Dying *Leptospira* : 1396 × 1000 × 1000 = 1.396 × 10^9 leptospires/ml

Fig. 5 Enumeration of two *Leptospira* cultures, a viable and a dying one. (**a**) FCM fluorescence (FL1/FL3) profiles. (**b**) Recorded number of events in the "Live" gate for each culture automatically calculated by flow cytometer software. (**c**) Obtention of enumeration titer for each culture diluted at 1:1000 prior to FCM analysis

3.6 Flow Cytometer Shutdown

1. Shut down the flow cytometer according to manufacturer's protocol. The process including running decontamination, cleaning, and deionized water solutions takes approximately 30 min.

2. The flow cytometer automatically powers off after an internal washing cycle.

4 Notes

1. If possible, use sterile material to avoid analysis by the flow cytometer of debris or contaminants which could generate important background noise.

2. Other combination of viability fluorescent dyes may be used with the following specifications: the first dye should stain all cells (e.g., other SYTO stains, SYBR-II, DAPI, or thiazole orange) or only living cells (e.g., CFDA), while the second should only mark dying and dead cells (e.g., TO-PRO-3 iodide). Their excitation wavelengths should correspond to a flow cytometer laser emission wavelength, and their emission wavelengths to two different flow cytometer detectors. Avoid spillover between the two dyes' emission spectra as it would require implementing flow cytometer compensation.

3. Other flow cytometers can be used. Check if laser emission and detectors' wavelengths correspond to the desired fluorochromes. Here, we describe the use of the benchtop BD Accuri™ C6 flow cytometer.

4. These dilution factors were determined to obtain a speed of analysis compatible with our flow cytometer performance. According to Accuri™ C6 manufacturer, the system can accommodate an upper limit of 10,000 events per second, but it is recommended to acquire samples at a rate of 2500 events per second or less to ensure the best data resolution. We experimentally determined that enumeration is statistically linear up to 7000 events per second corresponding to 30,000 events per microliter [9], but we routinely perform dilutions to reach a speed of analysis of approximately 250 events per second or 1000 events per microliter for maximal data resolution.

5. Perform the dilution first and then the staining. Staining will not be optimal otherwise.

6. Do not incubate bacteria with PI for an extended period of time (>10 min) since the dye is toxic and slowly kills bacteria. It would result in an overestimation of dead cells.

7. Dyes are sensitive to light. Although 3 min of incubation exposed to light will not result in any significant degradation

of the dyes, it is important to keep the stock solution in the dark to retain maximum efficiency.

8. Slow speed of analysis usually results in more accurate acquisitions. Changing speed leads to a population shift. Thus, always use the same speed of analysis along the experiments.

9. It can be necessary to wash the flow cytometer with decontamination, cleaning, and deionized water solution between analyses of stained and unstained samples, since fluorescent dyes tend to adhere to the pipes.

10. Contrarily to eukaryotic cells or bacteria such as *Escherichia coli* [10], *Leptospira* cannot be differentiated from background noise on FSC or SSC parameters because they are too thin to be distinguished based on morphological parameters.

11. When performing the first stained bacteria analysis of the day, FCM profile can be diffuse at first. It can be necessary to delete and restart the run after 15–30 s to obtain a more defined and accurate profile.

12. As an indication, the timescale for *Leptospira* enumeration is approximately 6 min: 2 min of sample preparation, 3 min of incubation with the dyes, 1 min of analysis by FCM.

13. Be aware that percentage of events contained in the "Live Leptospires" gate strictly depends on the fixed threshold and cannot be translated to an absolute percentage of bacterial viability. Indeed, as dying and dead cells' fluorescence is low, a part—large or small depending on FL1 threshold value—of the corresponding events are under the threshold and therefore not considered for percentage calculation. However, this percentage still provides valuable indications. High percentages (>70%) usually correspond to viable bacteria, whereas lower percentages imply dying ones. Besides, it is relevant to compare viability of different suspensions based on this percentage when analyzed with the same threshold.

14. FCM enumeration technique is highly reproducible, with no observed operator effect, and a repeatability variance, corresponding to the variability of the technique itself, is very low. Sensitivity of *Leptospira* enumeration by FCM is statistically determined to be 10^4 leptospires/mL. A bias in the measures is observed for titers as low as 10^3 leptospires/mL, leading to a slight overestimation of the titers, but this overestimation can be acceptable for most applications [9].

References

1. Picardeau M (2015) Genomics, proteomics, and genetics of leptospira. Curr Top Microbiol Immunol 387:43–63. https://doi.org/10.1007/978-3-662-45059-8_4

2. Murray GL, King AM, Srikram A, Sermswan RW, Adler B (2010) Use of luminescent *Leptospira interrogans* for enumeration in biological assays. J Clin Microbiol

48:2037–2042. https://doi.org/10.1128/JCM.02541-09

3. Nervig RM, Sebring RW, Scheevel KF (1986) The enumeration of *Leptospira interrogans* serovar pomona by a bioluminescence ATP assay. J Biol Stand 14:21–24

4. Casanovas-Massana A, Pedra GG, Wunder EA Jr, Diggle PJ, Begon M, Ko AI (2018) Quantification of *Leptospira interrogans* survival in soil and water microcosms. Appl Environ Microbiol 84(13). https://doi.org/10.1128/AEM.00507-18

5. Humberd CM, Murray CK, Stuart SK, Reeb BA, Hospenthal DR (2005) Enumerating leptospires using the coulter counter. Am J Trop Med Hyg 73:962–963

6. Steen HB (2004) Flow cytometer for measurement of the light scattering of viral and other submicroscopic particles. Cytometry A 57A:94–99. https://doi.org/10.1002/cyto.a.10115

7. Bakker RG, Li C, Miller MR, Cunningham C, Charon NW (2007) Identification of specific chemoattractants and genetic complementation of a Borrelia burgdorferi chemotaxis mutant: flow cytometry-based capillary tube chemotaxis assay. Appl Environ Microbiol 73:1180–1188. https://doi.org/10.1128/AEM.01913-06

8. Orth R, O'Brien-Simpson N, Dashper S, Walsh K, Reynolds E (2010) An efficient method for enumerating oral spirochetes using flow cytometry. J Microbiol Methods 80:123–128. https://doi.org/10.1016/j.mimet.2009.11.006

9. Fontana C, Crussard S, Simon-Dufay N, Pialot D, Bomchil N, Reyes J (2017) Use of flow cytometry for rapid and accurate enumeration of live pathogenic Leptospira strains. J Microbiol Methods 132:34–40. https://doi.org/10.1016/j.mimet.2016.10.013

10. Gatza E, Peña PV, Srienc F, Overton T, Lavarreda CA, Rogers CE (2012) Bioprocess monitoring with the BD Accuri™ C6 flow cytometer. BD Biosciences White Paper

Chapter 5

The Single-Step Method of RNA Purification Applied to *Leptospira*

Crispin Zavala-Alvarado and Nadia Benaroudj

Abstract

Establishing a rapid method to obtain pure and intact RNA molecules has revolutionized the field of RNA biology, enabling laboratories to routinely perform RNA analysis such as Northern blot, reverse transcriptase quantitative PCR, and RNA sequencing. Here, we describe an application of the effective single-step method of RNA extraction (or guanidinium thiocyanate-phenol-chloroform extraction) applied to *Leptospira* species. This method is based on the powerful ability of guanidinium thiocyanate to inactivate RNases and on the different solubilities of RNA and DNA in acidic phenol. This method allows one to reproducibly obtain total RNAs with high yield and integrity, as determined by capillary electrophoresis, suitable for the RNA sequencing technology.

Key words Spirochetes, *Leptospira*, RNA, Guanidinium thiocyanate, Phenol-chloroform extraction, RIN, RNA-Seq, RT-PCR

1 Introduction

Efficient acquisition of pure and intact RNA molecule is a prerequisite for numerous analytical techniques such as reverse transcriptase quantitative PCR (RT-qPCR), Northern blotting, microarray analysis, and RNA sequencing (RNA-Seq). Particularly powerful are RNA-Seq technologies that allow for profiling and quantification of RNA. Knowing which gene is expressed and how genes are regulated in a particular condition provides scientists with a comprehensive knowledge of the physiological state of cells. The pioneering transcriptomic studies performed in the 1990s have used hybridization-based microarray technology [1]. Since the development of affordable, high-throughput sequencing technologies, transcriptomes are determined by RNA-Seq [2].

Leptospira spp. are microorganisms with remarkable adaptation capacities allowing survival in different ecological niches. Pathogenic strains disseminate in the blood of infected hosts, can persist intracellularly in macrophages, colonize different animal tissues

Nobuo Koizumi and Mathieu Picardeau (eds.), *Leptospira spp.: Methods and Protocols*, Methods in Molecular Biology, vol. 2134,
https://doi.org/10.1007/978-1-0716-0459-5_5, © Springer Science+Business Media, LLC, part of Springer Nature 2020

(including the kidney, liver, and brain), and are shed in the environment (soil and water) through the urine of infected hosts [3]. Knowledge of the molecular basis of *Leptospira* pathogenicity is very limited compared to other bacteria, mainly due to the lack of genetic tools available for manipulation of leptospiral genome. Inactivating a gene by allelic exchange in pathogenic *Leptospira* strain is feasible but very inefficient. To study the function of a given leptospiral gene, scientists usually rely on random transposon insertion mutants [4, 5]. The transcriptomic approach is therefore instrumental not only in identifying cellular pathways involved in one particular physiological condition but also to speculate gene function when mutants are not available. Effective RNA extraction has allowed several laboratories to perform transcriptomic studies in *Leptospira*, thereby leading to a better knowledge of bacterial adaptation to host osmotic stress [6], in the presence of serum [7], upon temperature changes [8, 9], and to the host environment [10, 11].

Different methods can be used to extract RNA from a biological sample. One method relies on the different solubilities of cellular components in organic solvents and RNA precipitation by alcohol. Another method is based on the ability of RNA to bind to specific adsorbing material, such as silica and cellulose matrixes, and is used in most commercial RNA purification kits. In a third method, RNA is separated on density gradient centrifugation, but this method is laborious and does not allow for simultaneous processing of multiple samples.

Here, we describe the method based on RNA extraction with an organic solvent and precipitation with alcohol currently applied to *Leptospira* strains and allowing for high yields of pure and intact RNA, compatible with the use of RNA-sequencing technology. In this protocol, harvested *Leptospira* are first lysed in TRIzol™. This reagent contains guanidinium isothiocyanate, a chaotropic agent which is very effective at inactivating endogenous RNases. It also contains low-pH phenol for separating DNA from RNA [12]. After adding chloroform to the samples and subsequent centrifugation, RNAs remain in the upper clear aqueous phase, while precipitated proteins and DNA remain in the interphase and lower organic phase, respectively. The RNA contained in the upper phase is transferred to a new tube and undergoes alcohol precipitation. The RNA pellet is then diluted in a suitable buffer. Traces of contaminating DNA are eliminated by DNase treatment. This guanidinium thiocyanate-phenol-chloroform extraction also known as the "single-step method" greatly improved and expedited RNA purification and has become the gold standard widely used for any type of biological samples [13].

RNA quantification and purity can be determined by absorbance measurement at 260 and 280 nm. A ratio A_{260}/A_{280} of at least 1.80 indicates an acceptable purity with low protein contamination, suitable for RT-PCR. For performing RNA-Seq, RNA preparation should be of the highest quality. The integrity of RNA (i.e., absence of RNA degradation) can be assessed by analyzing the RNA preparation by capillary electrophoresis using, for instance, the chip-based device of the Agilent BioAnalyzer. This analysis will provide with a RIN (RNA integrity number) value that represents an objective measurement of RNA integrity ranging from 10 (highly intact RNA) to 1 (completely degraded RNA) [14]. For RNA-Seq, a RIN value above 8 should be aimed.

The total RNAs obtained via this method are mostly ribosomal RNAs. Depending on the analysis method used downstream, depletion of ribosomal RNAs allowing enrichment of messenger RNAs might be necessary.

2 Materials

We have applied this protocol to pathogenic *Leptospira* (*L. interrogans* serovar Manilae strain L495) and saprophyte (*L. biflexa* serovar Patoc strain Patoc) strains cultivated in vitro in EMJH medium (*see* **Note 1**).

1. Albumin supplement: 10% (w/v) bovine serum albumin, 0.004% (w/v) zinc sulfate, 0.015% (w/v) magnesium chloride, 0.015% (w/v) calcium chloride, 0.1% (w/v) sodium pyruvate, 0.4% (w/v) glycerol, 1.25% (v/v) Tween 80, 0.0002% (w/v) vitamin B12, 0.05% (w/v) ferrous sulfate (added at the last moment) in sterile water for injection (WFI).

2. EMJH base: dissolve 2.3 g of Difco Leptospira Medium Base EMJH (Becton Dickinson) in 900 ml sterile WFI. Autoclave the solution.

3. EMJH medium: add 100 ml albumin supplement to 900 ml EMJH base. Adjust the pH to 7.5 and filter sterilize the solution.

4. Refrigerated centrifuge and rotor reaching $12,000 \times g$.

5. Water bath at 55 and 4 °C.

6. Vortexer.

7. Fume hood.

8. P1000, P200, P20, P10/2 micropipettes.

9. RNase-free barrier tips for pipettes.

10. 1–2 ml disposable serological plastic pipettes.

11. 1.5 ml RNase-free polypropylene microcentrifuge tubes (*see* **Note 2**).

12. 50 and 15 ml RNase-free polypropylene conical tubes (*see* **Note 2**).

13. Surface RNase decontaminant solution.

14. TRIzol™ reagent or other commercially available guanidinium thiocyanate-acidic phenol solution (*see* **Note 3**).

15. Chloroform.

16. Isopropanol.

17. 75% ethanol in RNase-free water (*see* **Note 4**).

18. RNase-free H_2O.

19. DNase treatment kit (*see* **Note 5**).

20. UV spectrophotometer.

21. Tris-acetate-EDTA (TAE) running buffer (50×): 242 g of Tris base, 57.1 ml of glacial acetic acid, 100 ml of 0.5 M EDTA, pH 8.0. Adjust the volume to 1 L with distilled water (the pH should be around 8.5). Dilute the solution with ultrapure water to 1× for use.

22. Nucleic acid staining such as ethidium bromide (supplied in a dropper bottle at 625 μg/ml).

23. 1% agarose: 1 g of agarose in 100 ml of TAE running buffer. Add one drop (about 25 μg/40 μl) of ethidium bromide in 50 ml of the solution before agarose polymerization.

24. 6× gel loading buffer for nucleic acid: 10 mM Tris–HCl, pH 7.6, 60% glycerol, 60 mM EDTA, 0.03% bromophenol blue, 0.03% xylene cyanol FF. Mix 1 volume of the 6× gel loading buffer with 5 volume of RNA solution (containing 0.5–1 μg of RNA).

25. Gel equipment for nucleic acid electrophoresis.

26. Electrophoresis power supply.

27. UV transilluminator to visualize nucleic acids.

3   Methods

Great care should be taken to prevent RNA degradation by exogenous RNases. Gloves should be worn at all times and changed frequently. People with long hair should secure it. If possible, a designated laboratory space should be reserved exclusively for RNA extraction and manipulation (*see* **Note 6**). All the consumable materials (tips, tubes) and solutions should be RNase-free and protected from the dust. All the non-disposable materials that will be in contact with the RNA (pipettes, benches, centrifuge, gel equipment) should be washed with a surface RNase decontaminant solution (*see* **Note 7**). All the steps are performed at room temperature unless otherwise noted.

3.1 Cell Lysis

Optimally, the starting material should be in vitro-cultured *Leptospira* consisting of at least 10^9 cells. This corresponds to a 30 ml *Leptospira* culture at exponential phase (*see* **Note 8**).

1. Centrifuge the *Leptospira* cells in a 50 ml conical tube for 15 min at 3000 × *g* at 4 °C (*see* **Note 9**).

2. Resuspend the cell pellet in 1 ml of TRIzol™ reagent and transfer the suspension in a 1.5 ml polypropylene tube (*see* **Note 10**).

3. Vortex well to fully resuspend the pellet.

4. Flash freeze samples in liquid nitrogen and store them at −80 °C until further use (*see* **Note 11**).

3.2 RNA Extraction

1. Thaw the sample(s) at room temperature (*see* **Note 12**).

2. Add 260 μl chloroform, mix thoroughly by inversion for 15 s, and incubate for 10 min (*see* **Note 13**).

3. Centrifuge for 15 min at 12,000 × *g* at 4 °C. After the centrifugation, three phases are observed in the tube. The top clear aqueous phase contains RNA, the white ring at the interphase contains denatured precipitated proteins, and the bottom pink organic phase contains DNA.

4. Carefully, transfer the aqueous top layer containing RNA to a new clean 1.5 ml polypropylene tube (*see* **Note 14**).

5. Add 600 μl isopropanol to precipitate RNA. Mix thoroughly by gently inverting the tube. Incubate for 5–10 min at room temperature (*see* **Note 15**).

6. Centrifuge for 10 min at 12,000 × *g* at 4 °C and discard the supernatant (*see* **Note 16**).

7. Wash the RNA pellet by adding 1 ml of 75% ethanol (*see* **Note 17**).

8. Centrifuge for 5 min at 12,000 × *g* at 4 °C. Discard the supernatant (*see* **Note 18**).

9. Air-dry the RNA (*see* **Note 19**).

10. Resuspend the pellet in 40 μl RNase-free H_2O by pipetting up and down several times.

11. Incubate for 10 min in a water bath at 55 °C in order to enhance the resuspension of the pellet (*see* **Note 20**).

3.3 DNase Treatment

Here, we describe the DNase treatment using the Turbo DNA-free™ kit, but any other commercially available kit might work as well.

1. Add 5 μl of the 10× Turbo DNA-free™ buffer and 4 μl of RNase-free H_2O (provided in the kit) to 40 μl of the RNA suspension obtained in **step 11** in Subheading 3.2.

2. Add 1 µl of Turbo DNA-free™ DNase (at 2 U/µl). Mix by pipetting and incubate 30 min in a water bath at 37 °C (*see* **Notes 20** and **21**).

3. Add 5 µl of DNase inactivation reagent (provided in the kit), mix well by flicking the tube to disperse the inactivating reagent, and incubate for 5 min at room temperature. During this incubation, flick the tube to disperse the inactivating reagent each minute in order to increase the binding of the DNase to the reagent (*see* **Note 22**).

4. Centrifuge for 2 min at 10,000 × *g* and transfer the supernatant to a new clean polypropylene tube (*see* **Note 23**).

3.4 Assessing Quantity and Quality of RNA

The absorbance measurement at 260 nm allows the calculation of RNA concentration. An absorbance value of 1 corresponds to 40 µg/ml of RNA (for a spectrophotometer with 1 cm light path).

For RT-PCR, the quality of the RNA preparation can be assessed by electrophoresis on an agarose gel. When 0.5–1 µg of RNA are loaded on a 1% agarose gel, three main bands can be observed, the 23S, the 16S, and the 5S ribosomal RNAs (Fig. 1) as the total RNA preparation contains mainly ribosomal RNAs. Messenger RNAs can be sometimes visible as faint smear.

If the RNA is to be used in RNA-Seq, the integrity of RNA should be assessed by capillary electrophoresis (*see* **Note 24**). A typical electrophoresis pattern of high-quality RNA is shown in

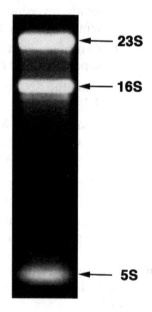

Fig. 1 Analysis of RNA preparation on agarose gel. 0.5 µg of total RNAs were loaded on a 1% agarose gel in 1×TAE. Nucleic acid was stained with ethidium bromide. The bands corresponding to 23S, 16S, and 5S ribosomal RNA are indicated

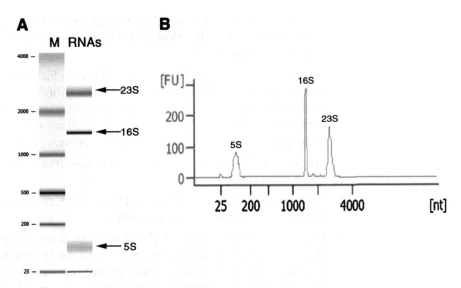

Fig. 2 Analysis of RNA preparation by capillary electrophoresis. 0.5 μg of total RNAs were analyzed on an Agilent RNA 6000 Nano chip with the 2100 Bioanalyzer. (**a**) Capillary electrophoresis gel-like image of the Agilent RNA 6000 ladder (lane M) and RNAs (lane RNAs). The migration position of 23S, 16S, and 5S rRNA is indicated at the right of the image. (**b**) Electropherogram trace of the RNA preparation. The peaks corresponding to different rRNAs are indicated. The RNA preparation displayed here has a RIN of 9.5. A degraded RNA preparation would display a decrease in the 23S and 16S rRNA signal and a concomitant increased baseline in the fast-migrating zone (before the position of the 16S rRNA peak)

Fig. 2. In this analysis, the abundant 23S and 16S rRNAs are well resolved, and the smaller peak corresponds to the 5S rRNA. Here, a RIN value of 9.5 was obtained, which indicates pure and non-degraded RNAs.

The yield of the purification method presented here can be up to 75 μg of RNA per 10^9 *Leptospira*, and RIN values of at least 8.5 are routinely obtained, which makes RNA obtained suitable for RNA-Seq analysis.

3.5 Storage

RNA can be stored at −20 °C for a short-term storage but −80 °C is preferential for a long-term storage. RNA samples could be aliquoted into several tubes to minimize freeze-thaw and reduce RNase degradation occurring upon accidental RNase contamination.

4 Notes

1. Only use autoclaved glassware dedicated for EMJH medium preparation. In order to avoid contaminating the glassware with components that could prevent growth of *Leptospira*, we rinse beforehand the glassware with sterile WFI, and all the chemical stock solutions are prepared with sterile WFI.

2. You do not need to use autoclaved tubes, but tubes exclusively reserved for RNA purification, and do not manipulate the tubes without wearing gloves.

3. We recommend using TRIzol™ from Thermo Fisher Scientific as it has proven to work optimally with this protocol. Another equivalent commercially available reagent might work with a comparable efficiency. Alternatively, home-made solutions can be prepared (*see* ref. 12, 13, 15), but the process is laborious.

4. Prepare the 75% ethanol solution in a RNase-free conical tube and discard any leftover.

5. We recommend using the Turbo DNA-free™ turbo kit from Invitrogen (Thermo Fisher Scientific).

6. If you do not have at your disposal an exclusive designated laboratory space for RNA extraction, it might be wise to perform RNA purification when there are not too much people in the laboratory, lowering air perturbation and the risk of dust movement and contamination. Regardless, the surface and any equipment used should be cleaned with a surface decontaminant DNA/RNase removing solution. We recommend using the RNase Away from Merck.

7. If possible, we recommend having a designated pipette set and gel equipment exclusively used for RNA purification.

8. We cultivate *Leptospira* in EMJH medium. It is possible to extract RNA from lower amounts of cells; however, the yield will be lower, and, in our experience, working with low amounts of cells leads to an RNA preparation with a lower integrity. It is not necessary to wash the cells before adding the guanidinium thiocyanate-acidic phenol solution.

9. The cells should be rapidly processed in the guanidinium thiocyanate-acidic phenol solution (TRIzol™) after harvesting them as rapid inactivation of endogenous RNases is essential for obtaining high-quality RNA.

10. Manipulation of the guanidinium thiocyanate-acidic phenol solution (TRIzol™) should be done under a fume hood, as the solution is highly volatile and toxic.

11. Even if RNA extraction is conducted right after cell resuspension in the guanidinium thiocyanate-acidic phenol solution (TRIzol™), samples should be frozen at −80 °C as freezing promotes cell lysis. Samples can be stored in the guanidinium thiocyanate-acidic phenol solution (TRIzol™) at −80 °C for at least several weeks. If you plan to analyze RNAs extracted from different biological samples, it is better to perform the RNA extraction of all samples at the same time.

12. In order to increase efficiency of cell lysis, up to three cycles of freezing/thawing can be applied to the samples. However, as

promptness is key to obtain high-quality RNAs, we avoid this especially when extracting RNAs for RNA-Seq.

13. This step should be performed under a fume hood as chloroform is highly volatile and toxic.

14. In order to prevent contamination with DNA and precipitated proteins, great care should be taken to avoid perturbating the three phases. It is better not to try to retrieve the totality of the upper phase to prevent carryover.

15. You can pause at this step, and store the samples in isopropanol at −20 °C until you are ready to proceed with the procedure, although we avoid this especially when extracting RNAs for RNA-Seq.

16. You should be able to see a white gel-like pellet containing RNA at the bottom of the tube. Great care should be taken when removing the supernatant as sometimes the RNA pellet does not tightly stick to the tube and tends to move on the tube wall.

17. You can pause at this step, and store the samples in 75% ethanol at −20 °C until you are ready to proceed with the procedure, although we avoid this especially when extracting RNAs for RNA-Seq.

18. As washing the RNA pellet with 75% ethanol will dissolve salts contained in the pellet, the aspect of the pellet will change. Very often, the pellet becomes smaller and less visible. Again, great care should be taken when removing the supernatant. You can use a micropipette or a syringe with a 22 G needle to remove most of the supernatant without disturbing the pellet.

19. The time for air-drying the RNA pellet will depend on the amount of ethanol left in the tube after removing the supernatant. You can put your samples under a fume hood or laminar follow cabinet to enhance the drying step. If the ethanol is properly removed, this step should take 5–10 min. You should avoid excessively drying the RNA pellet as it will decrease its solubility.

20. You should use a clean water bath. A heat block might work as well.

21. To enhance the DNase reaction, you can perform the reaction with 2 μl (4 U) of Turbo DNA-free™ DNase. You can also perform a two-step incubation with the enzyme. In a first step, 1 μl of Turbo DNA-free™ DNase are added and the sample is incubated for 30 min at 37 °C. Then, an additional 1 μl of Turbo DNA-free™ DNase is added to the sample, and the second incubation at 37 °C is conducted for 30 min.

22. The amount of DNase Inactivating Reagent to be added should be adjusted depending on the number of DNase units

used for the reaction. The manufacturer recommends using 5 μl of DNase Inactivating Reagent for 1 μl (2 U) of DNase.

23. It might be more comfortable to perform the DNase treatment in a 0.5 ml polypropylene tube as it will ease removal of the supernatant.

24. We recommend using the chip-based device of the Agilent BioAnalyzer as its software is the only one allowing for a RIN determination. This analysis is performed in transcriptomic facilities.

Acknowledgments

Crispin Zavala-Alvarado is part of the Pasteur-Paris University (PPU) International Ph.D. Program. Our laboratory has received funding from the European Union's Horizon 2020 research and innovation programme under the Marie Sklodowska-Curie grant agreement No. 665807 and from Fondation Etchebès-Fondation de France. We would like to thank Robert Anthony Gaultney for his suggestions on this manuscript.

References

1. Stears RL, Martinsky T, Schena M (2003) Trends in microarray analysis. Nat Med 9 (1):140–145

2. Wang Z, Gerstein M, Snyder M (2009) RNA-Seq: a revolutionary tool for transcriptomics. Nat Rev Genet 10:57–63

3. Picardeau M (2017) Virulence of the zoonotic agent of leptospirosis: still terra incognita? Nat Rev Microbiol 15:297–307

4. Bourhy P, Louvel H, Saint Girons I, Picardeau M (2005) Random insertional mutagenesis of *Leptospira interrogans*, the agent of Leptospirosis, using a mariner transposon. J Bacteriol 187(9):3255–3258

5. Murray GL, Morel V, Cerqueira GM, Croda J, Srikram A, Henry R et al (2009) Genome-wide transposon mutagenesis in pathogenic leptospira species. Infect Immun 77(2):810–816

6. Matsunaga J, Lo M, Bulach DM, Zuerner RL, Adler B, Haake DA (2007) Response of *Leptospira interrogans* to physiologic osmolarity: relevance in signaling the environment-to-host transition. Infect Immun 75(6):2864–2874

7. Patarakul K, Lo M, Adler B (2010) Global transcriptomic response of *Leptospira interrogans* serovar Copenhageni upon exposure to serum. BMC Microbiol 10(1):31–46

8. Qin J-H, Sheng Y-Y, Zhang Z-M, Shi Y-Z, He P, Hu B-Y et al (2006) Genome-wide transcriptional analysis of temperature shift in *L. interrogans* serovar lai strain 56601. BMC Microbiol 6(1):51–60

9. Lo M, Bulach DM, Powell DR, Haake DA, Matsunaga J, Paustian ML et al (2006) Effects of temperature on gene expression patterns in *Leptospira interrogans* serovar lai as assessed by whole-genome microarrays. Infect Immun 74 (10):5848–5859

10. Xue F, Dong H, Wu J, Wu Z, Hu W, Sun A et al (2010) Transcriptional responses of *Leptospira interrogans* to host innate immunity: significant changes in metabolism, oxygen tolerance, and outer membrane. PLoS Negl Trop Dis 4 (10):e857

11. Caimano MJ, Sivasankaran SK, Allard A, Hurley D, Hokamp K, Grassmann AA et al (2014) A model system for studying the transcriptomic and physiological changes associated with mammalian host-adaptation by *Leptospira interrogans* serovar copenhageni. PLoS Pathog 10(3):e1004004

12. Chomczynski P, Sacchi N (1987) Single-step method of RNA isolation by acid guanidinium thiocyanate-phenol-chloroform extraction. Anal Biochem 162(1):156–159

13. Chomczynski P, Sacchi N (2006) The single-step method of RNA isolation by acid guanidinium thiocyanate–phenol–chloroform extraction: twenty-something years on. Nat Protoc 1(2):581–585

14. Schroeder A, Mueller O, Stocker S, Salowsky R, Leiber M, Gassmann M et al (2006) The RIN: an RNA integrity number for assigning integrity values to RNA measurements. BMC Mol Biol 7(1):3–16

15. Kingston RE, Chomczynski P, Sacchi N (1996) Guanidine methods for total RNA preparation. Curr Protoc Mol Biol 36(1):4.2.1–4.2.9

Chapter 6

Purification of LPS from *Leptospira*

Delphine Bonhomme and Catherine Werts

Abstract

Leptospira species are one of the few spirochetes to possess a lipopolysaccharide (LPS) embedded in their outer membrane. Two protocols are currently available to extract and/or purify the leptospiral lipopolysaccharides: the rapid proteinase K method and the classical hot water/phenol extraction. The first method allows to get a quick overview of the LPS O antigen structure, whereas the second method is fitted to study the immunological properties of the leptospiral LPS. These two methods will be detailed in this chapter. Methodologies to assess the quality of the purification, such as the modified silver staining coloration, will also be reviewed. Both advantages and limitations of the different analyses will be described.

Key words *Leptospira* ssp., Lipopolysaccharide/endotoxin purification, Proteinase K, Hot water/phenol extraction, Silver staining

1 Introduction

Unlike most other spirochetes, *Leptospira* species possess surface-exposed lipopolysaccharides (LPS). These LPS, also referred to as "endotoxins," are one of the many Microbe-associated molecular patterns (MAMPs) that are recognized by the innate immune system and play an important role in the bacterial detection by the infected host. They are also essential components of the outer membrane structure and are known to play a key role in the leptospiral virulence [1].

Leptospiral LPS are composed of (1) a conserved lipidic part called lipid A and (2) a variable polysaccharide part, highly antigenic (Fig. 1). This part is composed of a succession of sugars divided between the core section and the O antigen (responsible for the antigenicity of the LPS). The lipid A is composed of a hexa-acylated lipidic part attached to a disaccharide and insures the docking of the LPS in the outer membrane. The lipid A is the part recognized by the innate immune system and often referred to as the "endotoxin." Unlike other bacteria, the lipid A of *L. interrogans* serovars Icterohaemorrhagiae, Lai, and Manilae possesses a peculiar

Nobuo Koizumi and Mathieu Picardeau (eds.), *Leptospira spp.: Methods and Protocols*, Methods in Molecular Biology, vol. 2134, https://doi.org/10.1007/978-1-0716-0459-5_6, © Springer Science+Business Media, LLC, part of Springer Nature 2020

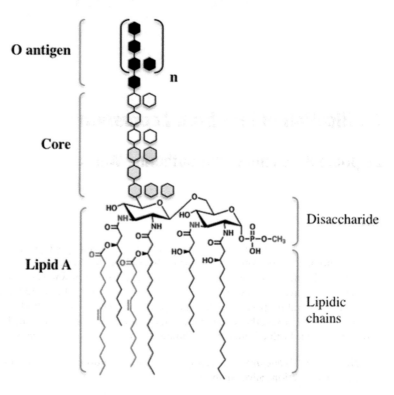

O antigen

Core

Lipid A

Disaccharide

Lipidic chains

Fig. 1 Peculiar structure of the LPS of *Leptospira interrogans*. (Adapted from Werts [3])

structure exhibiting (1) non-saturated fatty chains, (2) amide liaisons at the level of the disaccharide, and (3) a methylated 1-phosphate group and (4) lacks a 4′-phosphate group (Fig. 1) [2, 3].

As a consequence of their peculiar structural properties, the leptospiral LPS are recognized by the innate immune system in a particular manner. First of all, and unlike other bacteria, the leptospiral lipid A is not recognized by the human toll-like-receptor 4 (hTLR4) although it is recognized by the murine toll-like-receptor 4 (mTLR4) [4]. In addition, it has been shown that leptospiral LPS have the ability to activate the toll-like-receptor 2 (TLR2) [5], independently of the lipid A moiety [4]. Such activity could be attributed to the presence of lipoproteins that are known to activate TLR2 and often co-purify with the LPS.

Two methods for leptospiral LPS purification have been described: a rapid proteinase K purification and the classical hot water/phenol extraction [6]. The choice of the purification method depends on the intended use for the LPS. Indeed, the proteinase K purification is appropriate to detect differences in the O antigen. On the other hand, the hot water/phenol extraction is advised for studying the immunological properties of the LPS, since it requires purifying and separating the LPS from the rest of the bacterial membrane components. Both protocols will be detailed hereafter, and analysis methods will also be reviewed.

2 Materials

2.1 Specific Reagents for LPS Extraction and Analysis

1. Bacterial culturing medium Ellinghausen-McCullough-Johnson-Harris (EMJH): 1.0 g/L disodium phosphate, 0.3 g/L monopotassium phosphate, 1.0 g/L sodium chloride, 0.25 g/L ammonium chloride, 5 mg/L thiamine. EMJH medium is supplemented with 10 g/L serum bovine albumin, 4 mg/L heptahydrate zinc sulfate, 50 mg/L heptahydrate iron sulfate, 15 mg/L hexahydrate magnesium chloride, 15 mg/L dihydrate sodium chloride, 0.1 g/L sodium pyruvate, 0.12 mg/L B12 vitamin, 0.4 g/L glycerol, 1.25 g/L Tween 80.

2. Polypropylene centrifuge bottles of 250 mL.

3. Dulbecco's phosphate buffered saline (DPBS): 8 g/L sodium chloride, 200 mg/L potassium chloride, 1.44 g/L sodium phosphate dibasic, 240 mg/L monopotassium phosphate, pH 7.4.

4. Sodium dodecyl sulfate (SDS): 10% or more solution in H_2O.

5. Ammonium persulfate (APS): 10% solution in H_2O.

6. Precasted or hand-casted linear gradient gels for SDS-polyacrylamide gel electrophoresis (SDS-PAGE): 4% (light solution) to 15% (heavy solution) acrylamide/bisacrylamide (29.2:0.8), 0.375 M Tris–HCl pH 8.8, 5 mL/L of 10% APS, 0.5 mL/L N,N,N,N'-tetramethylethylenediamine.

7. Laemmli buffer: 4% SDS, 20% glycerol, 10% 2-mercaptoethanol, 0.004% bromophenol blue, 0.125 M Tris–HCl, pH 6.8.

8. SDS-PAGE running buffer: 25 mM Tris–HCl, 192 mM glycine, 0.1% SDS, pH 8.3.

9. Phenol (solid).

10. Endotoxin-free water.

11. Periodic acid (solid).

12. Silver nitrate: 20% solution in H_2O.

13. Citric acid (solid).

14. Proteinase K.

2.2 Specific Equipment for LPS Extraction and Analysis

1. Incubator with no agitation at 30 °C.

2. Freezer −80 °C.

3. Water bath at 68 °C located under a chemical hood.

4. Neoprene gloves.

Table 1
Composition and indicative volumes for preparation of silver stainning solutions

Name	Composition	Indicative volumes for one mini-gel	
Fixation solution	Ethanol 40% Acetic acid 5%	(1) Endotoxin-free water (2) Ethanol (>99%) (3) Acetic acid (>99%)	27.5 mL 20 mL 2.5 mL
Oxydation solution	Ethanol 40% Acetic acid 5% Periodic acid 3.5 mg/mL	(1) Endotoxin-free water (2) Ethanol (>99%) (3) Acetic acid (>99%) (4) Periodic acid (solid)	27.5 mL 20 mL 2.5 mL 175 mg
Staining solution	NaOH 0.2 N Ammonim hydroxide 1.4% Silver nitrate 0.7%	(1) NaOH 1 N (2) Ammonium hydroxide (>99%) (3) Silver nitrate (20%) (4) Endotoxin-free water	7 mL 500 µL 1.25 mL 26.25 mL
Revelation solution	Citric acid 0.2 mg/mL Formaldehyde 0.2%	(1) Endotoxin-free water (2) Citric acid (solid) (3) Formaldehyde	49.9 mL 10 mg 100 µL

5. 50 mL plastic tubes resistant to hot phenol.

6. Dialysis tubes of molecular weight cutoff: 12,000–14,000 Da.

7. Centrifuge: spinning up to 3345 × g.

8. Ultracentrifuge: spinning up to 100,000 × g.

9. 2 mL endotoxin-free glass tubes for lyophilization.

10. Lyophilization equipment: freeze dryer with temperature between −80 and −90 °C and pressure between 30 and 100 mTorr.

11. Precision balance.

2.3 Necessary Buffers for the Silver Staining Analysis

All the necessary buffers for the silver staining analysis are described in Table 1.

3 Methods

The purification protocol of the leptospiral LPS involves several toxic chemicals and takes a lot of time. It is recommended to read carefully **Notes 1–4** before starting the preparation.

3.1 Leptospira Cultures and Centrifugation

Leptospira strains are grown in liquid EMJH medium at 30 °C and with no agitation. Small volumes (1–2 mL) can be enough for the proteinase K protocol, but large volumes are used for the hot phenol procedure. For the latter, around 2–3 L of late exponential bacterial cultures are required to get enough LPS to be weighed. Such cultures can be obtained in the following manner:

1. Inoculate culture at 1×10^6 bacteria/mL to fresh EMJH medium and incubate them during 7–10 days (*see* **Note 5**).

2. Centrifuge the culture in 250 mL or 500 mL bottles at $6500 \times g$ for 25–30 min (*see* **Note 6**).

3. Resuspend the pellets in pre-warmed 37 °C endotoxin-free DPBS and pool them in a 50 mL plastic tube (*see* **Note 7**).

4. Centrifuge again to obtain a pellet that should be around 1–5×10^{12} bacteria.

5. Carefully discard the supernatant and store the pellet at −80 °C or use right away for LPS extraction.

3.2 Fast Extraction for O Antigen Analysis

The fast extraction protocol relies on the incubation of the leptospires with proteinase K to digest proteins and release the lipopolysaccharides from the bacterial membranes. This method is much faster than the classical hot/water extraction and requires less bacterial material, but does not allow the study of the immunological properties of the LPS because it is not separated from the rest of the bacterial cell wall components. Described by Murray et al. [1] for LPS mutant analyses, this technique has been widely used and modified. The protocol described hereafter is based on the original protocol modified by Eshghi and collaborators [7].

1. Resuspend the small bacterial pellet in endotoxin-free DPBS containing 0.1% SDS at the concentration of 10^9 bacteria/mL.

2. Sonicate the bacteria in ultra-sound bath for 45 s.

3. Incubate the lysate with proteinase K at a final concentration of 30 µg/mL overnight at room temperature.

4. Mix the lysate with Laemmli buffer, denature the sample at 99 °C for 10 min, and perform the silver staining analysis (described in Subheading 3.4.2) using the amount corresponding to 10^7 bacteria per well on the gel.

3.3 Classical Purification for Immunological Studies

3.3.1 Hot Water/Phenol Extraction

Like other bacterial LPS, leptospiral LPS can be purified by the classical hot water/phenol extraction described in the literature by Westphal et al. [6]. Nevertheless, LPS are usually extracted from the aqueous phase, whereas the leptospiral LPS are essentially obtained in the phenol phase because of their structural peculiarities (Fig. 2).

1. Resuspend the bacterial pellet in hot endotoxin-free water (temperature above 68 °C) (*see* **Note 8**). Add the same volume of pre-heated 90% phenol (68 °C) to the solution (*see* **Note 9**). For 5×10^{12} bacterial pellet, add 20 mL of water and 20 mL of 90% phenol.

2. Perform the extraction in a 50 mL endotoxin-free plastic tube by incubating the bacteria with the water/phenol mixture for

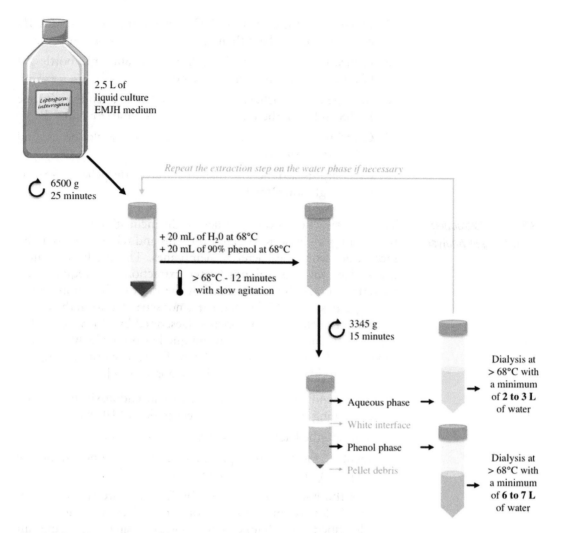

Fig. 2 Schematic protocol of the classical hot water/phenol extraction purification described by Westphal et al. [6] adapted for leptospiral LPS extraction

12 min at a temperature above 68 °C with slow agitation (hence insuring proper phenol/water miscibility).

3. Separate the different phases of the extraction by centrifuging at 3345 × *g* for 15 min at room temperature. Such centrifugation should result in obtaining three distinct phases of various densities. The lighter, whiter solution is the aqueous phase, whereas the heavier, yellower solution is the phenol phase in which the leptospiral LPS are expected. Both phases are separated by a white interface containing denatured proteins. Membrane fragments and a residual pellet of non-lysed bacteria might remain visible at the bottom of the tube (*see* **Note 10**).

4. Recover the aqueous and phenol phases without contamination by the white interface and the bacterial debris (*see* **Notes 11** and **12**).

3.3.2 Dialysis and Ultracentrifugation

1. In order to evacuate the phenol contaminant, perform dialysis of the two phases recovered from the extraction against large volumes of hot tap water (minimum 68 °C), using dialysis membranes with a molecular weight cutoff of 12,000–14,000 Da. Slow agitation is recommended for a more efficient dialysis (*see* **Note 13**). A minimum of 2–3 L should be use to dialyze the water phase and minimum of 6–7 L for the phenol phase.

2. Perform an ultracentrifugation of 3 h, $100,000 \times g$, 10 °C to remove nucleic acids and proteins from the LPS preparation.

3. Resuspend the white "flour-like" pellet obtained in large volume (25 mL minimum) of ultrapure endotoxin-free water. To ensure proper removal of the contaminants, it is important to vortex the LPS pellet thoroughly before processing to **step 4**.

4. Repeat **steps 2** and **3** three times.

5. Resuspend the pellet obtained after the last step of ultracentrifugation in 1 or 2 mL of endotoxin-free water. This LPS preparation can be used for quantitative and qualitative analysis and can be lyophilized for weighting.

3.4 "Crude" LPS Analysis

3.4.1 Quantification by Lyophilization

Up to now, the best way to quantify the amount of LPS extracted during the purification remains weighting after lyophilization. A 12 h lyophilization is in general sufficient to remove all the water from LPS purifications. However, this method allows the quantification of the LPS mass but not molarity, since the LPS is composed of a mixture of different molecules with various O antigen structures and sizes. Hence, this quantification (which is currently the only available one) is strongly biased, and such limit must be taken into consideration during experiments (*see* **Notes 14** and **15** for other methods).

1. Weight and tare an empty 2 mL endotoxin-free glass tube (*see* **Note 16**).

2. Put the resuspended LPS sample in the tube and freeze it for at least 3 h at −80 °C.

3. Remove the cap and replace it with parafilm in which small holes must be drilled. Conserve the cap in sterile environment.

4. Lyophilize the LPS for at least 12 h.

5. Recover the sample, close back the tube in a sterile environment, and weight the amount of LPS obtained (*see* **Note 17**).

6. Resuspend the LPS at concentration between 1 and 10 mg/mL for long-term storage at −20 or −80 °C.

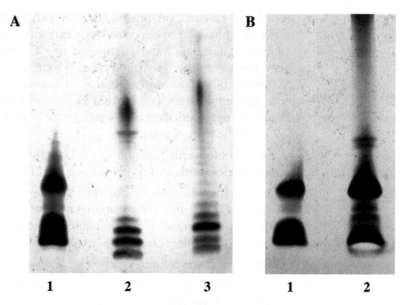

Fig. 3 Silver staining of SDS-PAGE (4–15%) of (**a**) various bacterial lipopolysaccharides, *Leptospira interrogans* serovar Icterohaemorrhagiae strain Verdun (lane 1), *Escherichia coli* O111:B4 (lane 2), and *Salmonella enterica* serotype Minnesota (lane 3), and (**b**) comparison between the classical hot water/phenol extraction (lane 1) and the fast proteinase K preparation (lane 2) of *Leptospira interrogans* serovar Icterohaemorrhagiae strain Verdun LPS

3.4.2 Qualitative Analysis by LPS Adapted Silver Staining

In order to visualize the LPS obtained after hot water/phenol extraction, it is recommended to perform a silver staining coloration of the LPS after SDS-PAGE analysis (Fig. 3). Tsai and Frasch have described a very sensitive silver staining method for LPS visualization [8]. This protocol includes a specific oxidation step with periodic acid that allows for a better coloration of the LPS (*see* **Note 18** for another method).

1. Perform the migration of the LPS samples boiled in Laemmli buffer (1:1 dilution) on SDS-PAGE using a gradient mini-gel 4–15%. Recommended amount of LPS per well is between 100 ng and 10 µg. Migrate at 80 V for 1–1.5 h (*see* **Note 19**).

2. Fix the LPS sample by incubating the gel in the fixation solution overnight at room temperature without agitation.

3. Perform the oxidation step by transferring the gel in oxidation solution for 5 min at room temperature and under slow agitation (40 rpm) (*see* **Note 20**).

4. Wash thoroughly the gel with ultrapure water (three times 15 min in 50 mL of water).

5. Prepare the staining solution: first, mix 7 mL of 1 N NaOH to 500 µL of ammonium hydroxide; then, add 1.25 mL of 20% silver nitrate and finally add 28.75 mL of ultrapure water (*see* **Note 21**). Incubate the gel in the coloration solution for

10 min at room temperature and under mild agitation (70 rpm).

6. Wash thoroughly the gel with ultrapure water (three times 10 min in 50 mL of water).

7. Incubate the gel in the revelation solution for 1–5 min at room temperature without agitation. Stop the reaction by discarding the revelation solution and replacing it with water as soon as the first bands of LPS become visible (*see* **Notes 22** and **23**).

4 Notes

1. The first dangerous step of the protocol is the centrifugation of the bacterial cultures. Indeed, the bacterial pellet is composed of around 10^{12} bacteria, and hence there is a high risk of contamination. Such step must be performed under a PSM to ensure the purity of the bacterial pellet and to prevent contamination of the experimenter. Precaution measures also include wearing safety lab coat and glasses and wearing two pairs of gloves (including one that properly covers the wrist).

2. The extraction protocol involves the use of large volumes of phenol, which is a highly toxic chemical component classified as a CMR (carcinogenic, mutagenic and reprotoxic) substance. Phenol is toxic by inhalation, toxic by ingestion, and toxic by skin absorption, corrosive, and mutagen. Extra precautions must be taken when manipulating:

   • Work exclusively under a functional chemical hood.

   • Wear a double pair of neoprene gloves.

   • Wear a cotton lab coat and safety glasses.

   In case of contact with the skin, eyes, or mouth, rinse immediately with water and consult a physician in emergency. In addition, all the phenol-contaminated wastes must be properly discarded in the CMR trashes, away from any heat-inducing material. Obviously, this note concerning the safety precautions linked to the usage of phenol is not exhaustive and one must consult the chemical data sheet associated with the product before use.

3. The silver staining analysis of the LPS requires the use of various chemicals that are toxic and corrosive. First of all, the periodic acid used to oxidize the gel is an oxidizer and is highly corrosive for the skin and eyes (always be reminded that acid must be poured in water and not the other way around). The silver nitrate solution presents the same harmful properties, but in addition, it is explosive. Once again, the safety precautions require wearing a cotton lab coat as well as gloves covering the

	MONDAY	TUESDAY	WEDNESDAY	THURSDAY	FRIDAY
Week 1	Inoculation of leptospiral cultures				
Week 2	Centrifugation of the cultures Freezing of the pellet at -80°C	Extraction of the LPS Dialysis over night	Dialysis Ultracentrifugations 1 to 2 x 3h	Ultracentrifugations 2 x 3h Storage at -80°C 3h & ON lyophilisation	Resuspension and storage at -20°C
Week 3	SDS-PAGE & ON fixation	Silver staining analysis			

Fig. 4 Example schedule of the whole process of the LPS extraction, from the bacterial culture inoculation to the silver staining analysis

wrist and safety glasses. The silver nitrate wastes must be evacuated in a dedicated appropriate container and must never be close to heating material. Once again, this note is not an exhaustive review of the necessary precautions associated with silver staining analysis, and consulting the data sheets associated with these chemicals is highly recommended.

4. To purify leptospiral LPS, one must consider the following schedule (Fig. 4):

- One to two weeks to expand the leptospiral culture.

- Half a day to perform the centrifugation and obtain the bacterial pellet.

- Three hours to perform the hot water/phenol extraction.

- One to two whole days of dialysis (timing can be reduced to less than 24 h by frequent change of the water or increasing the volumes used. Also, pre-heating the water that will be used is also a good way to gain time).

- Two days for the ultracentrifugations.

- One night for the lyophilization of the LPS.

- Two days for the silver staining analysis that required an overnight incubation.

5. The duration of the bacterial culture incubation depends on the growth rate of the strain. It is recommended to pellet the bacteria in late exponential phase.

6. The plastic bottles used for the centrifugation of the leptospiral cultures should be dedicated to such task and should not be used with any other bacteria and/or detergent to prevent any endotoxin contamination.

7. It is recommended to check before doing the purification that the 50 mL plastic tubes used are appropriate for hot phenol extraction and that there is no degradation of the plastic during the heating process.

8. The expected yield of the LPS extraction depends mainly on the quality of the pellet resuspension in hot water. This step must not be underestimated and should last at least 3–5 min depending on the size of the pellet.

9. In order to prepare the 90% phenol solution required for the extraction, the pure solid phenol must be heated to temperature above 68 °C and then mixed to endotoxin-free water. The solution must be kept in a water bath above 68 °C in order to ensure the proper water/phenol miscibility. Otherwise, the extraction will not be performed properly.

10. It is quite frequent that the two phases are inverted during the centrifugation and that the phenol phase is above the aqueous phase.

11. It is recommended to conserve all the phases obtained during the extraction to control the potential presence of LPS even in the aqueous phase.

12. If the aqueous phase is not completely transparent, and contaminated by debris, it is advised to repeat one more time the extraction step. Indicative volumes for second extraction: 10 mL of water and 10 mL of 90% phenol.

13. The dialysis tubes must be rehydrated at least 2 h prior to usage and should also be pre-warmed up to avoid breaking when thrown in hot water. They should also be washed with endotoxin-free water.

14. The limulus amebocyte lysate (LAL) test, which traditionally allows a precise quantification of endotoxins, is not useful when it comes to the quantification of the leptospiral LPS. Indeed, *Leptospira* LPS barely react with the lysate and usually exhibit negative or results 1000-fold lower than with the same amount of *E. coli* LPS [4].

15. In order to determine the number of active molecules present in the LPS sample, one possibility is to stimulate HEK293T-NFκB reporter cells transfected with the mTLR4 receptor. Indeed as described earlier, mTLR4 recognizes the lipid A part of the leptospiral LPS, and cells respond proportionally to the number of molecules present in the purification. This

indirect dosage does not allow a real quantification but can be valuable to compare several LPS purifications.

16. The glass tubes used for lyophilization should be detoxified at 256 °C for at least 4 h in order to get rid of any potential endotoxin traces that could contaminate the leptospiral LPS preparation.

17. On average, a 7-day-old stationary culture of 2.5 L that had been inoculated at around 10^6 leptospires/mL will give a $1–2 \times 10^{12}$ bacterial pellet. Such bacterial pellet should result in the extraction of 1–10 mg of LPS, depending on the strain and the quality of the extraction.

18. The bacterial lipopolysaccharides absorb light at the wavelength of 190 nm. However, the heterogeneity of the molecule makes it impossible to quantify the amount of LPS present in the preparation by measuring the absorbance. Despite these limitations, UV spectrum can be used to assess the quality of the extraction in terms of nucleic acid (260 nm) and protein (280 nm) contaminations.

19. Prefer the use of 10-well gels rather than 15-well gels, on which the migration of the LPS is often altered by the narrowness of the well.

20. Throughout the silver staining analysis, it is important to avoid touching the gel with metal component or with hands, because that will leave traces visible during the revelation of the coloration. Only touch the edges of the gel with wet gloves.

21. It is very important to respect the order of addition of the components of the solution. Also, note that a brown precipitate will form upon addition of the silver nitrate but should disappear under gentle homogenization.

22. Be careful not to over incubate the gel in the revelation solution; otherwise the background coloration will lower the quality of the visualization.

23. In general, LPS of various bacteria such as *Escherichia coli* are characterized by numerous bands of various sizes. However, in the case of *Leptospira interrogans* strain Verdun, the LPS only possess two major bands expected around 25 and 10 kDa, as well as several minor bands (Fig. 3a). Other strains of *L. interrogans* such as Manilae L495 and Copenhageni Fiocruz L1–130 have similar LPS with minor variations in the size and intensity of the bands. In addition, the differences in purity between the LPS extracted with the fast proteinase K method and the LPS purified with the classical method are visible on silver staining analysis (Fig. 3b).

References

1. Murray GL, Srikram A, Henry R, Hartskeerl RA, Sermswan RW, Adler B (2010) Mutations affecting *Leptospira interrogans* lipopolysaccharide attenuate virulence. Mol Microbiol 78 (3):701–709

2. Que-Gewirth NL, Ribeiro AA, Kalb SR, Cotter RJ, Bulach DM, Adler B et al (2004) A methylated phosphate group and four amide-linked acyl chains in *Leptospira interrogans* lipid A. The membrane anchor of an unusual lipopolysaccharide that activates TLR2. J Biol Chem 279(24):25420–25429

3. Werts C (2010) Leptospirosis: a Toll road from B lymphocytes. Chang Gung Med J 33 (6):591–601

4. Nahori MA, Fournie-Amazouz E, Que-Gewirth NS, Balloy V, Chignard M, Raetz CR et al (2005) Differential TLR recognition of leptospiral lipid A and lipopolysaccharide in murine and human cells. J Immunol 175 (9):6022–6031

5. Werts C, Tapping RI, Mathison JC, Chuang TH, Kravchenko V, Saint Girons I et al (2001) Leptospiral lipopolysaccharide activates cells through a TLR2-dependent mechanism. Nat Immunol 2(4):346–352

6. Westphal O, Lüderitz O, Bister F (1952) Über die Extraktion von Bakterien mit Phenol/Wasser. Z Naturforsch 7(3):148–155

7. Eshghi A, Henderson J, Trent MS, Picardeau M (2015) *Leptospira interrogans* lpxD homologue is required for thermal acclimatization and virulence. Infect Immun 83(11):4314–4321

8. Tsai CM, Frasch CE (1982) A sensitive silver stain for detecting lipopolysaccharides in polyacrylamide gels. Anal Biochem 119(1):115–119

Chapter 7

Isolation, Purification, and Characterization of Leptophages

Olivier Schiettekatte and Pascale Bourhy

Abstract

To date, only three bacteriophages of leptospires—leptophages—are known. Nonetheless, numerous prophages have been found in the genus, especially in the genomes of pathogenic species. Thus, some laboratories attempt to isolate leptophage particles from environmental samples or following mitomycin C induction of bacterial cultures. Here, we propose multiple procedures to isolate, purify, and characterize bacteriophages, based on protocols used for LE3 and LE4 characterization.

Key words Leptophages, Polyethylene glycol precipitation, Cesium chloride purification, TEM

1 Introduction

Three bacteriophages of leptospires have been isolated as of yet and are named "leptophages" [1]. They have been sequenced [2, 3] and morphologically characterized [1, 3]. In the last 30 years, no other leptophage has been isolated, but numerous prophages have been found in the genus, especially in the genomes of pathogenic species [3, 4], suggesting a role in pathogenicity acquisition [5]. Thus, some laboratories have attempted to isolate additional phage particles by mitomycin C inductions [6] or from environmental samples, using the method of Saint Girons et al. [1].

We propose, herein, a combination of methods for the extraction of leptophages from environmental samples and induced cultures as well as their isolation, purification, and phenotypic characterization. These protocols are based on classical bacteriophage purification methods—i.e., polyethylene glycol precipitation, plating, and TEM observations—considering *Leptospira* spp.-specific medium requirements and slow growth.

Nobuo Koizumi and Mathieu Picardeau (eds.), *Leptospira spp.: Methods and Protocols*, Methods in Molecular Biology, vol. 2134, https://doi.org/10.1007/978-1-0716-0459-5_7, © Springer Science+Business Media, LLC, part of Springer Nature 2020

2   Materials

2.1   Prophage Inductions

1. Incubator.
2. Ellinghausen-McCullough-Johnson-Harris (EMJH) medium: dissolve 2.3 g of Difco Leptospira Medium Base EMJH (Becton Dickinson) into 900 mL of distilled water, sterilize by autoclaving, and add 100 mL of Difco Leptospira Enrichment EMJH (Becton Dickinson).
3. 400 mL flasks.
4. Fresh *Leptospira* cultures.
5. Spectrophotometer.
6. Spectrophotometry cuvettes.
7. Mitomycin C: 1 mg/mL solution in sterile water.

2.2   Bacteriophage Purification

1. Polyethylene glycol 8000.
2. Sodium chloride.
3. Cesium chloride.
4. TN buffer: 10 mM Tris–HCl, pH 7.5, 100 mM NaCl.
5. Chloroform.
6. Centrifuge.
7. Ultracentrifuge.
8. Beckman Coulter rotor SW41.
9. 13 mL thin-walled polypropylene ultracentrifuge Coulter tubes.
10. Peristaltic pump and flexible plastic tubing.
11. Ultrafiltration cassette (100 kDa).
12. 3.5 kDa cutoff dialysis cassettes.
13. 0.22 μm dead-end filters.
14. Syringes (1–5 mL) with 18-G needles.

2.3   Double Agar Overlay Plaque Lysis Assay

1. EMJH medium supplemented with 0.6% and 1.2% agar.
2. Water bath (43 °C).
3. Petri dishes (90 × 15 mm).
4. Polypropylene Falcon tubes (15 mL).

2.4   Leptophage Observation

1. Electron microscope.
2. Microscopy carbon-coated copper grids.
3. Uranyl acetate: 4% solution in water.
4. Glutaraldehyde: 25% solution in water (stock solution) and 2% solution (working solution).

5. PHEM buffer (2×): 60 mM PIPES, 25 mM HEPES, pH 7.0, 10 mM EGTA, 4 mM MgSO$_4$.

6. Filter paper.

2.5 Leptophage Kinetics

1. Incubator.

2. EMJH medium.

3. 50 mL flasks.

4. Fresh *Leptospira* cultures.

5. Petroff-Hausser counting chamber.

6. TN buffer.

7. Centrifuge.

8. 0.22 μm dead-end filter.

3 Methods

3.1 Prophage Inductions in Leptospira spp.

1. Grow the *Leptospira* sp. bacterial host in EMJH flasks at 30 °C (we suggest 400 mL).

2. Measure the optical density of the culture daily at a wavelength of 420 nm.

3. When the optical density of the culture reaches 0.1 (3–7 days depending on the species), divide the culture in two identical volumes. Induce one culture with 0.05 μg/mL final concentration of mitomycin C.

4. Measure the optical density of the culture daily at a wavelength of 420 nm. If a significant difference (>50%) in optical density is observed between the control and the induced culture, this is potentially due to prophage expression.

5. Bacteriophages or induced particles are stored at 4 °C.

3.2 Bacteriophage Extraction and Purification

Two protocols exist for leptophage extraction. Tangential flow filtration is available to extract viruses from huge volumes of environmental water (*see* **Note 1**), but, due to high density of EMJH, it is not practical for the extraction of viruses from leptospiral lysates. In this case, phage recovery is facilitated by polyethylene glycol precipitation (*see* **Note 2**). If working with bacterial lysate, the supernatant should be recovered after pelleting cells and other debris at 6000 × *g* for 45 min.

3.2.1 Tangential Flow Filtration

1. Filter up to 2 L of environmental water with a 0.22 μm dead-end filter, and recover the filtrate.

2. Concentrate this sample up to 400 times with a 100 kDa tangential flow filtration cassette (*see* Fig. 1 for assembly).

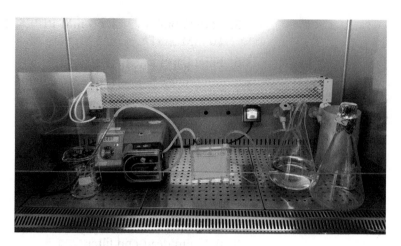

Fig. 1 Tangential flow filtration assembly

3. Add 10% final volume of chloroform to eliminate remnant cell contaminants and mix gently.

4. Centrifuge at $1700 \times g$ for 30 min at 4 °C to pellet chloroform-insoluble material.

5. Filter the supernatant with a 0.22 μm dead-end filter.

3.2.2 Polyethylene Glycol Precipitation

1. Filter 200 mL of leptospiral lysate (*see* Subheading 3.1) with a 0.22 μm dead-end filter.

2. Add 8% polyethylene glycol 8000 and 0.3 M NaCl (final concentrations).

3. Mix thoroughly to dissolve polyethylene glycol.

4. Incubate overnight at 4 °C.

5. Centrifuge at $24,000 \times g$ for 45 min at 4 °C.

6. Solubilize the pellet in a minimal volume of TN buffer for higher yields, and we recommend 5 mL.

3.2.3 Phage Amplification

1. Grow *Leptospira* sp. bacterial host in EMJH flasks at 30 °C (we suggest 400 mL).

2. When the optical density of the culture reaches 0.2, divide the culture in two identical volumes. Induce one culture with 5 mL of previously extracted phage (*see* Subheading 3.2.1 or 3.2.2) and the other culture with the same volume of TN buffer (control culture).

3. Measure the optical density of the culture daily at a wavelength of 420 nm. If a significant difference (>50%) in optical density is observed between the control and the induced culture, this is potentially due to phage amplification.

4. Repeat the PEG extraction and/or isolate phages with double agar overlay plaque lysis assay (*see* Subheading 3.3).

3.2.4 Cesium Chloride
Gradient
Ultracentrifugation

A purification step by cesium chloride gradient ultracentrifugation is recommended for particle characterization (TEM or mass spectrometry) but should be avoided prior to infection of bacteria because cesium chloride is toxic to leptospires.

1. Add 45% final concentration cesium chloride to sample.

2. Ultracentrifuge between 110,000 and 260,000 × g for 24 h at 10 °C in an ultracentrifuge tube.

3. Recover formed bands with a syringe (*see* **Note 3**).

4. Dialyze the sample with TN buffer with a 3.5 kDa cassette.

3.3 Double Agar
Overlay Plaque
Lysis Assay

3.3.1 Phage Isolation and
Counting

1. Prepare Petri dishes containing 20 mL EMJH with 1.2% agar and wait for complete solidification.

2. Prepare an equivalent number of Falcon tubes containing 5 mL EMJH with 0.6% agar, stored in 43 °C water bath to prevent solidification before plating.

3. Incubate 500 μL of exponential phase *Leptospira* sp. culture with 10 μL of different dilutions of an extract of leptophages, at 30 °C for 15 min.

4. Mix these co-cultures with 5 mL EMJH with 0.6% agar, and pour immediately over 1.2% agar EMJH plates. Work efficiently, as the overlays harden quickly.

5. Incubate at 30 °C and check every 2 days for the appearance of lysis plaques (*see* Fig. 2). The number of plaques formed is proportional to the bacteriophage concentration in the extract, considering the dilution used.

3.3.2 Fast Dropping

1. Prepare Petri dishes containing 20 mL EMJH with 1.2% agar and allow for complete solidification.

Fig. 2 LE4 leptophage plaque lysis on *Leptospira biflexa* overlay

2. Add 5 mL EMJH with 0.6% agar as an overlay and wait for complete solidification. Store dishes at 4 °C up to 1 month.

3. Spread 1 mL of exponential phase *Leptospira* sp. culture on a room temperature overlay medium.

4. Incubate at 30 °C for 2 h.

5. Remove any remaining liquid with a micropipette.

6. Leave the dish open under a sterile biosafety cabinet up to 15 min to allow the small volume of remnant liquid to dry.

7. (a) For screening: Drop 10 µL of each purified phage sample to test for clearing.

   (b) For raw counting: Drop 10 µL of serial dilutions of the bacteriophage extract to titer.

8. Incubate at 30 °C and check every 2 days for the appearance of lysis plaques. The number of plaques is proportional to the concentration of bacteriophages in the extract, considering the dilution used (*see* **Note 5**).

3.4 Leptophage Transmission Electron Microscopy

3.4.1 Induced Particle Morphological Characterization

1. Deposit a carbon-coated copper grid on a 15 µL drop of phage sample for 15 min.

2. Deposit the grid on a 30 µL drop of 4% uranyl acetate solution for 3 s for washing and staining.

3. Deposit the grid on another 30 µL drop of 4% uranyl acetate solution for 8 s for staining.

4. Dry the grid with a filter paper; take care to remove any remnant liquid.

5. Observe the grid with a transmission electron microscope at a magnification of 23,000 for screening and 49,000 for characterization (*see* Fig. 3a).

3.4.2 Infected Bacteria Observation

1. Deposit a carbon-coated copper grid on a 15 µL drop of phage-infected *Leptospira* culture (we suggest a MOI of 10) at the desired infection time, for 15 min.

2. Deposit the grid on a 15 µL drop of 2% glutaraldehyde in PHEM buffer for 10 min.

3. Deposit the grid successively on four 30 µL drops of distilled water for washing.

4. Deposit the grid on a 30 µL drop of 4% uranyl acetate solution for 3 s for washing and staining.

5. Deposit the grid on another 30 µL drop of 4% uranyl acetate solution for 8 s for staining.

6. Dry the grid with a filter paper; take care to remove any remnant liquid.

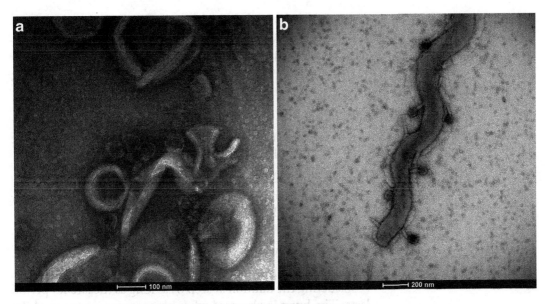

Fig. 3 LE4 bacteriophages (**a**) in a lysate and (**b**) with the bacterial host

7. Observe the grid with a transmission electron microscope at a magnification of 23,000 for screening and 49,000 for characterization (*see* Fig. 3b).

3.5 Leptophage Kinetics

3.5.1 Phage Adsorption Assays

1. Grow *Leptospira* sp. bacterial host in EMJH flask at 30 °C.

2. When the culture reaches exponential phase, dilute the bacteria to 10^8/mL in EMJH.

3. Infect this culture with a final concentration of 10^4 bacteriophages per mL (MOI of 0.0001) or with the same volume of buffer (control culture).

4. Prepare a control dilution of the bacteriophage at 10^4 bacteriophages per mL (without bacteria).

5. Sample 1 mL of the co-culture every 30 min for up to 6 h, and filter immediately with a 0.22 μm dead-end filter. Sample 1 mL of control culture and control dilution at times 0 and 6 h, and filter with a 0.22 μm dead-end filter.

6. Count the lytic bacteriophages from this sample with the double agar overlay plaque lysis assay (*see* Subheading 3.3).

7. The number of phages found in the non-infected control culture should stay at 0, and the concentration of phages in the control dilution should remain close to 10^4/mL and corresponds to the percent lost in the filter.

8. In the experimental co-culture, the number of recovered phages should decrease until the first burst. This decrease corresponds to the proportion of phages adsorbed on host cells (and retained in the 0.22 μm filter).

3.5.2 One-Step Growth Kinetics

1. Grow *Leptospira* sp. bacterial host in EMJH flask at 30 °C.

2. When the culture reaches exponential phase, standardize dilute with EMJH to 10^8 bacteria/mL in a final volume of 10 mL.

3. Infect this culture with a final concentration of 10^7 bacteriophages per mL (MOI of 0.1).

4. Incubate at 30 °C for 30 min.

5. Centrifuge the mixture at $6000 \times g$ for 30 min to remove non-adsorbed phages.

6. Resuspend the supernatant in 100 mL EMJH.

7. Each 30 min, sample 1 mL of each culture or control, and filter immediately with a 0.22 μm dead-end filter.

8. Count the lytic bacteriophages with the double agar overlay plaque lysis assays.

9. The number of phages should remain approximately 10^6/mL (considering the dilution at **step 6**) as long as the latency period lasts. The burst is observed when this number grows, and the percent increase corresponds to the number of phages produced per infected cell (burst size).

4 Notes

1. To date, leptophages have only been isolated from water sources contaminated with *Leptospira* species [1]. But soils [7] and organs of infected animals should also be considered.

2. For environmental samples, ultrafiltration can handle larger volumes than PEG purification with higher yields.

3. If no band is found, or if bacteriophages are not recovered from the bands, a systematic sampling and TEM observation of different layers of the gradient may be necessary.

4. The double agar overlay plaque lysis assay is required to isolate phage clones, but samples can be pretested in a fresh liquid culture as a preliminary assay.

5. Fast dropping is a quick counting or screening method, especially when testing numerous samples, but can only be used to elucidate the order of magnitude of the virus concentration.

Acknowledgments

This work was part of the Ph.D. thesis of O.S. who received financial support from "Université Paris Diderot" and "Sorbonne Paris Cité." We thank Soizick Lucas-Staat and Nicolas Dufour for technical assistance during leptophage extractions, Olivier Gorgette for technical assistance during the TEM observations, and Robert A. Gaultney for proofreading the manuscript.

References

1. Saint Girons I, Margarita D, Amouriaux P, Baranton G (1990) First isolation of bacteriophages for a spirochaete: potential genetic tools for Leptospira. Res Microbiol 141(9):1131–1138. https://doi.org/10.1016/0923-2508(90)90086-6

2. Bourhy P, Frangeul L, Couve E, Glaser P, Saint Girons I, Picardeau M (2005) Complete nucleotide sequence of the LE1 prophage from the spirochete *Leptospira biflexa* and characterization of its replication and partition functions. J Bacteriol 187(12):3931–3940. https://doi.org/10.1128/JB.187.12.3931-3940.2005

3. Schiettekatte O, Vincent AT, Malosse C, Lechat P, Chamot-Rooke J, Veyrier FJ, Picardeau M, Bourhy P (2018) Characterization of LE3 and LE4, the only lytic phages known to infect the spirochete Leptospira. Sci Rep 8(1):11781. https://doi.org/10.1038/s41598-018-29983-6

4. Fouts DE, Matthias MA, Adhikarla H, Adler B, Amorim-Santos L, Berg DE, Bulach D et al (2016) What makes a bacterial species pathogenic?: comparative genomic analysis of the genus Leptospira. PLoS Negl Trop Dis 10(2):e0004403

5. Picardeau M (2017) Virulence of the zoonotic agent of leptospirosis: still terra incognita? Nat Rev Microbiol 15:297–307. https://doi.org/10.1038/nrmicro.2017.5

6. Zhu W-N, Huang L-L, Zeng L-B, Zhuang X-R, Chen C-Y, Wang Y-Z, Qin J-H, Zhu Y-Z, Guo X-K (2014) Isolation and characterization of two novel plasmids from pathogenic *Leptospira interrogans* serogroup canicola serovar canicola strain Gui44. PLoS Negl Trop Dis 8(8):e3103. https://doi.org/10.1371/journal.pntd.0003103

7. Thibeaux R, Geroult S, Benezech C, Chabaud S, Soupé-Gilbert M-E, Girault D, Bierque E, Goarant C (2017) Seeking the environmental source of Leptospirosis reveals durable bacterial viability in river soils. PLoS Negl Trop Dis 11(2):e0005414. https://doi.org/10.1371/journal.pntd.0005414

Chapter 8

Creating a Library of Random Transposon Mutants in *Leptospira*

Christopher J. Pappas, Hui Xu, and Md A. Motaleb

Abstract

Generation of a random transposon mutant library is advantageous in *Leptospira* as site-directed mutagenesis remains a challenge, especially in pathogenic species. This procedure is typically completed by transformation of *Leptospira* with a *Himar1* containing plasmid via conjugation with *Escherichia coli* as a donor cell. Here we describe the methodology to generate random transposon mutants in the saprophyte *Leptospira biflexa* via conjugation of plasmid pSW29T-TKS2 harbored in *E. coli* β2163. Determination of transposon insertion site by semi-random nested PCR will also be described. A similar methodology may be employed to generate Tn mutants of pathogenic *Leptospira* species.

Key words Spirochete, *Leptospira*, *Himar1*, Transposon library, Mutagenesis, Conjugation, Nested PCR

1 Introduction

The genus *Leptospira* belongs to the phylum *Spirochaetes*, which consists of saprophytic species (e.g., *L. biflexa*), intermediate species (e.g., *L. licerasiae)*, and pathogenic species (e.g., *L. interrogans*), of which the latter are primarily responsible for the disease leptospirosis [1–3]. Leptospirosis is an emerging zoonotic disease of global importance, with over 1.7 million cases of severe disease occurring annually [4]. To facilitate our understanding of the biology of *Leptospira*, such as virulence factors, host adaptation mechanisms, or its physiology, effective genetic manipulation tools are required [1, 5–9].

Studies of *Leptospira* genes completed in the 1990s were primarily focused on functional complementation of *Escherichia coli* mutants. This led to the identification and characterization of *Leptospira* genes *recA*, *rfb*, *asd*, and *trpE* [10–13]. Later, electroporation and conjugation were both successfully applied to introduce DNA (by suicide or replicative plasmids) into saprophytic and pathogenic *Leptospira* species with selective antibiotic markers of

Nobuo Koizumi and Mathieu Picardeau (eds.), *Leptospira spp.: Methods and Protocols*, Methods in Molecular Biology, vol. 2134, https://doi.org/10.1007/978-1-0716-0459-5_8, © Springer Science+Business Media, LLC, part of Springer Nature 2020

kanamycin, spectinomycin, or gentamicin [14–21]. However, site-directed mutagenesis studies of *Leptospira* have remained very limited, such as to *flaB*, *recA*, and *hemH* genes in *L. biflexa* [22–24]; *metY*, *metX*, *metW*, and *trpE* in *L. meyeri* [15, 25]; and to a limited number of genes in pathogenic *L. interrogans* species [26].

Alternatively, transposon-based mutagenesis is a powerful tool for the study of bacterial physiology and pathogenesis through random DNA insertions into a chromosome [27]. *Himar1* transposon of the *mariner* family, which does not require host cofactors to randomly insert into a TA dinucleotide within a chromosome, has been used in a variety of prokaryotes, including the random Tn library generation of *Borrelia burgdorferi*, the spirochete that is the etiologic agent of Lyme disease [28–34]. Recently, *Himar1* has been adapted for use with *Leptospira* to create a large number of Tn mutants within saprophytic as well as pathogenic species, in which the latter has proven more challenging to generate [16–18]. Previous research has developed *Himar1* insertion into *Leptospira* by conjugation [17]. As an example, conjugative plasmid pCjTKS1 can deliver *Himar1* more efficiently than electroporation, with a DNA transfer frequency ranging from 1×10^{-6} to 8.5×10^{-8} transconjugants/recipient cell in *L. biflexa* and *L. interrogans*, respectively [17].

Here, we describe the methodology to generate random transposon mutants in the saprophyte *L. biflexa* via conjugation through *E. coli* β2163 as a donor cell harboring pCjTKS1-derived conjugative plasmid pSW29T-TKS2. Extensive libraries of saprophytic and pathogenic *Leptospira* Tn mutants created by this method could be used for studying the phenotypes of diverse Tn gene mutations including but not limited to amino acid biosynthesis, motility, iron acquisition [16, 35], virulence [36, 37], and sensitivity to ethidium bromide [38], among others. Additionally, the generation of Tn libraries could foster and facilitate collaborations between laboratories investigating different aspects of *Leptospira* physiology through sharing the Tn mutant libraries. For example, collaborations can valuably reduce the substantial resource and labor expenditure associated with the generation, sequencing, and upkeep of the library. Further, such collaborations foster a positive research and learning environment between laboratories, ultimately supporting research to transcend institutional borders.

2 Materials

Use only sterilized/autoclaved glassware for all procedures. Set aside specific glassware for EMJH preparation and *Leptospira* work. Do not use general glassware supplies as these may be contaminated with other chemicals to which *Leptospira* are sensitive.

2.1 Preparation for Conjugation

1. EMJH liquid medium: Prepare 100 mL of filter-sterilized bovine serum albumin (BSA) supplement which consists of pre-autoclaved ultrapure (18 MΩ at 25 °C) water supplemented with 10% fraction V BSA, 0.01% calcium chloride, 0.01% magnesium sulfate, 0.004% zinc sulfate, 0.0003% copper sulfate, 0.005% ferrous sulfate, 0.00002% vitamin B12, 1.25% Tween-80, and pH adjusted to 7.4 with 2 N NaOH. Add 100 mL of BSA supplement to 890 mL of autoclaved EMJH base, which consists of 886.6 mL of ultrapure water supplemented with 2.3 g of Difco Leptospira Medium Base EMJH (Becton Dickinson), 3.4 mL of 100% glycerol stock, and 10 mL of heat-inactivated rabbit serum [39, 40] (*see* **Notes 1–3**).

2. EMJH solid medium: 100 mL BSA supplement pre-warmed to 50 °C is added to 890 mL of warm EMJH base (886.6 mL of ultrapure water supplemented with 2.3 g of Difco Leptospira Medium Base EMJH (Becton Dickinson), 3.4 mL of 100% glycerol stock, and 1% noble agar prior to autoclaving) and 10 mL of heat-inactivated rabbit serum. Solution of 20–30 mL is poured into 100 mm × 15 mm Petri plates (*see* **Note 4**).

3. Luria-Bertani (LB) liquid medium: 10% tryptone, 5% yeast extract, 10% sodium chloride in distilled water. Autoclave prior to use.

4. Diaminopimelic acid (DAP) solution: 100 mM DAP in autoclaved ultrapure water, filter-sterilized.

5. Kanamycin solution: 50 mg/mL of kanamycin in autoclaved ultrapure water, filter-sterilized.

6. *Leptospira* biflexa serovar Patoc strain Patoc1 (ATCC 23582/ Paris) as the recipient cell (*see* **Notes 5** and **6**).

7. *E. coli* strain β2163 as the donor cell.

8. Plasmid pSW29T-TKS2 or other (*see* **Note 7**).

9. Biosafety cabinet type 2A (minimum) with UV disinfection cycle.

10. Dark-field microscope.

11. Spectrophotometer with detection available at 420 nm.

2.2 Conjugation

1. Glass filter apparatus consisting of 125 mL filter flask, filter holder base with frit and stopper 25 mm diameter, 15 mL glass funnel 25 mm diameter, clamp for 25 mm diameter filter holder (*see* **Note 8**).

2. Filter forceps, blunt end.

3. Membrane filter, hydrophilic PVDF type size 0.1 μm pore size, 25 mm diameter (*see* **Note 9**).

4. Aspiration system: in-line or bioreagent vacuum pump and tubing (*see* **Notes 10** and **11**).

5. Glass beads for cell culture plating, cleaned and autoclaved.

6. 15 mL conical tubes.

2.3 Isolating Clones in Solid Media and Long-Term Storage

1. Sterile mixed cellulose ester syringe filters, 0.22 μm pore size, 33 mm diameter (*see* **Note 12**).

2. 5 mL syringes without needle attached.

3. Glycerol solution: 60% glycerol in ultrapure water, autoclaved.

4. −80 °C and/or −160 °C ultralow temperature freezer(s).

5. 15 mL conical tubes.

2.4 Nested PCR Using a Semi-random Primer to Determine Transposon Insertion Site

1. Genomic DNA purification kit.

2. UV/vis spectrophotometer.

3. Standard PCR reagents.

4. Primers (Table 1).

5. Thermal cycler.

6. 10× Tris/borate/EDTA (TBE) running buffer solution: 0.89 M Tris base, 0.89 M boric acid, 20 mM EDTA in distilled water, pH 8.3.

7. Agarose gel: 0.8% agarose in 1× TBE solution.

8. 6× DNA loading dye solution: 40% sucrose, 0.25% xylene cyanol FF, 0.25% bromophenol blue in distilled water.

9. Nucleic acid staining solution.

10. Electrophoresis power supply.

11. DNA molecular weight marker.

12. Nucleic acid gel documentation system.

13. Capillary DNA sequencer or DNA sequencing outsourced to a commercial sequencing company.

3 Methods

All procedures should be performed within a biosafety cabinet type 2A to reduce risk of contamination of transconjugants and for biosafety. The random transposon mutagenesis protocol and conjugation plasmids were originally described and generously donated by Dr. Bourhy and Dr. Picardeau, respectively [16, 17, 41].

Table 1
Primers used for Tn insertion site determination

Primer name	Sequence (5′ → 3′)	Purpose	Expected size
VTK1-TKS2	GATCATGATATCGATTACAAGG	Transposon-specific sequence	258 bp
VTK2-TKS2	GTTAATATTTTGTTAAAATTCGCG		
LEPBIa0024F	AATATAAAACTCTCGATTCGCTTAACA	Amplification of LepBIa0024 in *L. biflexa*	500 bp
LEPBIa0024R	TCACAATTGCTTGTGTCAACC		
Deg1	GGCCACGCGTCGACTAG TACNNNNNNNNNNGATAT	Semi-random degenerate primer for first round of nested PCR with 5′ Tag	N/A
Deg2	GGCCACGCGTCGACTAG TACNNNNNNNNNNTCTT		N/A
TnK1	CTTGTCATCGTCATCCTTG	Transposon-specific primer for first round of nested PCR	N/A
TnK2	GTGGCTTTATTGATCTTGGG	Transposon-specific primer for first round of nested PCR	N/A
TnKN1	CGTCATGGTCTTTGTAGTCTATGG	Transposon-specific primer for second round of nested PCR	Varies
TnKN2	TGGGGATCAAGCCTGATTGGG	Transposon-specific primer for second round of nested PCR	Varies
Tag	GGCCACGCGTCGACTAGTAC	Tag sequence of semi-random degenerate primers for second round of nested PCR	Varies

3.1 Preparation for Conjugation

1. One to two weeks prior to conjugation, aliquot *L. biflexa* into 25 mL of fresh EMJH liquid media in a 50 mL conical, and incubate at 30 °C with or without agitation until bacteria reach mid-late exponential growth (2–3×10^8 cells/mL) (*see* **Note 13**).

2. One day prior to conjugation, aliquot *E. coli* β2163 harboring pSW29T-TKS2 from freezer culture into 5 mL of fresh LB liquid medium adulterated with 0.3 mM DAP and 50 µg/mL of kanamycin. Incubate overnight at 30 °C (*see* **Note 14**).

3. The morning of conjugation, transfer 0.5 mL of the *E. coli* incubated overnight into 4.5 mL of fresh EMJH liquid medium adulterated with 0.3 mM DAP (without kanamycin) (*see* **Notes 15–17**).

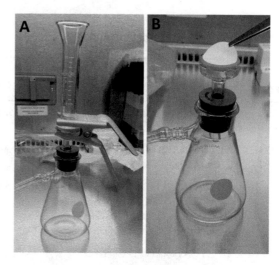

Fig. 1 (**a**) Complete filtration apparatus with membrane between frit and funnel. (**b**) Demonstration of applying a membrane filter to the frit and stopper of the filtration apparatus. Ensure aspiration is applied to system to hold membrane in place and secure funnel to top with clamp

4. Incubate *E. coli* at 37 °C with or without agitation until it reaches OD_{420} of 0.2–0.5, which correlates to mid-exponential growth (typically 2–4 h) (*see* **Note 18**).

5. Check *E. coli* and *Leptospira* by dark-field microscopy to ensure both samples demonstrate good motility, are not contaminated, and are in mid-exponential growth (e.g., approximately $2–3 \times 10^8$ cells/mL) before proceeding to conjugation.

3.2 Conjugation: Day 1

1. Set up filtration apparatus within the biosafety cabinet (Fig. 1a). Apply a vacuum to the apparatus by using an in-line or vacuum pump aspiration unit (*see* **Notes 19** and **20**).

2. Sandwich membrane filter between filter holder base with frit and stopper and glass funnel. Hold this system in place using the clamp to create a temporary seal between the glass pieces (Fig. 1b).

3. Add 1:10 ratio of *E. coli* containing EMJH media to *Leptospira* containing EMJH media by placing both into the filtration apparatus via the 15 mL funnel (e.g., 0.5 mL *E. coli* to 4.5 mL of *Leptospira*) (*see* **Notes 21** and **22**).

4. Allow all media to flow through membrane filter into filtration flask, depositing *E. coli* and *Leptospira* on the upper facing surface of the membrane filter (*see* **Note 23**).

5. Once flow-through is complete and no bubbles/liquid remains on membrane filter surface, remove clamp and funnel.

6. Remove membrane filter with deposited bacteria from glass frit filter using forceps (Fig. 2) (*see* **Note 24**).

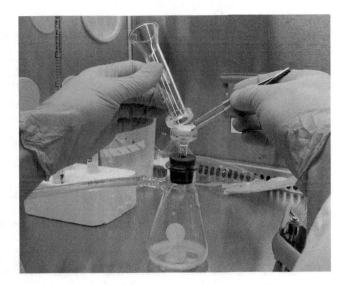

Fig. 2 Removing membrane from glass frit filter with stopper. Note how 15 mL funnel is placed at 45° angle to frit for purchase while removing membrane filter so that membrane filter doesn't jump due to static or vacuum pressure. (*Photo by Natalia Lizardo, Manhattanville College*)

Fig. 3 How to position seven membrane filters on one EMJH solid media plate. Bacteria deposited on membrane filter are visible as a circular sheen in the middle of each membrane filter. No bubbles are present between membrane filters and solid media to allow transfer of constituents to bacteria during conjugation

7. Place membrane filter on an EMJH media plate with 0.3 mM DAP (without kanamycin) with the bacteria-containing side facing up/away from the media (Fig. 3) (*see* **Notes 25** and **26**).

8. Repeat this filtration procedure with additional media and membrane filters until *Leptospira*-containing EMJH media is exhausted (*see* **Note 27**).

Fig. 4 Right: Three conjugation plates wrapped in Parafilm (end of conjugation day 1). Left: aluminum foil wrapped conjugation plates to be incubated for 10–50 days (end of conjugation day 2). Incubator is set to 30 °C

9. Once membrane filters have been placed, Parafilm Petri plate and incubate for approximately 16–20 h (overnight) at 30 °C to allow conjugation to occur (Fig. 4) (*see* **Note 28**).

3.3 Conjugation: Day 2

1. The next day, pre-warm to 30 °C a sufficient supply of EMJH liquid media (without kanamycin, without DAP) and EMJH plates (with 40 µg/mL of kanamycin) for plating (*see* **Note 29**).

2. Transfer a pair of membrane filters deposited with bacteria into a 15 mL conical tube with forceps (*see* **Note 30**).

3. Add 0.8 mL of pre-warmed EMJH (without kanamycin, without DAP) to the 15 mL conical tube. Using a handheld 1 mL serological pipette as a tool, move each membrane filter one at a time into the media.

4. Using the 1 mL serological pipette, create a gentle plunging/up-and-down movement of the membrane filter into the media to transfer the bacteria from the membrane filter to the EMJH liquid media followed by briefly vortexing (Fig. 5) (*see* **Note 31**).

5. Repeat this procedure with additional pairs of membrane filters into different 15 mL conical tubes until membrane filter supply is exhausted (e.g., if you started with ten membrane filters, you should now have five 15 mL conical tubes with membrane filters).

6. Incubate 15 mL conical tubes for 1 h at 30 °C with agitation.

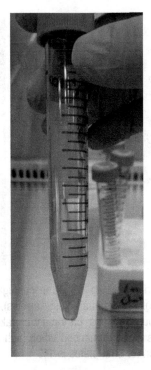

Fig. 5 15 mL conical with two membrane filters in 0.8 mL of EMJH liquid medium (without kanamycin, without DAP). This will be incubated for 1 h at 30 °C with agitation

7. After 1 h, transfer liquid EMJH media from one tube equally between at least three EMJH plates with 40 µg/mL of kanamycin (i.e., approximately 250–300 µL liquid media per plate). Agitate to uniformly spread bacteria across plate using glass beads.

8. Remove beads from plates.

9. To avoid desiccation, wrap plates in aluminum foil and place at 30 °C for at least 7–10 days (Fig. 4). After this time period, periodically check plates for colonies under aseptic conditions for up to for 50 days. *Leptospira* colonies may appear on the plates as small, medium, or large translucent circles within the solid media (Fig. 6) (*see* **Note 32**).

3.4 Isolating Clones from Solid Media and Long-Term Storage

1. Once colonies began to appear on plates, isolate them from solid media by using a 1 mL filter pipette tip attached to a micropipettor by piercing the media where the colony is present (i.e., a stab). The stab will collect a cylindrical column of agar with the transconjugant *Leptospira* within it (Fig. 6) (*see* **Notes 33–35**).

2. Transfer this isolated colony/stab into 3–5 mL of pre-warmed EMJH liquid media containing 40 µg/mL of kanamycin within 15 mL conical tubes.

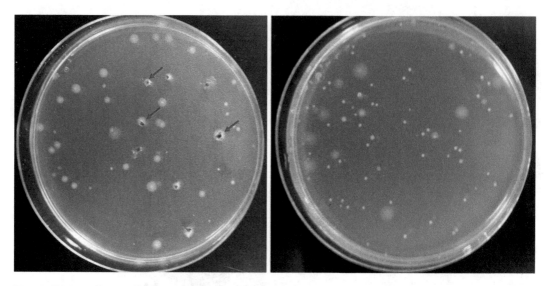

Fig. 6 EMJH solid medium plates with transconjugants. Left: Plate without 1% rabbit serum showing transconjugants with eight colony stabs (e.g., arrows). Right: Plate enriched with 1% rabbit serum showing transconjugants of various motilities. Care must be taken to ensure colony stabs are clonal; plating on more than three EMJH plates is recommended when high transformation efficiency/high density is anticipated. Noble agar concentration for both plates is 1%

3. Incubate the tubes for 5–28 days (growth rate is dependent upon which gene has been disrupted) at 30 °C with or without agitation until *Leptospira* growth is observed as turbid media.

4. Once samples become turbid, distribute a 10 μL aliquot onto a microscope slide to confirm *Leptospira* culture is not contaminated by dark-field microscopy (*see* **Note 36**).

5. If not contaminated and growth is robust and at mid- to late-exponential growth, sample can be labeled (e.g., Tn1, Tn2, etc.) and used for downstream applications and long-term storage at −80 °C or −160 °C (*see* **Notes 37–39**).

3.5 Nested PCR Using a Semi-random Primer to Determine Transposon Insertion Site

To determine transposon inserted site, nested PCR using a semi-random primer is completed as previously described [18, 42].

1. Purify genomic DNA from 1–2 mL of non-contaminated Tn *Leptospira* sample (*see* **Note 40**).

2. *Optional*: Use purified genomic DNA to confirm the sample is the correct *Leptospira* species and/or harboring the transposon by running PCR reactions for a species-specific gene (e.g., LEPBIa0024F and R for *L. biflexa*) or PCR for a transposon-specific primer set (e.g., VTK1-TKS2 and VTK2-TKS2) (Table 1).

3. Set up a mix for round 1 of nested PCR by using semi-random primers with the following reagents and quantities per reaction tube (total volume = 24.55 μL): 2.5 μL of 10× Taq polymerase

buffer without MgCl$_2$, 1.5 μL of 25 mM MgCl$_2$, 0.5 μL of 25 mM dNTPs, 0.6 μL of 10 μM oligo 1 (TnK1 or TnK2), 0.2 μL of 100 μM oligo 2 (Deg1 or Deg2), 0.25 μL of Taq DNA polymerase, 17 μL of nuclease-free water, 2 μL of template DNA (40 ng total) (*see* **Note 41**).

4. Run round 1 with the following reaction times and temperatures: Step 1, denature at 95 °C for 5 min; step 2, denature at 95 °C for 15 s, anneal at 45 °C for 1 min, extend at 72 °C for 2 min, and repeat step 2 39 additional times; and step 3, extend at 72 °C for 10 min (*see* **Note 42**).

5. Set up a mix for round 2 of nested PCR by using semi-random primers with the following reagents and quantities per reaction tube (total volume = 25 μL): 2.5 μL of 10× Taq polymerase buffer without MgCl$_2$, 1.5 μL of 25 mM MgCl$_2$, 0.5 μL of 25 mM dNTPs, 0.25 μL of 100 μM oligo 3 (TnKN1 or TnKN2), 0.25 μL of 100 μM oligo 4 (Tag), 0.2 μL of Taq DNA polymerase, 19 μL of nuclease-free water, 0.8 μL of PCR round 1 DNA (*see* **Note 43**).

6. Run round 2 with the following reaction times and temperatures: Step 1, denature at 95 °C for 5 min; step 2, denature at 95 °C for 15 s, anneal at 53 °C for 30 s, extend at 72 °C for 2 min, and repeat step 2 34 additional times; and step 3, extend at 72 °C for 10 min.

7. Run nested PCR products on a 1× TBE gel which will visualize multiple bands typically ranging from 100 to 1000 bp in length (Fig. 7) (*see* **Notes 44** and **45**).

L. biflexa Tn Mutants

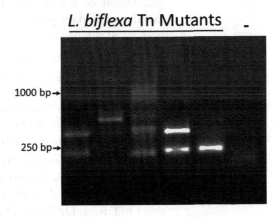

Fig. 7 Example of semi-random primer nested results after round 2. Lanes 1–5: 5 *L. biflexa* Tn mutants. Lane 6: negative control. For this experiment, round 1 primers used were TnK1 and Deg1. Round 2 primers used were TnKN1 and Tag. Electrophoresis completed using a 0.8× TBE gel adulterated with RedSafe, visualized with a Bio-Rad Gel Doc EZ System

8. Send round 2 nested PCR product for Sanger sequencing analysis using oligo 3 as the sequencing primer (i.e., either TnKN1 or TnKN2) (*see* **Note 46**).

9. Use sequencing results to run a BLAST analysis to determine transposon insertion site (*see* **Notes 47** and **48**).

10. Confirmation of transposon insertion site is conducted by running standard PCR with primers which flank approximately 100 bp proximal and distal to insertion site, which should increase PCR product corresponding to size of transposon insertion.

4 Notes

1. Ensure EMJH base has cooled to at least 50 °C before adding BSA supplement or rabbit serum.

2. Can also add 0.0001% superoxide dismutase (SOD), 1% lactalbumin hydrolysate, and 0.004% sodium pyruvate to further enrich media and improve growth of more sensitive Tn mutants.

3. Saprophytic *Leptospira* which may be present in municipal water are thin enough to fit through a 0.22 μm filter membrane. Therefore, we autoclave our ultrapure water before addition of BSA and other growth factors to ensure no microorganisms contaminate the medium.

4. Please bear in mind BSA enrichment supplement should not be warmed above 50–55 °C as BSA can denature.

5. A low passage, verified, or sequenced strain should be used for generation of the transposon mutant library. This helps to avoid a secondary phenotype as a result of SNPs that may be present in the parental strain [43].

6. Preliminary studies in our lab of transformation efficiency of *Himar1* via conjugation with the saprophytic strains *L. meyeri* and *L. yanagawae* did not result in a high yield of transconjugants. Therefore, these strains are not recommended until expertise of this protocol is completed first with the highly transformable strain *L. biflexa*.

7. pSW29T-TKS2 contains the *Himar1* random transposon system which confers kanamycin resistance for transconjugant selection along with a c9 hyperactive transposase, Ori R6K and OriT RP4. OriT and c9 are located outside of the transposable element. Other plasmids carrying the *Himar1* system are also useful for random transposon mutagenesis, such as pAL-614 [44], which confers spectinomycin resistance. However, kanamycin resistance-conferring transposons are

preferred for Tn library generation as spectinomycin resistance can occur spontaneously in *Leptospira* [19].

8. We recommend the Millipore filter apparatus system as it works well with 25 mm diameter filter membranes. These items are 125 mL filter flask item no. 1002505, filter holder base with frit and stopper 25 mm size item no. 1002502, 15 mL glass funnel 25 mm size item no. 1002514, and clamp for 25 mm filter holder item no. 1002503.

9. We recommend Millipore Durapore membrane filter item no. VVLP02500. Be certain to use a 0.1 μm pore size filter, as *Leptospira* may pass through 0.22 μm pores.

10. We use a Welch 2546B pump (which features an inlet water trap) which is attached to an intermediate 0.2 μm Whatman Polydisc in-line filter between flask and vacuum pump.

11. We recommend buying silicone tubing that can be autoclaved, since this decreases risk of contamination outside of the filtration system as well as decreases risk of equipment contamination.

12. We have found that some filters are less effective for filtering *Leptospira*. We recommend Millipore Millex-GS filters.

13. This growth can be from resurrection or from low-passage *L. biflexa*. *L. biflexa* can be grown at 30 °C or room temperature (18 °C) to allow flexibility of conjugation experiment to your work schedule. If grown at room temperature, we recommend to move *L. biflexa* to 30 °C at least 1 day prior to conjugation to increase growth rate.

14. Some laboratories prefer to aliquot *E. coli* β2163 harboring pSW29T-TKS2 from freezer culture the morning of conjugation directly into EMJH media, bypassing the LB liquid medium growth step. This approach also works; however, it typically has a delayed growth (4–6 h for exponential growth vs. 2–4 h when grown from LB liquid medium). Neither approach noticeably reduces transformation efficiency.

15. Dilution of *E. coli* can range from 1:10 to 1:100, with the 1:100 conferring less salts into EMJH media used for conjugation, which may be advantageous. In our lab, either dilution has yielded no notable difference in transformation efficiency.

16. This EMJH media should not contain kanamycin, since *Leptospira* will still be sensitive to kanamycin until transconjugation occurs.

17. If you are planning on completing many conjugation rounds, scale up with several conicals of *E. coli* in EMJH.

18. Do not allow *E. coli* (or *Leptospira*) to reach stationary phase as this will decrease transformation efficiency.

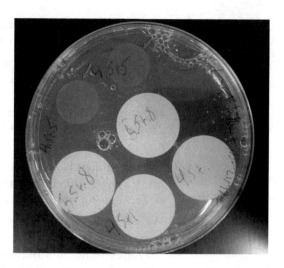

Fig. 8 Plate showing four membrane filters, as well as two circular cloudy patches (arrows) where membrane filters were removed. These cloudy patches are *Leptospira* that have traversed through the membrane filter during the 18-h incubation period into the EMJH solid media

19. To prevent tipping of filtration apparatus which would spill media, we recommend attaching a clamp system to the filter flask.

20. For a pump system, we find a vacuum pressure of 20–40 cmHg works well. With a lower vacuum pressure, the filtration step runs slowly. With higher vacuum pressure, the system may create bubbles in the flow-through which can transfer into the vacuum tubing.

21. Vary amounts since transformation efficiency will vary by density of *E. coli* and *Leptospira* (err on the side of more *Leptospira*). We typically vary between 3 and 8 mL of *Leptospira* and vary the dilution ratio between 1:2 and 1:10. It is recommended to write down the volume of each added on the back of the EMJH plate (e.g., Fig. 8) where membrane filters are placed, since on conjugation day 2, two membrane filters will be paired for the plating step. By doing this, it will help optimize volume vs. cell density for your lab.

22. Biosafety: While *L. biflexa* is considered non-infectious, if this procedure is applied to pathogenic *Leptospira*, it is essential to use prudent practices to avoid aspiration and contamination of *Leptospira*, especially if using a vacuum pump. Risk of infection in the laboratory by *Leptospira* is considered higher than average [45]. At least 67 laboratory-acquired infections and ten deaths have been reported for pathogenic *Leptospira* [46]. This is typically by direct or indirect contact with cultures or from the urine of an infected animal. Therefore, it is imperative to use an in-line filter when using this filtration technique with

pathogenic *Leptospira*, to clean all tubing by chemical or auto-clave sterilization methods regularly, and to wear relevant personal protective equipment.

23. Depending on concentration of bacteria and volume, this can take between 2 and 25 min to complete. Patience is key since disturbing the setup can result in bacteria bypassing the membrane filter if funnel is removed before all media has completed flow-through.

24. The membrane filter can carry static cling and as a result may jump as it is being removed from glass frit filter. To avoid the membrane filter from becoming contaminated by falling to the cabinet surface, we place the funnel at a 45° angle on the glass frit filter surface (Fig. 2). This gives purchase to the membrane filter as it is removed with forceps and reduces risk of static jump.

25. Ensure no bubbles remain between membrane filter and EMJH plate to optimize transfer of nutrients to bacteria.

26. If spaced properly, up to seven membrane filters can fit on one EMJH plate (Fig. 3).

27. We usually prepare at least 50 mL of *Leptospira*-containing EMJH media disbursed between two 50 mL conicals. This gives us sufficient volume to complete between seven and nine conjugations (i.e., create seven to nine membrane filters containing *E. coli* and *Leptospira*) on average.

28. Clean glassware and tubing used for conjugation day 1 by chemical disinfection, rinse thoroughly with ultrapure water, and allow to dry. Place in autoclave sleeves and autoclave for future use.

29. For every two membrane filters (pairs), 0.8 mL of EMJH liquid and three EMJH solid media plates are required.

30. When transferring, you may notice a cloudy circular patch on the EMJH plate where the membrane filter was located. This is *Leptospira* that has migrated through the membrane filter overnight. While there will still be sufficient *Leptospira* on the membrane, this is one reason why using a 0.1 μm filter pore and not exceeding a 20-h incubation period for conjugation to occur on EMJH solid media are best practices (Fig. 8).

31. Place the residual liquid media (i.e., ~10 μL) in the serological pipette onto a microscope slide, and visualize by dark-field microscopy. *E. coli* and *Leptospira* should be visualized in close proximity and in certain instances appearing attached (i.e., undergoing conjugation). This will allow you to verify amount of bacteria present and their motility. This will aid you in determining the likelihood that generation of transconju-gants was successful (Fig. 9).

Fig. 9 Microscopy of *E. coli* β2163 and *L. biflexa* potentially undergoing conjugation. In instances where *E. coli* and *Leptospira* are overlapped (e.g., arrows), cells are monitored to determine if those move together. If so, *E. coli* will appear bound to cell membrane of *Leptospira* (visualized as *E. coli* traversing longitudinally along the length of the leptospire). (Image captured with Zeiss Axio Lab A.1 (EC PLAN-Neofluar 20×/0.5 NA) with attached Axiocam 105 color CMOS camera)

32. Since *Leptospira* transconjugants may be motile, it is important to check periodically to ensure colonies remain clonal.

33. Only a small fraction of the colony is required for growth in EMJH (approx. 10–50 μL of volume of the agar containing the colony or one stab into solid media).

34. Some colonies will appear faintly translucent as they spread in the media. To aid in removal, we typically use a black sheet of paper placed below the Petri plate to more easily visualize the colonies during removal.

35. Continue to monitor plates for up to 50 days, since some mutants may grow slowly or be motility-deficient which will delay the appearance of a colony. If a plate develops mold, discard the plate according to your standard biohazard disposal protocol.

36. If *Leptospira* culture is contaminated with *E. coli* or another microbe which may occasionally occur, the contaminating bacteria can be removed by syringe filtration. Take a clean 15 mL conical and place 1–3 mL of fresh EMJH liquid media into it. Filter into this fresh media the contaminated sample by using a 5 mL syringe outfitted with 0.22 μm syringe filter. *Leptospira* should transit through this filter while other microbes will not. Allow *Leptospira* to grow for a minimum of 5–7 days at 30 °C with or without agitation. Likelihood of

recovery is best if *Leptospira* concentration was at least 10^7 cells/mL prior to filtration.

37. Once transposon is *cis*-inserted into genome, it is not spontaneously lost. Therefore, kanamycin selection is not necessary after initial growth in EMJH liquid media with 40 µg/mL of kanamycin.

38. It is typically considered that the concentration of Tween-80 and glycerol present in EMJH liquid media is sufficient for long-term storage in cryogenic vials without adding additional cryoprotectant. Some laboratories use 5% DMSO as a cryoprotectant [47]. Our laboratory typically makes up two freezer tubes for each Tn mutant for long-term storage at −80 °C: one with 30% glycerol added and one without. We have not noticed any difference in growth between those stored with and without glycerol during resurrection.

39. It is highly recommended to share the Tn library with a collaborator. This will provide an off-site backup of your Tn library in case of catastrophic equipment failure.

40. Spot PCR is less reliable for this procedure; we recommend using a genomic DNA purification kit or an automated genomic DNA purification system, such as PureLink or Maxwell 16, respectively.

41. We typically run at least two different reactions at once to save on time (e.g., TnK1 with Deg1, TnK2, and Deg1). One will typically amplify. If nested PCR does not yield multiple bands after round 2, we will try the other combinations.

42. Round 1 does not typically generate visible bands by gel electrophoresis. Therefore, it is not recommended to attempt visualization by electrophoresis from round 1 product.

43. Match TnK1 to TnKN1 and TnK2 to TnKN2.

44. We use RedSafe in lieu of ethidium bromide as an intercalating fluorescent dye and get excellent and well-resolved visualization of bands.

45. The multiple bands visualized by electrophoresis do not infer multiple insertion sites. Instead, since oligo 2 is a semi-random/degenerate primer [41], it is annealing to several locations upstream of the transposon insertion site, resulting in the appearance of multiple bands (Fig. 7). Previous research confirmed multiple banding is due to single-site transposon insertion [16]. Therefore, the PCR product can be sent for sequencing without single-band gel extraction.

46. Since some companies use enzymatic purification of PCR products which may impact multiband PCR products, we occasionally complete PCR cleanup by column purification before sending products for sequencing. With Genewiz, we have

found this is unnecessary as we still get good sequencing data, even if quality becomes low from enzymatic purification.

47. We typically use either NCBI BLAST (https://blast.ncbi.nlm. nih.gov/Blast.cgi) or Genoscope (http://www.genoscope.cns. fr/agc/microscope/home/index.php). Either website yields similar results. One advantage of using Genoscope is that many bioinformatic programs are also available within the program for initial characterization of the disrupted gene that is identified by BLAST analysis.

48. Transposon may have inserted in same or reverse orientation as disrupted gene since the FRT sites in pSW29T-TKS2 are not in tandem. This allows the transposon to become inverted.

Acknowledgments

This work was supported by NIH grant 1R01AI132818 (M.A.M.) and the Manhattanville College Faculty Research Fund (C.J.P.).

References

1. Haake DA, Levett PN (2015) Leptospirosis in humans. Curr Top Microbiol Immunol 387:65–97. https://doi.org/10.1007/978-3-662-45059-8_5

2. Adler B, de la Pena Moctezuma A (2010) Leptospira and leptospirosis. Vet Microbiol 140:287–296. https://doi.org/10.1016/j.vetmic.2009.03.012

3. Chiriboga J, Barragan V, Arroyo G, Sosa A, Birdsell DN, España K, Mora A, Espín E, Mejía ME, Morales M, Pinargote C, Gonzalez M, Hartskeerl R, Keim P, Bretas G, Eisenberg JNS, Trueba G (2015) High prevalence of intermediate *Leptospira* spp. DNA in febrile humans from urban and rural ecuador. Emerg Infect Dis 21:2141–2147. https://doi.org/10.3201/eid2112.140659

4. Costa F, Hagan JE, Calcagno J, Kane M, Torgerson P, Martinez-Silveira MS, Stein C, Abela-Ridder B, Ko AI (2015) Global morbidity and mortality of leptospirosis: a systematic review. PLoS Negl Trop Dis 9:e0003898. https://doi.org/10.1371/journal.pntd.0003898

5. Levett PN (2001) Leptospirosis. Clin Microbiol Rev 14:296–326. https://doi.org/10.1128/CMR.14.2.296-326.2001

6. Evangelista KV, Coburn J (2010) Leptospira as an emerging pathogen: a review of its biology, pathogenesis and host immune responses. Future Microbiol 5(9):1413–1425

7. De Brito T, da Silva AMG, Abreu PAE (2018) Pathology and pathogenesis of human leptospirosis: a commented review. Rev Inst Med Trop Sao Paulo 60:e23

8. Guernier V, Goarant C, Benschop J, Lau CL (2018) A systematic review of human and animal leptospirosis in the Pacific Islands reveals pathogen and reservoir diversity. PLoS Negl Trop Dis 12(5):e0006503. https://doi.org/10.1371/journal.pntd.0006503

9. Wang S, Stobart Gallagher MA, Dunn N (2020) Leptospirosis (Weil Disease). In: *StatPearls*. Treasure Island (FL): StatPearls Publishing

10. Yelton DB, Cohen RA (1986) Analysis of cloned DNA from *Leptospira biflexa* serovar patoc which complements a deletion of the *Escherichia coli* trpE gene. J Bacteriol 165 (1):41–46. https://doi.org/10.1128/jb.165.1.41-46.1986

11. Stamm LV, Parrish EA, Gherardini FC (1991) Cloning of the recA gene from a free-living leptospire and distribution of RecA-like protein among spirochetes. Appl Environ Microbiol 57 (1):183–189

12. Baril C, Richaud C, Fournie E, Baranton G, Saint GI (1992) Cloning of dapD, aroD and asd of *Leptospira interrogans* serovar icterohaemorrhagiae, and nucleotide sequence of the asd gene. J Gen Microbiol 138(1):47–53

13. Mitchison M, Bulach DM, Vinh T, Rajakumar K, Faine S, Adler B (1997) Identification and characterization of the dTDP-rhamnose biosynthesis and transfer genes of the lipopolysaccharide-related rfb locus in *Leptospira interrogans* serovar Copenhageni. J Bacteriol 179(4):1262–1267. https://doi.org/10.1128/jb.179.4.1262-1267.1997

14. Girons IS, Bourhy P, Ottone C, Picardeau M, Yelton D, Hendrix RW, Glaser P, Charon N (2000) The LE1 bacteriophage replicates as a plasmid within *Leptospira biflexa*: construction of an *L. biflexa-Escherichia coli* shuttle vector. J Bacteriol 182:5700–5705

15. Bauby H, Saint Girons I, Picardeau M (2003) Construction and complementation of the first auxotrophic mutant in the spirochaete *Leptospira meyeri*. Microbiology 149(Pt 3):689–693. https://doi.org/10.1099/mic.0.26065-0

16. Bourhy P, Louvel H, Saint Girons I, Picardeau M (2005) Random insertional mutagenesis of *Leptospira interrogans*, the agent of leptospirosis, using a mariner transposon. J Bacteriol 187:3255–3258. 187/9/3255 [pii]

17. Picardeau M (2008) Conjugative transfer between *Escherichia coli* and *Leptospira* spp. as a new genetic tool. Appl Environ Microbiol 74:319–322. AEM.02172-07 [pii]

18. Murray GL, Morel V, Cerqueira GM, Croda J, Srikram A, Henry R, Ko AI, Dellagostin OA, Bulach DM, Sermswan RW, Adler B, Picardeau M (2009) Genome-wide transposon mutagenesis in pathogenic Leptospira species. Infect Immun 77:810–816. https://doi.org/10.1128/IAI.01293-08

19. Poggi D, De Giuseppe PO, Picardeau M (2010) Antibiotic resistance markers for genetic manipulations of *Leptospira* spp. Appl Environ Microbiol 76(14):4882–4885. https://doi.org/10.1128/AEM.00775-10

20. Pappas CJ, Benaroudj N, Picardeau M (2015) A replicative plasmid vector allows efficient complementation of pathogenic Leptospira strains. Appl Environ Microbiol 81:3176–3181. https://doi.org/10.1128/AEM.00173-15

21. Zhu W, Wang J, Zhu Y, Tang B, Zhang Y, He P, Zhang Y, Liu B, Guo X, Zhao G, Qin J (2015) Identification of three extrachromosomal replicons in Leptospira pathogenic strain and development of new shuttle vectors. BMC Genomics 16:90. https://doi.org/10.1186/s12864-015-1321-y

22. Picardeau M, Brenot A, Saint Girons I (2001) First evidence for gene replacement in *Leptospira* spp. Inactivation of L biflexa flaB results in non-motile mutants deficient in endoflagella.

23. Mol Microbiol 40(1):189–199. https://doi.org/10.1046/j.1365-2958.2001.02374.x

24. Kameni APT, Couture-Tosi E, Saint-Girons I, Picardeau M (2002) Inactivation of the spirochete recA gene results in a mutant with low viability and irregular nucleoid morphology. J Bacteriol 184(2):452–458. https://doi.org/10.1128/JB.184.2.452-458.2002

24. Guégan R, Camadro JM, Saint Girons I, Picardeau M (2003) *Leptospira* spp. possess a complete haem biosynthetic pathway and are able to use exogenous haem sources. Mol Microbiol 49(3):745–754. https://doi.org/10.1046/j.1365-2958.2003.03589.x

25. Picardeau M, Bauby H, Saint Girons I (2003) Genetic evidence for the existence of two pathways for the biosynthesis of methionine in the *Leptospira* spp. FEMS Microbiol Lett 225(2):257–262. https://doi.org/10.1016/S0378-1097(03)00529-9

26. Croda J, Figueira CP, Wunder EA, Santos CS, Reis MG, Ko AI, Picardeau M (2008) Targeted mutagenesis in pathogenic Leptospira species: disruption of the LigB gene does not affect virulence in animal models of leptospirosis. Infect Immun 76(12):5826–5833. https://doi.org/10.1128/IAI.00989-08

27. Craig NL (1997) Target site selection in transposition. Annu Rev Biochem 66:437–474. https://doi.org/10.1146/annurev.biochem.66.1.437

28. Rubin EJ, Akerley BJ, Novik VN, Lampe DJ, Husson RN, Mekalanos JJ (1999) In vivo transposition of mariner-based elements in enteric bacteria and mycobacteria. Proc Natl Acad Sci U S A 96(4):1645–1650

29. Zhang JK, Pritchett MA, Lampe DJ, Robertson HM, Metcalf WW (2000) In vivo transposon mutagenesis of the methanogenic archaeon *Methanosarcina acetivorans* C2A using a modified version of the insect mariner-family transposable element Himar1. Proc Natl Acad Sci U S A 97(17):9665–9670. https://doi.org/10.1073/pnas.160272597

30. Ashour J, Hondalus MK (2003) Phenotypic mutants of the intracellular actinomycete *Rhodococcus equi* created by in vivo Himar1 transposon mutagenesis. J Bacteriol 185:2644–2652. https://doi.org/10.1128/JB.185.8.2644-2652.2003

31. Lamichhane G, Zignol M, Blades NJ, Geiman DE, Dougherty A, Grosset J, Broman KW, Bishai WR (2003) A postgenomic method for predicting essential genes at subsaturation levels of mutagenesis: application to *Mycobacterium tuberculosis*. Proc Natl Acad Sci U S A 100(12):7213–7218. https://doi.org/10.1073/pnas.1231432100

32. Stewart PE, Hoff J, Fischer E, Krum JG, Rosa PA (2004) Genome-wide transposon mutagenesis of *Borrelia burgdorferi* for identification of phenotypic mutants. Appl Environ Microbiol 70(10):5973–5979. https://doi.org/10.1128/AEM.70.10.5973-5979.2004

33. Maier TM, Pechous R, Casey M, Zahrt TC, Frank DW (2006) In vivo Himar1-based transposon mutagenesis of *Francisella tularensis*. Appl Environ Microbiol 72:1878–1885. https://doi.org/10.1128/AEM.72.3.1878-1885.2006

34. Lin T, Gao L, Zhang C, Odeh E, Jacobs MB, Coutte L, Chaconas G, Philipp MT, Norris SJ (2012) Analysis of an ordered, comprehensive STM mutant library in infectious *Borrelia burgdorferi*: insights into the genes required for mouse infectivity. PLoS One 7(10):e47532. https://doi.org/10.1371/journal.pone.0047532

35. Louvel H, Bommezzadri S, Zidane N, Boursaux-Eude C, Creno S, Magnier A, Rouy Z, Médigue C, Saint Girons I, Bouchier C, Picardeau M (2006) Comparative and functional genomic analyses of iron transport and regulation in *Leptospira* spp. J Bacteriol 188:7893–7904. https://doi.org/10.1128/JB.00711-06

36. Ristow P, Bourhy P, da Cruz McBride FW, Figueira CP, Huerre M, Ave P, Girons IS, Ko AI, Picardeau M (2007) The OmpA-like protein Loa22 is essential for leptospiral virulence. PLoS Pathog 3:e97. 07-PLPA-RA-0228 [pii]

37. Lourdault K, Cerqueira GM, Wunder EA, Picardeau M (2011) Inactivation of clpB in the pathogen *Leptospira interrogans* reduces virulence and resistance to stress conditions. Infect Immun 79(9):3711–3717. https://doi.org/10.1128/iai.05168-11

38. Petrosova H, Picardeau M (2014) Screening of a *Leptospira biflexa* mutant library to identify genes involved in ethidium bromide tolerance. Appl Environ Microbiol 80:6091–6103. https://doi.org/10.1128/AEM.01619-14

39. Ellinghausen HC Jr, McCullough WG (1965) Nutrition of *Leptospira pomona* and growth of 13 other serotypes: fractionation of oleic albumin complex and a medium of bovine albumin and polysorbate 80. Am J Vet Res 26:45–51

40. Johnson RC, Harris VG (1967) Differentiation of pathogenic and saprophytic letospires. I. Growth at low temperatures. J Bacteriol 94:27–31

41. Slamti L, Picardeau M (2012) Construction of a library of random mutants in the spirochete *Leptospira biflexa* using a mariner transposon. Methods Mol Biol 859:169–176. https://doi.org/10.1007/978-1-61779-603-6_9

42. Beeman RW, Stauth DM (1997) Rapid cloning of insect transposon insertion junctions using "universal" PCR. Insect Mol Biol 6(1):83–88. https://doi.org/10.1046/j.1365-2583.1997.00159.x

43. Fontana C, Lambert A, Benaroudj N, Gasparini D, Gorgette O, Cachet N, Bomchil N, Picardeau M (2016) Analysis of a spontaneous non-motile and avirulent mutant shows that FliM is required for full endoflagella assembly in *Leptospira interrogans*. PLoS One 11(4):e0152916. https://doi.org/10.1371/journal.pone.0152916

44. Eshghi A, Becam J, Lambert A, Sismeiro O, Dillies MA, Jagla B, Wunder EA Jr, Ko AI, Coppee JY, Goarant C, Picardeau M (2014) A putative regulatory genetic locus modulates virulence in the pathogen *Leptospira interrogans*. Infect Immun 82:2542–2552. https://doi.org/10.1128/IAI.01803-14

45. Committee on Hazardous Biological Substances in, Laboratory NRC (1989) Biosafety in the laboratory: prudent practices for handling and disposal of infectious materials. In: Biosafety in the laboratory: prudent practices for handling and disposal of infectious materials. National Academies Press, Washington, DC

46. Sewell DL (1995) Laboratory-associated infections and biosafety. Clin Microbiol Rev 8 (3):389–405

47. Cameron CE (2015) Leptospiral structure, physiology, and metabolism. Curr Top Microbiol Immunol 387:21–41. https://doi.org/10.1007/978-3-662-45059-8_3

Chapter 9

Transposon Sequencing in *Leptospira interrogans*

Kristel Lourdault and James Matsunaga

Abstract

Our limited understanding of the relationship of genotype to phenotype in the spirochete *Leptospira interrogans* stems from the inefficiency of the genetic tools available to manipulate the pathogen. The recent development of random transposon mutagenesis in *L. interrogans* has allowed the creation of large libraries of mutants, permitting the identification of several genes involved in certain functions such as virulence. However, the process of phenotypically screening individual mutants in the library remains time- and labor-intensive. Here, we describe a transposon sequencing technique (Tn-Seq), which combines random transposon mutagenesis with high-throughput sequencing for screening *L. interrogans* mutants more rapidly with fewer resources than traditional methods.

Key words *Leptospira*, High-throughput sequencing, Virulence factors, Transposon, Genomic library, Fitness

1 Introduction

The contribution of most leptospiral genes to the characteristics of *Leptospira interrogans* remains unknown due to the poor performance of the tools currently used to genetically manipulate the spirochete [1]. The development of random transposon mutagenesis in *L. interrogans* a decade ago opened the door to the identification of gene function at a genome-wide scale [2]. However, the process of individually screening mutants for altered phenotypes is laborious, particularly for identification of virulence genes, which requires a large number of animals.

To facilitate the identification of virulence factors, we recently applied transposon sequencing (Tn-Seq) to *L. interrogans* [3–6]. Tn-Seq, which combines random transposon mutagenesis with high-throughput sequencing, is the perfect tool to screen the fitness of a large number of mutants under a given condition. Briefly, mutants are pooled (input pool) and grown under a condition of interest to study in vitro fitness. To examine the fitness of the mutants in vivo, the pools are inoculated into animals and later

Nobuo Koizumi and Mathieu Picardeau (eds.), *Leptospira spp.: Methods and Protocols*, Methods in Molecular Biology, vol. 2134,
https://doi.org/10.1007/978-1-0716-0459-5_9, © Springer Science+Business Media, LLC, part of Springer Nature 2020

recovered from different tissues. DNA extracted from cultures or tissues is then sheared by sonication. DNA fragments are processed by two PCR steps that amplify sequences across one end of the transposon insertion site. The resulting genomic library is then analyzed by Illumina sequencing.

Here, we describe the steps to prepare genomic libraries for high-throughput sequencing. The sequencing reactions generate FASTQ files, which can be analyzed to determine the counts for each mutant in the pool [7].

2   Materials

2.1   Preparation of the Pool of Mutants and In Vitro or In Vivo Testing

1. *Leptospira Himar1* mariner transposon mutants (*see* **Note 1**).

2. EMJH basal medium [8, 9]: Dissolve 1 g Na_2HPO_4, 0.3 g KH_2PO_4, and 1 g NaCl in 998 ml sterile distilled water. Add the following stock solutions prepared in sterile distilled water: 1 ml 10% (v/v) glycerol and 1 ml 25% (w/v) NH_4Cl. Measure the pH, which should be 7.4. Sterilize by autoclaving.

3. EMJH supplement: Dissolve 10 g bovine serum albumin (BSA) in 50 ml sterile distilled water with slow and constant stirring. Once the BSA is dissolved, add the following stock solutions prepared in sterile distilled water (stock solutions should be filtered a 0.22 µM filter): 1 ml thiamine (stock 5 mg/ml thiamine chloride), 1 ml calcium chloride (stock 10 mg/ml $CaCl_2\cdot 2H_2O$), 1 ml magnesium chloride (stock 10 mg/ml $MgCl_2\cdot 6H_2O$), 1 ml zinc sulfate (stock 4 mg/ml $ZnSO_4\cdot 7H_2O$), 0.1 ml manganese sulfate (stock 3 mg/ml $MnSO_4\cdot H_2O$), 10 ml ferrous sulfate (stock 5 mg/ml $FeSO_4\cdot 7H_2O$), 1 ml vitamin B_{12} (stock 0.2 mg/ml vitamin B_{12}), 12.5 ml Tween 80 (make fresh, 10 ml Tween 80 in 90 ml distilled water). Adjust pH to 7.4 with NaOH, and then bring volume to 100 ml with sterile distilled water. Sterilize by filtration through a 0.22 µM filter.

4. EMJH liquid medium: Add 100 ml EMJH supplement to 900 ml EMJH basal medium. Add 1 g lactalbumin hydrolysate, 100 µl superoxide dismutase (stock 10 mg/ml superoxide dismutase in PBS, store aliquots at −20 °C), and 0.04 g sodium pyruvate. Filter sterilize (0.22 µM filter). Aliquot into smaller volumes and store at 4 °C.

5. Kanamycin: 50 mg/ml solution in water.

6. 50 ml centrifuge tubes.

7. 125 ml screw-capped Erlenmeyer flasks.

8. Shaking incubator.

9. Dark-field microscope.

10. Petroff-Hausser chamber.

11. Female Golden Syrian hamsters, 3- to 4-weeks-old.

12. 1 ml insulin syringes U-100 with 26 G × ½″ needle.

13. 3 ml syringes with 25 G × 5/8″ needle.

14. 5 ml EDTA blood collection tubes.

15. Precision balance (to weigh animals).

16. Isoflurane and isoflurane vaporizer.

17. Scalpels.

18. Scissors.

19. Forceps.

20. 1.5 ml microcentrifuge tubes.

21. 1 ml cryotubes.

22. Phosphate buffered saline (PBS): 1.8 mM KH_2PO_4, 137 mM NaCl, 10 mM Na_2HPO_4, 2.7 mM KCl, pH 7.2. Autoclave and then store at room temperature (RT).

23. Tissue homogenizer (Omni Bead Ruptor).

2.2 Preparation of Genomic Libraries

1. DNA extraction kit.

2. Sonicator with cup horn.

3. Tris-acetate-EDTA (TAE) buffer (50×): 2 M Tris base, 1 M acetic acid, and 50 mM disodium EDTA.

4. Ethidium bromide: 10 mg/ml solution in water.

5. Agarose gel: 2% gel in 1× TAE buffer with 0.5 μg/ml of ethidium bromide (5 μl of 10 mg/ml ethidium bromide in 100 ml gel).

6. Microvolume spectrophotometer.

7. dCTP + ddCTP: 9.5 mM dCTP, 0.5 mM ddCTP solution in water.

8. Terminal deoxynucleotidyl transferase (TdT): 30 U/μl.

9. Thermal cycler.

10. *Pfu* DNA polymerase and master mix.

11. PCR purification kit.

12. Oligonucleotides (Tables 1 and 2).

13. Fluorometer.

3 Methods

3.1 Assembly of the Pool of Mutants (Input Pool)

1. Grow each transposon insertion mutant individually in EMJH supplemented with 50 μg/ml kanamycin at 30 °C under 100 rpm agitation to a density of 10^8 leptospires/ml.

Table 1
Oligonucleotide primers for semi-random PCR [10]

Oligonucleotide primers for semi-random PCR [10]	
Name	Sequence (5′–3′)
Tnk1	CTTGTCATCGTCATCCTTG
Tnk2	GTGGCTTTATTGATCTTGGG
Deg1[a]	GGCCACGCGTCGACTAGTACNNNNNNNNNNGATAT
Deg2[a]	GGCCACGCGTCGACTAGTACNNNNNNNNNNTCTT
TnkN1	CGTCATGGTCTTTGTAGTCTATGG
TnKN2	TGGGGATCAAGCCTGATTGGG
Tag	GGCCACGCGTCGACTAGTAC

[a]"NNNNNNNNNN" indicates a random nucleotide at each of the ten positions

Table 2
Oligonucleotide primers for Illumina sequencing [5, 6]

Oligonucleotide primers for Illumina sequencing [5, 6]	
Name	Sequence (5′–3′)
TnkN3	CGGGGAAGAACAGTATGTCGAGCTATTTTTTGACTTACTGGGGATCAAGCC TGATTGGG
olj376	GTGACTGGAGTTCAGACGTGTGCTCTTCCGATCTGGGGGGGGGGGGGGGGGG
pMargent2	AATGATACGGCGACCACCGAGATCTACACTCTTTCCGGGGACTTA TCAGCCAACCTGTTA
IP[a]	CAAGCAGAAGACGGCATACGAGAT**XXXXXX**GTGACTGGAGTTCAGACGTG TGCTCTTCCGATCT
IP seq	GATCGGAAGAGCACACGTCTGAACTCCAGTCAC
pMargent3	ACACTCTTTCCGGGGACTTATCAGCCAACCTGTTA

IP indexing primer
[a]**XXXXXX** refers to a specific barcode that differs for each primer

2. Determine the cell density of each culture with the Petroff-Hausser chamber under a dark-field microscope (*see* **Note 2**).

3. Dilute each culture to the same cell density and pool equal volumes together (input pool) (*see* **Notes 3** and **4**).

3.2 Fitness Testing in Hamsters

All animal experiments must be approved by the local Institutional Animal Care and Use Committee and must be performed under biosafety level 2 (BSL2) containment.

1. Inject each animal intraperitoneally (*see* **Note 5**) with 1 ml of the input pool with an insulin syringe U-100 with 26 G × ½″ needle (*see* **Note 6**).

2. Monitor animals daily until they are terminated at a predetermined date. Weigh animals and look for endpoint criteria, which include loss of appetite, ruffled fur, abnormal gait, prostration, and loss of more than 10% of the animal's maximum weight.

3. Euthanize animals by isoflurane inhalation (4–5%) in an induction chamber followed by thoracotomy.

4. Collect blood by cardiac puncture with a 3 ml syringe with a 25 G × 5/8″ needle. Immediately transfer blood into an EDTA blood collection tube, and mix well by inversion (five to six times) (*see* **Note 7**).

5. Harvest tissues and store them in cryotubes at −80 °C for use in **step 1** of Subheading 3.4 (*see* **Note 8**).

3.3 Fitness Testing In Vitro

1. Collect 10 ml of the input pool, and spin it down for 20 min at $3200 \times g$ in a 50 ml centrifuge tube. Store the pellet at −80 °C until use in **step 4** of Subheading 3.4.

2. Inoculate three 125 ml screw-capped Erlenmeyer flasks containing 25 ml EMJH with 1 ml of the input pool. Incubate the cultures under the desired conditions (*see* **Note 9**).

3. After a predetermined time or when cultures reach density $\approx 1 \times 10^8$/ml, spin down each culture for 20 min at $3220 \times g$ in 50 ml centrifuge tubes (*see* **Note 10**).

4. Store cell pellets at −80 °C for use in **step 2** of Subheading 3.4.

3.4 DNA Extraction

1. For in vitro cultures, skip to **step 2**. To prepare animal tissues for DNA extraction, dice 50–80 mg of tissue, and transfer it into a 1.5 ml microcentrifuge tube with 500 µl PBS. Homogenize tissue with the Omni Bead Ruptor for 1 min at five movements per second. Determine the volume corresponding to 25 mg of tissue (*see* **Note 11**), and transfer it into a 1.5 ml microcentrifuge tube.

2. Extract the DNA with a DNA extraction kit following the manufacturer's instructions. Elute DNA with 100 µl elution buffer.

3. For blood, extract DNA from 100 µl blood collected in an EDTA blood collection tube using a DNA extraction kit. Elute the DNA with 100 µl elution buffer.

4. For the input pool, extract DNA from pellet using a DNA extraction kit following the manufacturer's instructions. Elute the DNA with 100 µl elution buffer.

3.5 Shearing DNA

1. Transfer 50 μl of extracted DNA into a 1.5 ml microcentrifuge tube.

2. Place the microcentrifuge tubes in the rack of the cup horn's sonicator. Add ice-cold water into the cup horn.

3. Run the sonicator at 80% intensity with cycles of 10 s on pulse and 5 s off pulse for a total of 3 min (*see* **Note 12**).

4. Check the size of the DNA fragments by running 2.5 μl of sheared DNA on a 2% agarose gel (*see* **Note 13**).

3.6 Addition of the C-Tail

1. Determine the DNA concentration using a spectrophotometer, and calculate the volume corresponding to 500 ng of DNA (*see* **Note 14**).

2. Set up the reaction as described in Table 3 to add C-tails.

3. Run the following program on the thermal cycler: 1 h at 37 °C then 20 min at 75 °C.

4. Clean samples with a mini-PCR purification kit, and elute the DNA with 12 μl of buffer.

3.7 Nested PCR

1. Prepare PCR mixes for the first PCR round (Table 4).

2. Run the following program on the thermal cycler: 95 °C for 2 min; 24 cycles: 95 °C for 30 s, 60 °C for 30 s, 72 °C for 2 min; 72 °C for 2 min.

3. Set up the reaction for the second round of PCR (Table 5). An indexing primer with a different barcode sequence should be used for each replicate.

4. Run the following program on the thermal cycler: 95 °C for 2 min; 18 cycles: 95 °C for 30 s, 60 °C for 30 s, 72 °C for 2 min; 72 °C for 2 min.

5. Confirm the size of DNA fragments by running 3 μl of sheared DNA on a 2% agarose gel (*see* **Note 15**).

6. Purify the PCR products with a PCR purification kit following the manufacturer's instructions. Elute the DNA with 30 μl elution buffer.

3.8 Preparation of the Genomic Libraries for Sequencing

1. Measure the DNA concentration using a fluorometer.

2. Calculate the volume of each library corresponding to 15 or 20 ng, and pool all libraries together (*see* **Note 16**).

3. Sequence genomic libraries as 64 bp single-end reads with the standard commercial sequencing primer (IP seq) and the custom primer pMargent3 (*see* **Note 17**). Run "sequencing-by-synthesis" reactions on a next-generation sequencing platform.

Table 3
Addition of the C-tail's reaction

Reagent	One reaction
Sheared DNA	500 ng
Water	Up to 14.5 μl
5× TdT buffer	4 μl
(9.5 mM dCTP + 0.5 mM ddCTP) mix	1 μl
TdT (30 U/μl)	0.5 μl
Total	*20 μl*

Table 4
PCR mixes for the first PCR round

Reagent	One reaction	
	Library reaction	Control reaction
Master mix, 2×	12.5 μl	12.5 μl
Primer 1, 30 μM (TnkN3)	0.5 μl	–
Primer 2, 30 μM (olj376)	1.5 μl	1.5
Water	7.5 μl	8 μl
C-tailed DNA	3 μl	3 μl
Total	*25 μl*	*25 μl*

Table 5
PCR mixes for the second PCR round

Reagent	One reaction	
	Library reaction	Control reaction
Master mix, 2×	25 μl	12.5 μl
Primer 1, 30 μM (pMargent2)	0.5 μl	0.25 μl
Primer 2, 30 μM (indexing primer)	0.5 μl	0.25 μl
Water	23 μl	11.5 μl
PCR products from PCR #1	1 μl	0.5 μl
Total	*50 μl*	*25 μl*

4 Notes

1. The protocol assumes that insertion mutants were generated with a transposon carrying a Kan^R marker. Transposon mutants can be generated from saprophytic [10] or pathogenic strains [2]. The creation of the library of mutants is described in Chapter 8 of the book. The Tn-Seq protocol can be applied to both strains.

2. To determine the density of *Leptospira* using a Petroff-Hausser chamber, dilute the culture 1:100 in EMJH. Place coverslip over the slide, and put 10 μl of diluted bacterial suspension under the coverslip. Count bacterial cells located within the big center square (corresponding to 25 small squares) under a dark-field microscope at 400× magnification. Calculate the density of *Leptospira* using the following equation: cells/ml = number of cells × dilution factor × 50,000.

3. The bacterial cell density of the inoculum or input pool depends on the goals of the experiment. For animal experiments, the challenge dose, the strain's ED_{50}, the route of inoculation, the animal model, and the length of the experiment need to be considered to determine the appropriate input pool cell density. The kinetics of dissemination of *Leptospira* is affected by all of these characteristics. For example, after intraperitoneal challenge with the Fiocruz L1-130 strain of *L. interrogans*, bacteria can be detected in blood after an hour when the inoculum is 10^8 leptospires, but when the dose is only 250 leptospires, bacteria are not detected until the fifth day of infection. The route of challenge also influences the kinetics of dissemination. With the same dose of *L. interrogans*, bacteria are detected in the kidney 1 day after intraperitoneal challenge but 4 days after conjunctival challenge [11]. The number of *Leptospira* cells in the inoculum required to generate a lethal infection also depends on the animal model.

   The starting density and the length of the in vitro experiment depend on the conditions being examined. For in vitro testing, the main criterion to consider is the properties of the strain itself. Saprophytic and pathogenic strains have very different in vitro doubling times, 4.5 and 14–18 h, respectively [12]. Because of these differences, the density of the input pool and the length of the experiment need to be determined for each strain. To detect subtle effects on fitness, a longer period of incubation of the culture and a lower starting density may be necessary so that a larger number of cell doublings occur.

4. When possible, transposon insertion mutants unaffected by the condition to be examined should be included. Mutants whose fitness is known to be affected by the condition under study

should also be included. For fitness testing in animals, the input pool should include previously described mutants that are attenuated in their fitness such as *loa22* [6, 13] and mutants with wild-type fitness in tissues such as *ligB* [6, 14] or *flaA1* [6, 15].

5. Although intraperitoneal inoculation is the most common challenge route for studies of *Leptospira* infection [16], it does not reflect the natural route of infection. Routes into or through the skin (subcutaneous or intradermal inoculation), respectively [17], or through mucous membranes (ocular inoculation) [11], which simulate the natural entry of the bacteria, can be considered. The choice of challenge route depends on the purpose and the design of the experiment. The volume of inoculum needs to be adapted to the route of challenge. For intraperitoneal challenge, the maximum volume that can be injected is 1 ml, whereas only 10 µl can be used for ocular inoculation [11]. Keep in mind that the more natural routes of inoculation have not been examined for bottleneck effects, which lead to stochastic rather than deterministic changes in measured fitness levels.

6. The number of animals to use for each experiment depends on the number of mutants in the input pool. This number can be calculated by power analysis, which may require a pilot experiment [7].

7. DNA must be extracted from the blood immediately after its collection.

8. Selection of the organs to be harvested depends on the goals of the experiment. Tissues that are the main sites of disease pathology, such as kidney, liver, and lung, are often collected. However it has been shown that *Leptospira* can disseminate to and colonize many other organs [11].

9. The exact medium composition and incubation scheme depend on the conditions being examined. The number of replicates to be tested depends on the conditions being examined and the number of mutants in the pool. A pilot experiment with a small number of mutants may be necessary to calculate the appropriate number of replicates from a power analysis.

10. Carefully remove the supernatant; pellets are loose due to the low-speed centrifugation.

11. Calculate the volume of tissue to use for DNA extraction with the following formula: volume in µl = (500 µl × 25 mg)/(mg of diced tissue). For example, 250 µl homogenate will be used from a 50 mg tissue resuspended in 500 µl PBS.

12. Wear earmuffs to protect hearing.

13. Sheared DNA is characterized by a smear located between 200 and 600 bp.

14. Calculate the volume corresponding to 500 ng of DNA with the following equation: volume in µl = (500 ng)/(DNA concentration in ng/µl).

15. The library should show a smear in which most of the PCR products are between 200 and 600 bp and no amplification for the control reaction.

16. Calculate the volume corresponding to 15 or 20 ng with the following equation: volume in µl = (DNA concentration of library in ng/ml)/(15 or 20 ng).

17. DNA concentration and volume required for sequencing vary depending on the sequencing platform.

References

1. Picardeau M (2018) Toolbox of molecular techniques for studying *Leptospira* spp. Curr Top Microbiol Immunol 415:141–162. https://doi.org/10.1007/82_2017_45

2. Murray GL, Morel V, Cerqueira GM, Croda J, Srikram A, Henry R, Ko AI, Dellagostin OA, Bulach DM, Sermswan RW, Adler B, Picardeau M (2009) Genome-wide transposon mutagenesis in pathogenic *Leptospira* species. Infect Immun 77(2):810–816. https://doi.org/10.1128/IAI.01293-08

3. van Opijnen T, Bodi KL, Camilli A (2009) Tn-seq: high-throughput parallel sequencing for fitness and genetic interaction studies in microorganisms. Nat Methods 6 (10):767–772. https://doi.org/10.1038/nmeth.1377

4. van Opijnen T, Dw L, Camilli A (2015) Genome-wide fitness and genetic interactions determined by Tn-seq, a high-throughput massively parallel sequencing method for microorganisms. Curr Protoc Microbiol 36:1E.3.1–1E.324. https://doi.org/10.1002/9780471729259.mc01e03s36

5. Troy EB, Lin T, Gao L, Lazinski DW, Camilli A, Norris SJ, Hu LT (2013) Understanding barriers to *Borrelia burgdorferi* dissemination during infection using massively parallel sequencing. Infect Immun 81 (7):2347–2357. https://doi.org/10.1128/IAI.00266-13

6. Lourdault K, Matsunaga J, Haake DA (2016) High-throughput parallel sequencing to measure fitness of *Leptospira interrogans* transposon insertion mutants during acute infection. PLoS Negl Trop Dis 10(11):e0005117. https://doi.org/10.1371/journal.pntd.0005117

7. Lourdault K, Matsunaga J, Evangelista KV, Haake DA (2017) High-throughput parallel sequencing to measure fitness of *Leptospira interrogans* transposon insertion mutants during golden Syrian hamster infection. J Vis Exp (130):56442. https://doi.org/10.3791/56442

8. Ellinghausen HC Jr, McCullough WG (1965) Nutrition of *Leptospira pomona* and growth of 13 other serotypes: fractionation of oleic albumin complex and a medium of bovine albumin and polysorbate 80. Am J Vet Res 26:45–51

9. Johnson RC, Harris VG (1967) Differentiation of pathogenic and saprophytic letospires. I. Growth at low temperatures. J Bacteriol 94(1):27–31

10. Slamti L, Picardeau M (2012) Construction of a library of random mutants in the spirochete *Leptospira biflexa* using a mariner transposon. Methods Mol Biol 859:169–176. https://doi.org/10.1007/978-1-61779-603-6_9

11. Wunder EA Jr, Figueira CP, Santos GR, Lourdault K, Matthias MA, Vinetz JM, Ramos E, Haake DA, Picardeau M, Dos Reis MG, Ko AI (2016) Real-time PCR reveals rapid dissemination of *Leptospira interrogans* after intraperitoneal and conjunctival inoculation of hamsters. Infect Immun 84 (7):2105–2115. https://doi.org/10.1128/iai.00094-16

12. Cameron CE (2015) Leptospiral structure, physiology, and metabolism. Curr Top Microbiol Immunol 387:21–41. https://doi.org/10.1007/978-3-662-45059-8_3

13. Ristow P, Bourhy P, da Cruz McBride FW, Figueira CP, Huerre M, Ave P, Girons IS, Ko AI, Picardeau M (2007) The OmpA-like protein Loa22 is essential for leptospiral virulence. PLoS Pathog 3(7):e97. https://doi.org/10.1371/journal.ppat.0030097

14. Croda J, Figueira CP, Wunder EA Jr, Santos CS, Reis MG, Ko AI, Picardeau M (2008) Targeted mutagenesis in pathogenic *Leptospira* species: disruption of the LigB gene does not affect virulence in animal models of leptospirosis. Infect Immun 76(12):5826–5833. https://doi.org/10.1128/IAI.00989-08

15. Lambert A, Picardeau M, Haake DA, Sermswan RW, Srikram A, Adler B, Murray GA (2012) FlaA proteins in *Leptospira interrogans* are essential for motility and virulence but are not required for formation of the flagellum sheath. Infect Immun 80(6):2019–2025. https://doi.org/10.1128/IAI.00131-12

16. Haake DA (2006) Hamster model of leptospirosis. Curr Protoc Microbiol 12:2. https://doi.org/10.1002/9780471729259.mc12e02s02

17. Coutinho ML, Matsunaga J, Wang LC, de la Pena Moctezuma A, Lewis MS, Babbitt JT, Aleixo JA, Haake DA (2014) Kinetics of *Leptospira interrogans* infection in hamsters after intradermal and subcutaneous challenge. PLoS Negl Trop Dis 8(11):e3307. https://doi.org/10.1371/journal.pntd.0003307

Chapter 10

Specific Gene Silencing in *Leptospira biflexa* by RNA-Guided Catalytically Inactive Cas9 (dCas9)

Luis Guilherme Virgílio Fernandes and Ana Lucia Tabet Oller Nascimento

Abstract

Easy, practical, and affordable gene silencing techniques are constantly progressing, and genetic tools such as TALEs, RNAi, and CRISPR/Cas9 have emerged as new techniques for understanding the basic biology and virulence mechanisms of pathogenic organisms, including bacteria. Here, we describe one-step targeted gene silencing in *Leptospira biflexa* by using plasmids expressing catalytically inactive *Streptococcus pyogenes* Cas9 (dCas9) and a single-guide RNA (sgRNA) capable of pairing to the coding strand of a desired gene.

Key words Single-guide RNA, dCas9, Gene silencing, *Leptospira*, Electroporation

1 Introduction

The development of gene inactivation or silencing techniques is a pivotal step for understanding the basic biology and virulence of pathogenic microorganisms. In *Leptospira* spp., these aspects have remained unexplored mainly due to the lack of genetic tools for easy and efficient genetic manipulation in spirochetes. Distinct types of CRISPR (*c*lustered *r*egularly *i*nterspaced *s*hort *p*alindromic *r*epeat)/Cas systems have been found in the genome of most *Archaea* and *Bacteria*, playing an important role in their immunity against phages and plasmids [1, 2]. The type II CRISPR/Cas system of *Streptococcus pyogenes* has been widely used in several cell types for mutagenesis [3–7]. The enzyme Cas9 is the RNA-guided DNA endonuclease capable of recognizing genomic sites and causing double-strand breaks (DSB).

Eukaryotic cells can repair DSBs introduced by Cas9 by directly ligating DNA broken ends in the absence of a repair template [8, 9]. However, Cas9-induced DSB in the chromosome of most prokaryotes, including *L. biflexa*, is lethal to the cells in the absence of a template for recombination [10–12].

Nobuo Koizumi and Mathieu Picardeau (eds.), *Leptospira spp.: Methods and Protocols*, Methods in Molecular Biology, vol. 2134, https://doi.org/10.1007/978-1-0716-0459-5_10, © Springer Science+Business Media, LLC, part of Springer Nature 2020

It is possible to overcome this problem using the catalytic null mutant of the Cas9 RNA-guided nuclease ("dead" Cas9 or dCas9), which induces gene silencing by steric hindrance of RNA polymerase, rather than disruption, thereby blocking transcription initiation (when dCas9 protein is bound to the promoter) or elongation (when dCas9 is bound to the coding region) [13]. Based on our previous results [12], when a single-guide RNA (sgRNA) is designed to hybridize to the coding strand of the DNA, full gene silencing is expected. We have created dCas9 and sgRNA-expressing plasmid and have successfully obtained gene silencing using this novel tool.

2   Material

2.1   Bacterial Strains and Media

1. *Escherichia coli* strain π1 [14].

2. Thymidine solution: 100 mM solution in water. Filter the solution through a 0.22 filter for sterilization, and keep the final solution refrigerated (4 °C).

3. Spectinomycin: 100 mg/mL spectinomycin in water.

4. Sodium pyruvate: 10 mg/mL in water.

5. Luria-Bertani (LB) medium: 1% NaCl, 0.5% yeast extract, 1% tryptone, pH to 7.0. Sterilize by autoclaving for 20 min at 121 °C. For solid medium, include 1.5% bacteriological agar before autoclaving.

6. *L. biflexa* serovar Patoc strain Patoc1 (*see* **Note 1**). Leptospiral cells are grown in liquid EMJH medium with gentle shaking (80–100 rpm) at 30 °C or on solid medium at the same temperature.

7. Genomic DNA from *L. interrogans* serovar Copenhageni strain Fiocruz L1-130.

8. Liquid EMJH medium: Dissolve 2.3 g of Difco™ Leptospira Medium Base EMJH (Becton Dickinson) in 900 mL Milli-Q H$_2$O, autoclave the solution (to avoid contamination), and after cooling, add 100 mL Difco™ Leptospira Enrichment EMJH (Becton Dickinson). Make sure the glass containers are properly washed and sterile. The final medium can be stored up to 2 months at 4 °C. It is important to avoid excess handling and always to check for possible contamination.

9. 2× EMJH: Dissolve 0.46 g of Difco™ Leptospira Medium Base EMJH (Becton Dickinson) in 80 mL of Milli-Q water, and sterilize by autoclaving. After cooling, add 20 mL of Difco™ Leptospira Enrichment EMJH (Becton Dickinson). This can be stored up to 2 months at 4 °C.

10. 2.4% noble agar: Dissolve 2.4 g of noble agar in Milli-Q H$_2$O and adjust the volume to 100 mL. Autoclave the solution. This can be stored up to 2 months at 4 °C.

11. Solid EMJH plates: 100 mL of 2× EMJH, 100 mL of 2.4% noble agar. When preparing the plates, warm the agar solution in a microwave (with the cap of the agar bottle loosened before microwaving), and place both solutions in water bath at 56 °C. Next, mix 100 mL of each solution and add 200 μL of sodium pyruvate (10 mg/mL) and 80 μL of spectinomycin stock solution (*see* **Note 2**). Normally, 20 mL of medium are poured into plates (10 cm diameter).

2.2 Molecular Biology

1. Thermocycler.

2. Plasmid pMaOri.dCas9 [12].

3. Microtubes, 0.2 mL.

4. PCR reagents (high-fidelity polymerases are preferred) and PCR product purification kit.

5. UV transilluminator.

6. Dry-heat bath (65 °C).

7. T4 DNA ligase and ligation buffer (ATP-containing).

8. *E. coli* strain π1.

9. Petri dishes with LB medium plus 40 μg/mL spectinomycin and 0.3 mM dT.

10. Primers pMaOri2F and R:

    pMaOri2F: 5′ ACGCAATGTATCGATACCGAC 3′.

    pMaOri2R: 5′ ATAGGTGAAGTAGGCCCACCC 3′.

    These are flanking primers for the *Xma*I site we routinely use for sgRNA inclusion into pMaOri.dCas9.

11. Primer for amplifying *lipL32* promoter (restriction site not included):

    p32F: 5′ GAACAAGAAAGAGTCAGAG 3′.

    p32R: 5′ AAAATCACGGTATGAACTTAG 3′.

12. Tris-acetate-EDTA (TAE) buffer (10×): 400 mM Tris, 200 mM acetic acid, 10 mM EDTA. Store at room temperature.

13. 1% agarose gel: 1% agarose in 1× TAE buffer.

14. *Xma*I restriction enzyme.

2.3 Transformation

1. Plasmid pMaOri.dCas9, empty and containing sgRNA cassette (Fig. 1).

2. *L. biflexa* serovar Patoc strain Patoc1.

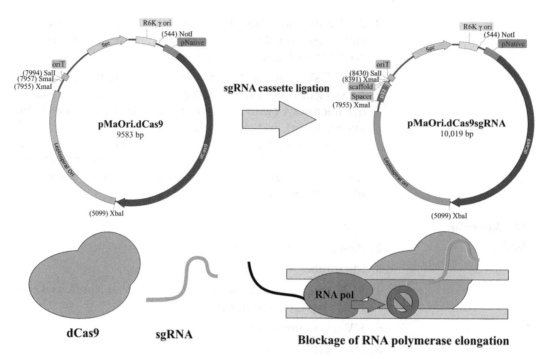

Fig. 1 Expressed dCas9 and sgRNA complex blocks RNA elongation. The plasmid pMaOri.dCas9 contains the R6k γ origin of replication, requiring the Pi protein for replication. dCas9 transcription is driven by the native *S. pyogenes* promoter (pNative). Unique *Xma*I, *Sma*I, and *Sal*I restriction sites can be used for sgRNA cassette ligation. Plasmids with pMaOri backbone can be used to transform *Leptospira* spp. by electroporation or conjugation, due to *oriT*. After sgRNA cassette ligation (here represented in *Xma*I site), recombinant pMaOri. dCas9sgRNA plasmid is able to code for both dCas9 and sgRNA (directed by *lipL32*, p32, promoter), conferring a one-step gene silencing tool. dCas9 is positioned in genomic targets by Watson-Crick base pairing between spacer sequence and DNA coding strand, provoking a blockage of RNA polymerase elongation, abolishing transcription, and consequently, affecting protein levels

3. Liquid EMJH medium (with or without spectinomycin, depending on the step).

4. Solid EMJH plates (with or without spectinomycin, depending on the step).

5. Sterile water (autoclaved or 0.22 μm filtered).

6. 30 °C shaker and incubator.

7. Spectrophotometer (420 nm).

8. Sterile 50 mL polypropylene centrifuge tubes.

9. 0.2 cm electroporation cuvettes.

10. Electroporator.

2.4 SDS-
Polyacrylamide Gel
Electrophoresis
(SDS-PAGE)
and Immunoblotting

1. Thirty percent acrylamide/bis-acrylamide solution (29:1): Wearing gloves and mask, weigh out 29 g of acrylamide and 1 g of bis-acrylamide, and mix the two reagents in water at a final volume of 100 mL. Shake the mixture until dissolved and filter the solution using a 0.22 μm filter. Store at 4 °C, in a

bottle wrapped with aluminum foil. Avoid skin contact with unpolymerized acrylamide.

2. Ammonium persulfate: 10% solution in water.

3. Sodium dodecyl sulfate (SDS): 10% solution in water.

4. N,N,N,N-Tetramethylethylenediamine (TEMED): store at 4 °C.

5. Resolving gel buffer: 1.5 M Tris–HCl, pH 8.8. Transfer 181.7 g Tris to a graduated cylinder or a glass beaker and then add about 900 mL of water and mix. Adjust the pH with HCl and add more water to a final volume of 1 L. Autoclave the solution (121 °C, 20 min) and store at room temperature.

6. Stacking gel buffer: 1 M Tris–HCl, pH 6.8. Dissolve 121.14 g Tris in 900 mL water, adjust pH to 6.8 with HCl, and then complete the volume to 1 L. Autoclave and store at room temperature.

7. SDS-PAGE sample buffer (5×): 250 mM Tris–HCl, pH 6.8, 0.1% SDS, 50% glycerol, 6.7% β-mercaptoethanol, 0.5 mg/mL bromophenol blue. Mix 7.5 mL of 1 M Tris–HCl, pH 8.0, 3 mL of 10% SDS, 15 mL of glycerol, and 15 mg of bromophenol blue, and complete the volume to 28 mL. Heat the solution at 65 °C until the dye is completely dissolved, and then add 2 mL β-mercaptoethanol.

8. Tris-glycine buffer (5×): 125 mM Tris–HCl, pH 8.3, 1.25 M glycine, 0.5% SDS. Keep the solution at room temperature.

9. Glass plates, gel caster, and combs.

10. Nitrocellulose membranes.

11. Transfer buffer (5×): Mix 625 mL of 5× Tris-glycine buffer, 18.5 mL of 10% SDS, 356.5 mL of water.

12. Phosphate buffered saline (PBS, 10×): 1.37 M NaCl, 27 mM KCl, 100 mM Na_2HPO_4, 20 mM KH_2PO_4, pH 7.4. Autoclave and store at room temperature.

13. PBS-Tween 20 (PBS-T): 1× PBS, 0.05% Tween 20. Since this detergent is viscous, pipetting should be done slowly.

14. Blocking buffer: PBS-T, 10% dry milk, or 5% bovine serum albumin (BSA). Weigh out 10 g of skimmed dry milk or 5 g of BSA and dissolve the powder in 100 mL of PBS-T.

15. Plastic container.

16. Ponceau S.

17. Semidry transfer system.

18. Filter paper.

19. Chemiluminescent substrate.

20. Imager system set for chemiluminescence.

---

3   Methods

3.1   Selection of Protospacer and Plasmid Ligation

In this section, we describe the steps for selecting appropriate protospacers for constructing and expressing sgRNA and further ligation into pMaOri.dCas9. This sgRNA will be recognized by the expressed dCas9 and will direct this protein to genomic targets based on Watson-Crick base pairing. This complex, bound to the coding strand of the desired gene, will hamper RNA polymerase elongation and, therefore, block transcription (Fig. 1). Both dCas9 and sgRNA are expressed in the same plasmids, allowing one-step gene silencing in *L. biflexa*. All cloning procedures are performed in the dT auxotrophic *E. coli* strain π1 [14], due to pMaOri [15] origin of replication, R6K-gamma (Fig. 1)

1. The backbone of sgRNA cassette ligation is the plasmid pMaOri.dCas9 [12] (Fig. 1). This plasmid allows transformation of *Leptospira* spp. by electroporation and conjugation [15]. In our experience, the *Xma*I site has been successfully used for inclusion of the sgRNA sequence.

2. Select the coding sequence of the desired gene (https://www.ncbi.nlm.nih.gov/genbank) and submit it to the CHOP-CHOP webserver (http://chopchop.cbu.uib.no/), selecting "Fasta Target" and setting to "CRISPR/Cas9." Cas9 from *S. pyogenes* recognizes the PAM (protospacer adjacent motif) NGG. The NGG trinucleotide must *not* be included in the sgRNA sequence.

3. Based on the output, select the protospacers with the best score (green arrow) and as close as possible to the 5′end of the coding region. It is very important to select a protospacer that is capable of pairing to the coding strand. In the output, these protospacers will be listed as minus strand. We recommend selecting also a protospacer to hybridize to the template strand, to serve as control. Around 30–50% gene silencing is expected in this situation, in contrast to 100% gene silencing when the sgRNA contains protospacer sequence pairing to the coding strand (*see* **Note 3**).

4. The 20 nt sequence selected referring to the protospacers should be fused to the *lipL32* promoter (at its 5′end) and sgRNA scaffold (3′end, including a dCas9 handle and intrinsic terminator) (Fig. 2). We use the promoter region comprising −334 to the TSS (transcription start site, according to Zhukova et al. [16]) and scaffold sequence: GTTTTAGAGCTA-

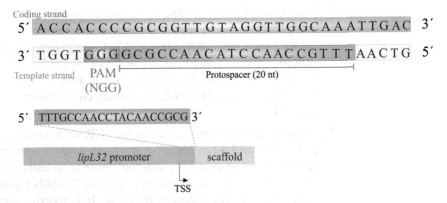

Fig. 2 Construction of sgRNA cassette. Protospacers within genomic targets are selected based on the presence of a 3'PAM (protospacer adjacent motif) composed of NGG. For full gene silencing, it is required pairing between sgRNA and coding strand. Thus, protospacer is selected in the template strand, and to its sequence, a 3'scaffold sequence is added. For sgRNA transcription, we have been using *lipL32* promoter, from −334 to its TSS

GAAATAGCAAGTTAAAATAAGGCTAGTCCGTTAT-CAACTTGAAAAAGTGGCACCGAGTCGGTGCTTTTTT. LipL32 is a constitutive, high-copy number protein from pathogenic strains, and its promoter has been shown to be functional in *L. biflexa*. Other functional promoters could be used to drive sgRNA transcription, but it is important to have the TSS well defined.

5. The sgRNA cassette can be obtained by ordering a synthetic gene or by PCR. With regard to PCR, we have succeeded in amplifying the *lipL32* promoter from genomic DNA of *L. interrogans* and including the 20 nt and scaffold sequence at the 5'end of the reverse primer. In both forward and reverse primer ends, include a *Xma*I restriction site (CCCGGG). The PCR mixture (50 μL) contains 1× PCR buffer, 2 mM MgCl$_2$ (some PCR buffers already contain magnesium), 0.2 mM dNTP mix, 0.5 μM forward and reverse primers (p32F and R), and 1 U DNA polymerase (*see* **Notes 4** and **5**).

6. Digest sgRNA cassette and pMaOri.dCas9 with *Xma*I enzyme. Purify the digested products directly from the reaction mixture, and perform the ligation with a 1:10 molar ratio plasmid to insert. First, mix plasmid, insert and water (11 μL reaction mixture), and heat at 65 °C for 10 min. After cooling, add the ligation buffer (normally 5×, add 3 μL) and T4 DNA ligase (1 μL, 1 U). The final volume of the reaction mixture is 15 μL, and the reaction is carried out for 1 h at room temperature. Alternatively, pMaOri.dCas9 digested with *Xma*I can be treated with calf intestinal alkaline phosphatase (CIP) to increase ligation efficiency.

7. Use the total reaction mixture to transform 50 μL of chemically competent *E. coli* π1 cells. Transformation steps include 30 min on ice (cells plus DNA), 90 s at 42 °C, and then 5 min on ice; 300 μL LB medium are added and cells are incubated at 37 °C for 1 h. Streak the bacteria on a LB agar plate containing 0.3 mM dT and 40 μg/mL spectinomycin. Incubate overnight at 37 °C.

8. Evaluate the recovered colonies by PCR with flanking primers pMaOri2F and R (*see* **Note 6**). Because only one restriction site is used, ligation efficiency may be lower than usual (when no CIP is used). We have observed around 20–50% ligation rate after checking the colonies by PCR. It is worth mentioning that, as the cloning is not directional, the insert can be ligated in either orientation.

9. Grow recombinant *E. coli* cells containing pMaOri. dCas9sgRNA in LB medium for plasmid purification.

10. Confirm sgRNA cassette ligation by sequencing and digestion of the recombinant plasmid with *Xma*I restriction enzyme.

3.2 Transformation

Due to the pMaOri backbone, the recombinant plasmid containing dCas9 and sgRNA cassette can be used to transform *Leptospira* spp. by conjugation and electroporation. Here, we describe briefly the electroporation steps. The detailed protocol for conjugation has been described by Picardeau [17]. We recommend, for each desired gene, the use of empty pMaOri.dCas9 (control) and this plasmid containing sgRNA for both the coding (full gene silencing) and template (partial) strand of DNA.

1. Each 10 mL of exponential-phase *L. biflexa* cells generates one aliquot of competent bacteria. Therefore, the final volume of culture will depend on the number of aliquots needed. Inoculate 1 volume of *L. biflexa* culture in 100 volumes of fresh liquid EMJH, and incubate with moderate shaking (100 rpm) at 30 °C until the culture reaches an OD_{420} between 0.3 and 0.5 (around 3 days). All procedures should be performed in a laminar flow cabinet, previously UV-irradiated to avoid contamination. Make sure all glass used for culture growth is properly washed.

2. Monitor the culture for possible contamination.

3. When the desired OD_{420} is reached, collect the bacteria by centrifugation (4000 × *g*, room temperature for 15 min) in 50 mL centrifuge tubes.

4. Discard the supernatant and resuspend the bacterial pellet with the same volume of sterile water by gently shaking the cells (avoid vortexing).

5. Repeat centrifugation, discard supernatant, and resuspend the resulting pellet in 1:100 volumes of water with respect to original culture volume. Aliquots of bacterial suspensions (100 μL) are stored in sterile 1.5 mL microtubes.

6. Centrifuge the microtubes (4000 × g, room temperature for 15 min), and completely remove the supernatant with pipet (*see* **Note 7**).

7. Resuspend the bacterial pellets with 100 μL of DNA solution in water (DNA volume varies according to the original concentration and final amount used). Incubate for 15 min at room temperature. We have performed transformation with DNA amounts ranging from 100 ng to 1 μg. However, even when the same amount of DNA is used, variations in transformation rates are observed between experiments.

8. Keep 0.2 cm electroporation cuvettes at −20 °C until right before the electroporation step. Transfer ~100 μL of cell suspensions (leptospires plus DNA) to the cuvettes; the electroporator should be set at 1.8 kV, 25 μF, and 200 Ω.

9. Electroporate the bacterial cells. The resulting pulse should render a time constant of 5–6 ms. After electroporation, add 1 mL of fresh liquid EMJH to the cells.

10. Transfer the cells to sterile 50 mL tubes and incubate with shaking overnight at 30 °C. This step allows the successful expression of the antibiotic resistance genes.

11. Spread 100–250 μL of solution into EMJH plates containing spectinomycin, and incubate the plates at 30 °C until colonies are visible, normally 7–14 days. Plates must be sealed with Parafilm or plastic film to avoid desiccation.

12. Validate the transformants by picking a few colonies from the plates. Since leptospires grow as subsurface colonies, make sure to "dig" below the medium surface. The collected colonies should have a bit of solid medium.

13. Resuspend the selected colony in 50 μL of fresh EMJH, and then place 5 μL of the suspension on glass slides for visualization of leptospires.

14. Transfer the remaining volume to 3–5 mL of liquid EMJH plus spectinomycin, and incubate with moderate shaking at 30 °C.

15. When growth is observed with the naked eye ("cloudy" appearance), confirm transformation by PCR using pMaOri2 primers. In our experience, it is not necessary to purify genomic DNA from the cultures (*see* **Note 6**).

16. *Leptospira* cells can be stored at −80 °C in EMJH medium plus 2.5% dimethyl sulfoxide (DMSO). Add 25 μL of DMSO to 1.5 mL microtubes and then 975 μL of leptospiral culture, mix gently, and then freeze immediately.

3.3 Gene Silencing Evaluation

The basic approach for assessing gene silencing in our laboratory is performing immunoblotting of the recovered recombinant *L. biflexa* cells. This strategy is based on the availability of antiserum against the protein encoded by the gene.

1. Collect the recombinant *L. biflexa* cells grown in liquid EMJH medium by centrifugation ($4000 \times g$, room temperature, 15 min). Wash the bacterial cell pellet twice with PBS (*see* **Note 8**).

2. Resuspend the cells in 100 μL of PBS and then check the OD_{420} of the samples, using PBS as the blank. After checking the densitometry reading, dilute all samples with PBS to the same OD, for normalization purposes. We use a final OD_{420} of 5–10. If your target protein is poorly expressed, we recommend using a higher cell density (*see* **Note 9**).

3. Add 25 μL of $5 \times$ SDS-PAGE buffer to the 100 μL of bacterial solutions, vortex briefly, and boil the normalized samples at 96 °C for 10 min.

4. Subject the samples to SDS-PAGE (percentage depending on the molecular weight of the target protein). For protein ranging from 20 to 60 kDa, we use 12% gels. For preparation of the separating gel, mix 4 mL of 30% polyacrylamide mixture, 3.3 mL of water, and 2.5 mL of 1.5 M Tris, pH 8.8, in a 50 mL tube. Add 100 μL of 10% SDS, 100 μL of 10% ammonium persulfate, and 10 μL of TEMED, mix gently, and pour gel in a 7.25 cm × 10 cm × 1-mm gel cassette. Normally, 7 mL is sufficient for the resolving gel. Overlay the gel with isopropanol or water, saving around 3 cm for the stacking gel.

5. For preparation of the stacking gel, mix 0.83 mL of 30% acrylamide mixture, 3.4 mL water, and 0.63 mL 1 M Tris, pH 6.8, in a 50 mL tube. Add 50 μL of 10% SDS, 50 μL of 10% ammonium persulfate, and 5 μL of TEMED. Insert the gel comb immediately, avoiding the formation of bubbles.

6. For the electrophoresis, include protein standards. Set the amperage at 25 mA per gel, and run the samples until the dye reaches the bottom of the gel.

7. As the electrophoresis is ending, prepare the system for protein transfer. Immerse three to four sheets of filter paper in $1 \times$ transfer buffer in a plastic container, then the nitrocellulose membrane, combining the elements on the surface of the semidry transfer system. When the run is over, carefully take the gel from the electrophoresis cassette, and let it sit in the transfer buffer for about 5 min. Next, place the gel over the nitrocellulose membrane and then cover it with three to four more sheets of filter paper, previously soaked in transfer buffer.

We recommend using a roll to remove any air bubbles from the "sandwich."

8. Set the transfer to 0.5 A for 30 min.

9. When finished, carefully remove the membrane and place it in a new plastic container. For evaluating the transfer efficiency, add 10 mL of Ponceau S solution and incubate the membrane for 5 min. Afterward, recover the solution and rinse the membrane with distilled water to decrease the background and visualize the transferred proteins. Mark every lane and standard proteins with a pencil to interpret the results. Ponceau S solutions can be reused several times. When loss of staining capacity is observed, discard the solution appropriately and use a fresh one. We recommend photographing the transferred protein to verify the normalization of the samples.

10. Wash the membrane with PBS-T until Ponceau S is removed. Next, add the blocking buffer (milk or BSA) to the membrane and incubate for 30 min at 37 °C.

11. Discard the blocking buffer and add the primary antibody solution. Dilute the polyclonal antiserum in blocking buffer, and incubate for 1 h at room temperature with moderate shaking (*see* **Notes 10** and **11**). We use 10 mL of solution per membrane. Dilution of the antiserum may vary. We start the experiments with 1:1000 dilution, and depending on the outcome, we adjust the concentration.

12. Wash the membrane three times with PBS-T and incubate with secondary antibody solution in blocking buffer. We use horseradish peroxidase-conjugated anti-mouse IgG (1:5000). Incubate at room temperature for 40 min with gentle shaking.

13. Wash the membranes five times with PBS-T and then add the chemiluminescent substrate for visualization of reactivity. For gene silencing, the expected results are intense signal in *L. biflexa* cells containing pMaOri.dCas9 alone, accompanied by a slightly reduced signal when dCas9 is expressed along the sgRNA pairing to the template strand and abolished signal when the sgRNA is capable of pairing to the coding strand. This result can be associated with measurable phenotypes, as described in **Note 12**. We recommend adding spectinomycin for maintaining the mutant cultures (*see* **Note 13**).

4   Notes

1. Even though our experiments were performed in the saprophyte *L. biflexa*, it is anticipated that our tool can be applied to pathogenic strains.

2. The addition of sodium pyruvate has been shown to hasten bacterial growth on plates.

3. If gene silencing is performed for an essential gene, colonies will not be recovered in plates. In our experience, depending on the gene, a partial silencing is tolerated by the cells. When an essential target gene is expected, designing sgRNA for both strands is critical for evaluating the results. In this situation, no colonies will be recovered when dCas9 is expressed along sgRNA pairing with the coding strand; when sgRNA is capable of pairing with the template strand, colonies will be recovered if the decrease in protein levels (30–50%) is tolerated by the cells.

4. For digesting the PCR fragments referring to the sgRNA cassette with *Xma*I, it is recommended that at least one more nucleotide be included at the 5′end of primers, right after the *Xma*I site. Some restriction enzymes need additional nucleotides to recognize their sites at the end of DNA.

5. Even when ordering a synthetic gene, we recommend re-amplifying only the sgRNA cassette. In our hands, digestion and ligation reactions are much more efficient if the DNA is purified directly from the reactions (PCR or digestion), with no agarose gel steps. We use an agarose gel only to verify DNA products, but never for product recovery. This strategy has greatly increased our ligation efficiency.

6. For PCR with pMaOri2 primers, we have adopted a protocol in which it is not necessary to purify genomic DNA. When performing colony PCR in *E. coli* to assess sgRNA ligation into pMaOri.dCas9, we add 1 μL of water in 0.2 mL PCR tubes, and with the aid of a pipette, we scrape gently the surface of the selected colony (marking below the plate the number or letter to identify the colony) and gently mix the PCR tube. Next, 14 μL of PCR mix are added to the samples. Positive colonies are then again scraped off the plates and grown in liquid medium. To evaluate *L. biflexa* recombinant cells, we perform PCR of the growth cultures: we centrifuge 100 μL of saturated culture, wash the pellet once in water, and resuspend it in 10 μL of water, where 1 μL of the cell suspension is used as the template for PCR. In both situations, an initial denaturation step of 5 min is established in PCR.

7. Additional washings and centrifugation steps with sterile water in 1.5 mL microtubes ensure the elimination of salt from the samples. Due to the small conical bottom of the tubes, leptospiral pellets, which are normally fragile, stay well sedimented, facilitating the aspiration of residual supernatant containing residual salt from the EMJH medium. After adopting this protocol, we drastically reduced the loss of cuvettes.

8. Any remaining albumin in the samples can turn them viscous. It is very important to wash the bacterial pellets thoroughly.

9. The cell concentration of leptospiral protein extracts can vary according to the abundance of the target protein. As a starting point, we recommend adjusting the OD_{420} of the cell suspension to 10. In some cases, as we experienced with FliG protein (low abundance), this was not enough, and therefore, we had to concentrate exponential-phase cells up to 300 times to prepare the extracts. We performed this by centrifuging 30 mL of the culture and resuspending the final pellet in 100 μL.

10. It is recommended to include an antiserum against another leptospiral protein as a loading control. For example, we have used anti-DnaK antiserum along with the antiserum against the target protein to show that leptospiral extracts are loaded into the lanes.

11. Avoid using shakers with circular rotation, since the contact of the solution with center of the membrane will be less efficient when using small volumes. If that is the only shaker available, we recommend using larger volumes. Using "up-and-down" shakers ensures that the whole membrane surface will be in contact with the solution, even with small volumes (6–10 mL per membrane).

12. Depending on the gene silenced, there is an expected phenotype to be observed, and this could work as a complementary result to Western blotting. For example, we have successfully silenced the β-galactosidase gene in *L. biflexa*, and since chromogenic substrates for this enzyme are routinely used, gene silencing was demonstrated by enzymatic assays. Also, when genes involved in motility are silenced, smaller colonies in plates can be observed.

13. Since our gene silencing tools are episomal, it is recommended to maintain selective pressure in in vitro cultures always. In our experience, the pMaOri backbone is quite stable, and β-galactosidase gene silencing in *L. biflexa* is maintained even after eight passages with no antibiotic addition. In the eighth passage, we observed a remaining 50% gene silencing, which was fully recovered after one passage in new medium with spectinomycin (100% silencing).

Acknowledgments

This work was financially supported by FAPESP (grants 2014/50981-0 and 2017/06731-8), CNPq (grants 301229/2017-1 and 441449/2014-0), and Fundação Butantan. Dr. A. Leyva (USA) provided English editing of the manuscript.

References

1. Barrangou R, Fremaux C, Deveau H, Richards M, Boyaval P, Moineau S, Romero DA, Horvath P (2007) CRISPR provides acquired resistance against viruses in prokaryotes. Science 315:1709–1712

2. Garneau JE, Dupuis M, Villion M, Romero DA, Barrangou R, Boyaval P, Fremaux C, Horvath P, Magadán AH, Moineau S (2010) The CRISPR/Cas bacterial immune system cleaves bacteriophage and plasmid DNA. Nature 468:67–71

3. Jiang Y, Chen B, Duan C, Sun B, Yang J, Yang S (2015) Multigene editing in the *Escherichia coli* genome via the CRISPR-Cas9 system. Appl Environ Microbiol 81:2506–2514

4. Shapiro RS, Chavez A, Collins JJ (2018) CRISPR-based genomic tools for the manipulation of genetically intractable microorganisms. Nat Rev Microbiol 16:333–339

5. Cobb RE, Wang Y, Zhao H (2015) High-efficiency multiplex genome editing of Streptomyces species using an engineered CRISPR/Cas system. ACS Synth Biol 4:723–728

6. DiCarlo JE, Norville JE, Mali P, Rios X, Aach J, Church GM (2013) Genome engineering in *Saccharomyces cerevisiae* using CRISPR-Cas systems. Nucleic Acids Res 41:4336–4343

7. Shan Q, Wang Y, Li J, Zhang Y, Chen K, Liang Z, Zhang K, Liu J, Xi JJ, Qiu JL, Gao C (2013) Targeted genome modification of crop plants using a CRISPR-Cas system. Nat Biotechnol 31:686–688

8. Matthews LA, Simmons LA (2014) Bacterial nonhomologous end joining requires teamwork. J Bacteriol 196:3363–3365

9. Weterings E, van Gent DC (2004) The mechanism of non-homologous end-joining: a synopsis of synapsis. DNA Repair (Amst) 3:1425–1435

10. Bikard D, Hatoum-Aslan A, Mucida D, Marraffini LA (2012) CRISPR interference can prevent natural transformation and virulence acquisition during in vivo bacterial infection. Cell Host Microbe 12:177–186

11. Cui L, Bikard D (2016) Consequences of Cas9 cleavage in the chromosome of *Escherichia coli*. Nucleic Acids Res 44:4243–4251

12. Fernandes LGV, Guaman LP, Vasconcellos SA, Heinemann MB, Picardeau M, Nascimento ALTO (2019) Gene silencing based on RNA-guided catalytically inactive Cas9 (dCas9): a new tool for genetic engineering in Leptospira. Sci Rep 9:1839

13. Qi LS, Larson MH, Gilbert LA, Doudna JA, Weissman JS, Arkin AP, Lim WA (2013) Repurposing CRISPR as an RNA-guided platform for sequence-specific control of gene expression. Cell 152:1173–1183

14. Demarre G, Guérout AM, Matsumoto-Mashimo C, Rowe-Magnus DA, Marlière P, Mazel D (2005) A new family of mobilizable suicide plasmids based on broad host range R388 plasmid (IncW) and RP4 plasmid (IncPalpha) conjugative machineries and their cognate *Escherichia coli* host strains. Res Microbiol 156:245–255

15. Pappas CJ, Benaroudj N, Picardeau M (2015) A replicative plasmid vector allows efficient complementation of pathogenic Leptospira strains. Appl Environ Microbiol 81:3176–3181

16. Zhukova A, Fernandes LG, Hugon P, Pappas CJ, Sismeiro O, Coppée JY, Becavin C, Malabat C, Eshghi A, Zhang JJ, Yang FX, Picardeau M (2017) Genome-wide transcriptional start site mapping and sRNA identification in the pathogen. Front Cell Infect Microbiol 7:10

17. Picardeau M (2008) Conjugative transfer between *Escherichia coli* and *Leptospira* spp. as a new genetic tool. Appl Environ Microbiol 74:319–322

Chapter 11

Leptospira spp. Toolbox for Chemotaxis Assay

Ambroise Lambert

Abstract

A toolbox for chemotaxis assay adapted to *Leptospira* spp. has emerged in the recent years: soft agar assay, capillary assay, and videomicroscopy tracking. Those methods allow to demonstrate chemotaxis defect, identify diverse chemoattractants, or decipher motile behavior quantitatively. These experiments have demonstrated a role of motility and potentially chemotaxis in leptospirosis pathogenesis. We describe extensively the methods and provide the key steps to use this toolbox.

Key words Chemotaxis, Leptospirosis, *Leptospira interrogans*, Capillary assay, Videomicroscopy, Motility

1 Introduction

Leptospira spp. are bacteria with saprophytic and pathogenic species belonging to the phylum of spirochetes. Pathogenic *Leptospira* are the causative agents of leptospirosis. The zoonotic disease mostly carried and disseminated by rodents represents over a million cases per year according to the most recent estimation [1]. Carriers are contaminated at the kidney level and disseminate the bacteria in the environment through their urine. Human infection mostly occurs by infection through contaminated soil or water during recreational activities or in poor sanitary conditions [2]. Pathogenic *Leptospira* strains are highly motile which enable them to quickly disseminate and reach target organs [3, 4]. The motility and possibly the chemotaxis of *Leptospira* spp. have been demonstrated to play a critical role in pathogenesis [5–9].

In recent years, cryotomography combined with genetics, videomicroscopy, and biochemistry has allowed to decipher the motility of *Leptospira* spp. [10, 11]. *Leptospira* spp. are thin (0.1–0.3 μm) and elongated (10–20 μm) helically shaped bacteria, while other spirochetes like *Borrelia burgdorferi* are flat wave shaped. This morphology combined with periplasmic flagella allows them to swim in viscous media where most peritrichous flagellated

Nobuo Koizumi and Mathieu Picardeau (eds.), *Leptospira spp.: Methods and Protocols*, Methods in Molecular Biology, vol. 2134, https://doi.org/10.1007/978-1-0716-0459-5_11, © Springer Science+Business Media, LLC, part of Springer Nature 2020

bacteria are stopped [12, 13]. Additionally, *Leptospira* spp. are atypical since they have only two endoflagella, each arising subterminally in the periplasm of the cell without overlapping at the cell center. The rotation of the motors in two different directions at each end is responsible for a forward thrust that allows the bacteria to translate [14]. However, the regulatory mechanism for coordination of flagellar motor remains unknown to date in *Leptospira* spp. Furthermore, chemotactic system regulating motility is very diverse with more than 26 chemotactic receptors in saprophytic strains and only 13 for most pathogenic strains [6]. On the other hand, motility and chemotaxis has been well characterized early in the 1960s with enterobacteria model like *Escherichia coli* with the use of capillary assay [15] or even videomicroscopy tracking [16]. We and other teams have applied and adapted those tools in different studies to decipher chemotaxis on *Leptospira* spp. bacteria model [6, 8, 17].

In this review I will describe the methods as a toolbox to understand chemotaxis, from the first intention assay soft agar plates that allow to characterize chemotaxis defect to chemoattractant identification assay like capillary assay, ending with quantitative analysis of motility with videomicroscopy tracking assay.

2 Materials

2.1 Soft Agar Assay

1. EMJH supplement [18]: Dissolve 10 g of bovine serum albumin in 50 ml of prewarmed distilled water (37 °C), slow agitation to prevent bubbles. Add 1 ml of $ZnSO_4$ (stock 0.04 g/L $ZnSO_4$ heptahydrated), 1 ml of $MgCl_2$ (stock 0.15 g/L $MgCl_2$ hexahydrated), 1 ml of $CaCl_2$ (stock 0.15 g/L $CaCl_2$ dihydrated), 1 ml of sodium pyruvate (stock 1 g/L sodium pyruvate), 2 ml of glycerol, 12.5 ml of Tween 80, and 1 ml of vitamin B12 (stock 2 mg/L vitamin B12). Add previously prepared solution of $FeSO_4$ at 0.5% and adjust volume to 100 ml and pH to 7.4.

2. EMJH basal medium: Dissolve 2.3 g of Difco Leptospira Medium Base EMJH (Becton Dickinson) in 900 ml of water.

3. EMJH medium: 100 ml of EMJH supplement, 900 ml of EMJH basal medium. Adjust pH to 7.4 if necessary and filter sterilize (0.22 μm filter), aliquot in flask of smaller volume, control sterility at 37 °C for 48 h, and then store at 4 °C.

4. Noble agar.

5. Square Petri dish with grid.

2.2   Capillary Assay

1. Motility buffer: 7 mM Na_2HPO_4, 2.2 mM KH_2PO_4, 17.1 mM NaCl, 4.7 mM NH_4Cl, 14.8 μM thiamine, 0.5% bovine serum albumin. Filter sterilize once pH adjusted to 7.4.

2. Petroff-Hausser chamber.

3. Hematocrit capillaries: 75 mm/60 μl, 0.95 mm diameter.

4. Hematocrit wax.

5. 96-Well storage plates, round bottom, 1.2 ml.

6. Automate of DNA purification.

7. DNA purification kit.

8. Real-time PCR detection system.

9. qPCR primers: use *lipL32* (forward primer (45F), 5′-AAG CAT TAC CGC TTG TGG TG-3′ reverse primer (286R) 5′-GAA CTC CCA TTT CAG CGA TT-3′) or *rpoB* primers (forward primer 5′-ATG ATG AGA CGG ATG ACT GC-3′, reverse primer 5′-CGA CGA AAC GTT TGA ACC AA-3′) to enumerate *L. interrogans* (pathogenic strain) or *L. biflexa* (saprophytic strain), respectively [19, 20].

10. qPCR SYBR green kit.

11. Methylcellulose (15 cP).

12. Microcentrifuge.

2.3   Videomicroscopy

1. Fast frame rate acquisition camera.

2. Dark-field microscope.

3. MATLAB 2011 or later version.

4. Micro-Manager 1.4.

5. FIJI, ImageJ software.

3   Methods

3.1   Soft Agar Assay

Soft agar assay is often used to assess motile behavior. This test is used as a first intention assay to evaluate chemotactic or motile behavior of *Leptospira* spp. and to provide information about the ability of *Leptospira* spp. to migrate once their nutrient source is depleted. However, it does not discriminate between motility and chemotaxis defect. Two types of soft agar assay have been developed: 0.5% agar plate and 0.3% agar plate. This strategy allows to identify intermediate chemotactic or motile phenotypes [5].

1. To prepare soft agar EMJH medium, add 300 mg or 500 mg of noble agar to 100 ml of freshly prepared EMJH medium for 0.3% agar plate and 0.5% agar plate, respectively. Dissolve carefully agar using a microwave. Cool down agar and media at 60 °C for 15 m, and pour 35 ml in square Petri dish with grid.

2. Grow *Leptospira* spp. to test until exponential phase in liquid EMJH medium; first evaluate motility in dark-field microscopy while counting using a Petroff-Hausser chamber. Then slow speed centrifugate bacteria in EMJH medium at 3000 × g for 10 m at room temperature. Resuspend the pellet in appropriate volume such that 10 µl contains 10^6 bacteria.

3. Once soft agar EMJH medium has solidified, cooled, and dried to room temperature, drill equally spaced holes in agar using the grid of the plate as a reference (used 1 ml pipette tip to drill the holes). Maximal capacity of square Petri dish can accommodate three rows and three columns of holes to perform optimally soft agar assay. Inoculate each hole with 10 µl of the previously prepared bacterial suspension to test.

4. Incubate plates for up to 15 days at 30 °C for most pathogenic *Leptospira* spp. and up to 7–10 days for the saprophytic model bacterium *Leptospira* biflexa.

5. Measure the diameters of the zone of spread, and repeat soft agar assays at least for three independent cultures of each bacterial strain to test. Always add the wild-type strain if you test mutant strain. Add for each plate assay a motility mutant like *ΔflaB* [21] as a negative control.

3.2 Capillary Assay

Capillary assay was first designed for bacteria by Julius Adler in 1967, with the first chemotaxis capillary assay in the model bacterium *E. coli* [15]. They provide a quantitative assessment of chemotactic response to specific compounds. First capillary assay for *Leptospira* spp. was designed in 1993 [22]. Additionally, another assay was adapted in *Borrelia burgdorferi* [23]. This assay is optimized for *Leptospira* spp. with higher throughput by combining it with quantitative PCR [6, 23], and it could be also combined with flow cytometry.

1. Grow exponential-phase *Leptospira* in EMJH liquid medium (10 ml). Suspension should be at $OD_{420nm} = 0.5$, which corresponds to approximately 5 × 10^8 bacteria/ml. Do a quick motility check under dark-field microscope, and most bacteria should be actively motile (95% of cells with actively gyrating ends). Harvest cells at room temperature at low centrifugation speed (3000 × g for 10 m).

2. On the same day prepare the motility buffer.

3. Resuspend harvested bacteria into motility buffer to 10^8 bacteria/ml, incubate overnight (18 h) at 30 °C to allow bacteria to recover motility, and deplete nutrient carried over during centrifugation.

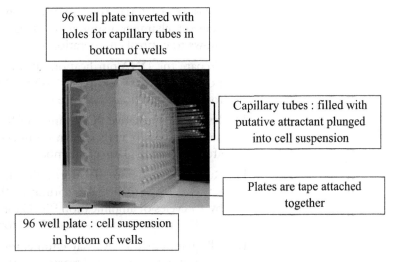

96 well plate inverted with
holes for capillary tubes in
bottom of wells

Capillary tubes : filled with
putative attractant plunged
into cell suspension

Plates are tape attached
together

96 well plate : cell suspension
in bottom of wells

Fig. 1 Chemotaxis chamber setup (picture annotated)

4. Measure motility after incubation using a Petroff-Hausser chamber, approximately 60% of cells should be translated, and the remaining cells should display gyrating ends.

5. Prepare chemotaxis chambers using two storage plates of 96 wells with 1.2 ml rounded bottom facing each other. Take one of the plates, and pierce holes at the bottom of wells through the plastic using a heated tweezer. The holes need to be large enough to easily allow capillaries to enter the chamber. Fill the other plate with 200 μl suspension of 2×10^7 bacteria in motility buffer. Invert the perforated plate and face it to the plate filled with bacterial suspension. Tape the two plates together.

6. Prepare four capillary tubes for each condition, and fill capillaries by plunging them in control or hypothetical chemoattractant solution (*see* **Note 1**). Once filled seal the end of the capillary with wax seal. Wipe carefully with low lint wipes all excess fluid remaining outside of the capillary's end immersed in attractant solution. Plunge capillaries carefully in chemotaxis chamber immersed in bacterial suspension.

7. Incubate the chemotaxis chamber from 30 m to 1 h at 30 °C (Fig. 1). Maintain the chemotaxis chamber with capillaries in horizontal position while in the incubator and after.

8. After incubation, remove capillaries and wipe out any excess fluid of bacteria suspension on the outside of capillary. Over a 15 ml conical tube, break the seal wax extremity and dispose of it (*see* **Note 2**). Centrifuge capillaries with liquid in conical tube for 1 m at $3000 \times g$. Approximately 50 μl of suspension should remain at the end of the procedure.

9. Measure bacteria concentration and motility in Petroff-Hausser chamber using 5 μl of each suspension. This step allows to control enumeration with quantitative PCR.

10. Prepare the DNA purification kit for automated DNA extraction (*see* **Note 3**), and pipette 40 μl of the capillary suspension for each extraction (*see* **Note 4**).

11. Prepare quantitative PCR using SYBR green mix. Prepare standard curves mix with known numbers of leptospires enumerated in Petroff-Hausser chamber.

12. Run PCR program: PCR conditions were as follows (ramp rates of 20 °C/s): initial denaturation 95 °C for 600 s; followed by 45 cycles of amplification 95 °C for 10 s, 57 °C for 8 s, and 72 °C for 10 s; and fusion at 95 °C for 360 s [23].

13. All quantitative PCR assay enumeration should be performed in duplicate with control reaction without template. Each assay is repeated in at least two independent experiments with three to four capillaries in each condition tested. Use enumeration values to quantify chemotactic response in fold change for capillary containing chemoattractant compared to motility buffer only containing capillary (a change of 2 or below is not considered significant).

3.3 Videomicroscopy Tracking Assay

Compared to previously mentioned methods, videomicroscopy tracking assay allows a direct observation of the motile behavior. We can track individual motile behavior, observe trajectories, pause, and increase in motility. We can quantify speed of bacteria and length of trajectories. It was first pioneered in 1972 in *E. coli* [16]. A dark-field videomicroscope with ImageJ analysis and MATLAB scripts can be used to track and adapt the method to *Leptospira* spp. [8].

1. Grow actively motile *Leptospira* spp. in EMJH liquid medium until exponential phase. Ensure that the bacteria remain motile daily using dark-field microscopy. Harvest cell using slow-speed centrifugation (3000 × g for 10 min).

2. Prepare motility buffer the day of the experiment. Prewarm 100 ml of motility buffer at 65 °C, add 1 g of methylcellulose (15 cP), and homogenize. Resuspend harvested bacteria in viscous motility buffer (1% methylcellulose).

3. Prepare sample for microscopy. Add four drops (2 mm size) of silicon grease on glass slides, and the drops should form a square of 15 mm. Deposit 0.8 μl of bacterial suspension in viscous motility buffer at the center of the four drops. Apply an 18 mm-squared cover glass on top of the drops until there is contact with the bacterial suspension drop.

4. Acquire 2 min video with ten frames per second. Use a dark-field microscope with 20× magnification. For each strain acquire at least ten independent 2 min videos from two independent culture batches. Make sure to dilute bacteria to a density of ten bacteria per field of view.

5. Perform image analysis using ImageJ to binarize video, decrease noise using background subtraction, and then export as .tiff file for quantification and trajectory analysis. Then, analyze movies using MATLAB scripts based from BACtrack software, a boundary detection algorithm is applied, trajectories using the center of mass between consecutive frames (http://www.rowland.harvard.edu/labs/bacteria/software/index.php). This software allows to extract the length of trajectories, average speed, and number of reversal (any change of direction of a total 180° ± 45°).

4 Notes

1. When filling capillaries leave enough space at the other end of capillaries to accommodate the wax that seals the capillary tube.

2. To extract capillary fluid, be su re to hold only hematocrit wax to avoid breaking the capillary in the middle and losing fluid.

3. For DNA extraction, automated DNA extraction is preferred but not mandatory for robust standardized results.

4. For DNA extraction from capillary fluid, avoid pipetting glass after centrifuging capillary tubes, as the latter can result in lower yield.

References

1. Costa F, Hagan JE, Calcagno J et al (2015) Global morbidity and mortality of leptospirosis: a systematic review. PLoS Negl Trop Dis 9: e0003898

2. McBride AJA, Athanazio DA, Reis MG, Ko AI (2005) Leptospirosis. Curr Opin Infect Dis 18:376–386

3. Faine S, Adler B, Bolin C, Perolat P (1999) Leptospira and leptospirosis. MediSci, Melbourne, Australia, p 259

4. Faine S, Vanderhoeden J (1964) Virulence-linked colonial and morphological variation in *Leptospira*. J Bacteriol 88:1493–1496

5. Lambert A, Picardeau M, Haake DA et al (2012) FlaA proteins in *Leptospira interrogans* are essential for motility and virulence but are not required for formation of the flagellum sheath. Infect Immun 80:2019–2025

6. Lambert A, Takahashi N, Charon NW, Picardeau M (2012) Chemotactic behavior of pathogenic and nonpathogenic *Leptospira* species. Appl Environ Microbiol 78:8467–8469

7. Eshghi A, Becam J, Lambert A et al (2014) A putative regulatory genetic locus modulates virulence in the pathogen *Leptospira interrogans*. Infect Immun 82:2542–2552

8. Lambert A, Wong JN, Picardeau M (2015) Gene inactivation of a chemotaxis operon in the pathogen *Leptospira interrogans*. FEMS Microbiol Lett 362:1–8

9. Fontana C, Lambert A, Benaroudj N et al (2016) Analysis of a spontaneous non-motile and avirulent mutant shows that FliM is required for full endoflagella assembly in *Leptospira interrogans*. PLoS One 11:e0152916

10. Wunder EA Jr, Figueira CP, Benaroudj N et al (2016) A novel flagellar sheath protein, FcpA, determines filament coiling, translational motility and virulence for the *Leptospira spirochete*. Mol Microbiol 101:457–470

11. Tahara H, Takabe K, Sasaki Y et al (2018) The mechanism of two-phase motility in the spirochete *Leptospira*: swimming and crawling. Sci Adv 4:eaar7975. https://doi.org/10.1126/sciadv.aar7975

12. Kaiser GE, Doetsch RN (1975) Letter: Enhanced translational motion of *Leptospira* in viscous environments. Nature 255:656–657

13. Greenberg EP, Canale-Parola E (1977) Relationship between cell coiling and motility of spirochetes in viscous environments. J Bacteriol 131:960–969

14. Goldstein SF, Charon NW (1988) Motility of the spirochete *Leptospira*. Cell Motil Cytoskeleton 9:101–110. https://doi.org/10.1002/cm.970090202

15. Adler J, Dahl MM (1967) A method for measuring the motility of bacteria and for comparing random and non-random motility. J Gen Microbiol 46:161–173. https://doi.org/10.1099/00221287-46-2-161

16. Berg HC, Brown DA (1972) Chemotaxis in *Escherichia coli* analysed by three-dimensional tracking. Nature 239:500–504

17. Affroze S, Islam MS, Takabe K et al (2016) Characterization of leptospiral chemoreceptors using a microscopic agar drop assay. Curr Microbiol 73:202–205. https://doi.org/10.1007/s00284-016-1049-1

18. Ellinghausen H Jr, McCullough W (1965) Nutrition Of *Leptospira pomona* and growth of 13 other serotypes: a serum-free medium employing oleic albumin complex. Am J Vet Res 26:39

19. Casanovas-Massana A, Pedra GG, Wunder EAJ et al (2018) Quantification of *Leptospira interrogans* survival in soil and water microcosms. Appl Environ Microbiol 84. https://doi.org/10.1128/AEM.00507-18

20. Lourdault K, Aviat F, Picardeau M (2009) Use of quantitative real-time PCR for studying the dissemination of *Leptospira interrogans* in the guinea pig infection model of leptospirosis. J Med Microbiol 58:648–655. https://doi.org/10.1099/jmm.0.008169-0

21. Picardeau M, Brenot A, Saint Girons I (2001) First evidence for gene replacement in *Leptospira* spp. Inactivation of *L. biflexa* flaB results in non-motile mutants deficient in endoflagella. Mol Microbiol 40:189–199

22. Yuri K, Takamoto Y, Okada M et al (1993) Chemotaxis of leptospires to hemoglobin in relation to virulence. Infect Immun 61:2270–2272

23. Li C, Bakker RG, Motaleb MA et al (2002) Asymmetrical flagellar rotation in *Borrelia burgdorferi* nonchemotactic mutants. Proc Natl Acad Sci U S A 99:6169–6174. https://doi.org/10.1073/pnas.092010499

Chapter 12

In Situ Structural Analysis of *Leptospira* spp. by Electron Cryotomography

Akihiro Kawamoto

Abstract

Spirochetes such as *Treponema*, *Borrelia*, and *Leptospira* species can rotate their bodies to swim in liquid environments by rotating periplasmic flagella or endoflagella, which are present inside the cell. Electron cryotomography (ECT) is an imaging technique that directly provides three-dimensional (3D) structures of cells and molecular complexes in their cellular environment at nanometer resolution. Here, I present a general protocol of ECT that covers the sample preparation, data collection, tilt series alignment, and tomographic reconstruction for visualization of intact periplasmic flagella in *Leptospira* spp. This protocol is capable of determining protein structures at resolutions high enough to visualize their individual domains and secondary structures in their cellular environment.

Key words Electron cryotomography (ECT), *Leptospira*, Bacterial flagellum

1 Introduction

Bacteria such as *Escherichia coli* and *Salmonella* actively swim in liquid environments by rotating long, helical, filamentous organelle called the bacterial flagellum [1]. At the base of each filament, there is a rotary motor powered by proton or other cation motive force across the cytoplasmic membrane. The bacterial flagellum is a large molecular assembly composed of nearly 30 different proteins that identify it as an elaborate biological nanomachine [2]. While the bacterial flagellum is generally present outside the cell body, the spirochetes such as *Treponema*, *Borrelia*, and *Leptospira* species have flagella in the periplasm of the cell [3–5]. The periplasmic flagella generate swimming motility in liquid environments and crawling motility on solid surfaces [6–8]. Interestingly, recent studies show that the periplasmic flagella determine the typical hook- and spiral-shaped cell ends of *Leptospira* [9–11]. Structural analysis of the periplasmic flagella is required for understanding their motility and cell shape determination mechanisms, but it is difficult to

Nobuo Koizumi and Mathieu Picardeau (eds.), *Leptospira spp.: Methods and Protocols*, Methods in Molecular Biology, vol. 2134, https://doi.org/10.1007/978-1-0716-0459-5_12, © Springer Science+Business Media, LLC, part of Springer Nature 2020

observe their structures inside the cell by conventional electron microscopy technique.

Here, I describe a general protocol for the in situ structural analysis of periplasmic flagella in *Leptospira* spp. by electron cryo-tomography (ECT). ECT is an imaging technique that directly provides 3D structures of cells and molecular complexes in their cellular environment at nanometer resolution. EM images are 2D projections of 3D objects. To obtain the structures of cells and molecules, the specimen grid is tilted incrementally around an axis perpendicular to the electron beam, e.g., from −60° to +60° with 2° increments, and an image is taken at each tilt angle. Each image represents a different view of the 3D objects on the grid. The images of a tilt series are aligned and are back-projected to generate a 3D image (tomogram) of the specimen. Using the protocols described here, the structures of periplasmic flagella are directly visualized whereby the interaction between the periplasmic flagella and outer membrane can be revealed [12]. In this chapter, I focus on sample preparation, data collection, tilt series image alignment, and tomographic reconstruction for visualization of intact periplasmic flagella in *Leptospira* spp.

2   Materials

Prepare all solutions using ultrapure water (resistance = 18 MΩ at 25 °C) and analytical grade reagents. Prepare all reagents at room temperature. Store all reagents at 4 °C. Particular attention should be paid to the adequate disposal of materials contaminated with *Leptospira*.

2.1 Bacterial Stain and Cell Culture

1. Bacterial strain: *Leptospira* biflexa serovar Patoc strain Patoc I.

2. EMJH basement: Dissolve 2.3 g of Difco™ Leptospira Medium Base EMJH (Becton Dickinson) in 800 mL of ultrapure water. Add 1 mL of glycerol (stock 10 g/100 mL glycerol) and 1 mL of sodium pyruvate (stock 10 g/100 mL sodium pyruvate).

3. EMJH supplement: Dissolve 10 g of bovine serum albumin in 100 mL of ultrapure water. Mix 1 mL of $ZnSO_4$ (stock 0.4 g/100 mL $ZnSO_4 \cdot 7H_2O$), 1 mL of $MgCl_2$ (stock 1.5 g/100 mL $MgCl_2 \cdot 6H_2O$), 1 mL of $CaCl_2$ (1.5 g $CaCl_2 \cdot 2H_2O$), 12.5 mL of Tween 80 (stock 10 g/100 mL Tween 80), 1 mL of cyanocobalamin (stock 0.02 g/100 mL cyanocobalamin), and 10 mL of $FeSO_4$ (0.05 g/10 mL $FeSO_4 \cdot 7H_2O$, prepare just before making EMJH supplement), and adjust to 198 mL with ultrapure water (*see* **Note 1**).

4. Liquid EMJH medium: 802 mL of EMJH basement, 198 mL of EMJH supplement, pH 7.4. Sterilize the medium by filtration with 0.22 μm pore size filter.

5. Phosphate buffered saline (PBS): 137 mM NaCl, 2.7 mM KCl, 10 mM Na_2HPO_4, 1.8 mM KH_2PO_4.

6. 10 nm colloidal gold particle solution.

2.2 Equipment and Electron Microscope

1. 30 °C incubator.

2. Sterile conical tubes of 15 mL capacity.

3. Microcentrifuge.

4. Tweezer.

5. Ion coater.

6. Quantifoil molybdenum 200 mesh R0.6/1.0 grid (Quantifoil Micro Tools, Jena, Germany).

7. Vitrobot Mark IV (Thermo Fisher Scientific).

8. Titan Krios FEG transmission electron microscope (Thermo Fisher Scientific).

9. FEI Falcon II 4k × 4k CMOS direct electron detector camera (Thermo Fisher Scientific).

2.3 Program for Data Collection and Structural Analysis by ECT

1. EPU software (Thermo Fisher Scientific).

2. Xplore3D software package (Thermo Fisher Scientific).

3. UCSF Chimera [13].

4. IMOD software package [14].

3 Methods

3.1 Sample Preparation

1. *Leptospira* cells are cultivated at 30 °C in EMJH medium for 7–8 days.

2. The 100 μL of pre-culture is added to 5 mL EMJH medium, and then the cells are incubated at 30 °C in EMJH medium for 3–4 days until a final concentration ~2 × 10^8 cells/mL.

3. The cells are collected by centrifugation (4600 × g, 2 min, 25 °C).

4. The pellet is suspended in 50 μL PBS.

3.2 ECT Sample Preparation

1. Quantifoil molybdenum 200 mesh R0.6/1.0 holey carbon grids are glow-discharged on a glass slide for 5 s.

2. A 3.0 μL of sample solution (*Leptospira* cells in PBS prepared in Subheading 3.1) containing 10 nm colloidal gold particles is applied onto the grid, and the grid is dried up at room temperature (*see* **Note 2**).

3. The pretreated grids are glow-discharged again on a glass slide for 20 s.

4. A 2.6 µL aliquot of the sample solution is applied onto the pretreated grid, and wait for 30 s to attach the *Leptospira* on the grid surface.

5. The grid is blotted by a filter paper for 4 s, maintained chamber at 22 °C and 100% relative humidity, and quickly frozen by rapidly plunging it into liquid ethane using Vitrobot Mark IV.

6. The grid is quickly transferred from the liquid ethane to liquid nitrogen.

7. The extra ethane solution is removed by a fan-shaped strip of a filter paper.

8. The grids are inserted into a Titan Krios FEG transmission electron microscope operated at 300 kV with a cryo-specimen stage cooled with liquid nitrogen.

3.3 Data Acquisition

1. After aligning the microscope, the grid is loaded into electron microscope.

2. Scan the grid at Atlas mode of EPU (*see* **Note 3**) to check the ice thickness and good position where potentially data can be collected (Fig. 1) (*see* **Notes 4** and **5**).

3. Go to the area where data will be collected and select the position of interest for data collection.

4. Check the eucentric height of interesting position by tilting the stage to ±30° and adjusting the stage Z position.

5. Save the X, Y, Z position of interesting area and then select another interesting area.

6. Set the "low-dose mode." Low-dose mode is to minimize beam damage on the area in which you want to collect data. Low-dose mode has three different modes, which are search, focus, and acquisition mode. First, set "search" mode at low magnification (e.g., 2900×) for centering the same position as the acquisition area. Generally, apply high defocus value and binning the images (binning 2× or 4×) is used to increase the signal to ratio (S/N) of images.

7. Set "acquisition" mode at an appropriate magnification, spot size, and beam intensity.

8. In acquisition mode, calculate the total dose, exposure time, angle range, and number of images in the tilt series (*see* **Note 6**).

9. Set "focus" mode at the same magnification, spot size, and beam setting as acquisition mode. Focus area is moved to carbon surface on the tilt axis.

10. Run the Xplore3D software package.

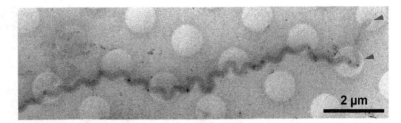

Fig. 1 Representative cryo-EM image of *Leptospira biflexa*. Target region is a cell tip area of the cell in the hole of grid. Red arrowheads indicate the region selected to collect a tilt series

3.4 Tilt Series Alignment and Tomographic Reconstruction Using IMOD

1. After data collection, the raw tilt series are saved as new stack file with the suffix ".st" using the *newstack* command of IMOD software.

2. Type "etomo" for starting the IMOD software graphical user interface (GUI).

3. Open the new stack file and select appropriate data type.

4. Click the "Scan Header" for reading pixel size and image rotation from header of new stack file. Type the diameter of fiducial gold maker.

5. Align tilt series based on cross-correlation. Click "Calculate Cross-Correlation" and "Generate Coarse Aligned Stack." Click "View Aligned Stack in 3dmod" for checking the result of coarse alignment.

6. Select ~20 gold fiducial makers and make seed model. Click "Track Seed Model" for tracing gold fiducial makers at each tilt angle. If you find any untracked or mis-tracked makers, you can manually fix them using "Fix Fiducial Model."

7. Click "Compute Alignment" for calculating fine alignment using seed model and "View/Edit Fiducial Model" for checking this result.

8. Type appropriate tomogram thickness and select "Use whole tomogram" with "Binning 2." Click "Create Whole Tomogram" and "Create Boundary Model" for selecting boundary lines and then save it. Finally, click "Create Final Alignment" for calculating final alignment parameter.

9. Click "Create Full Aligned Stack" for making aligned stack file (*see* **Note 7**).

10. Select "Weighted Back Projection" or "SIRT (simultaneous iterative reconstruction technique)" for generating final tomogram (*see* **Note 8**). Click "View Tomogram(s) in 3dmod" for checking the generated tomogram.

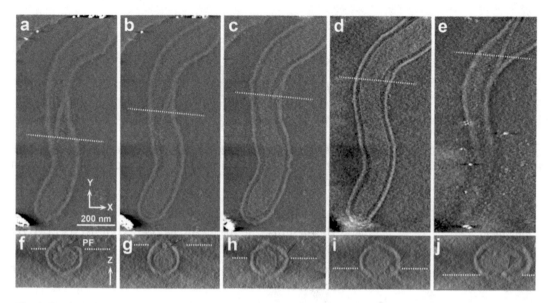

Fig. 2 Tomogram of the cell tip of *Leptospira biflexa*. (**a–e**) Slice images from tomogram of *Leptospira biflexa*. Each slice is a different *Z* height position as indicated white dotted line in panel (**f–j**). Cross-section images as indicated yellow dotted line in panel (**a–e**). Red arrowheads show periplasmic flagella (PF)

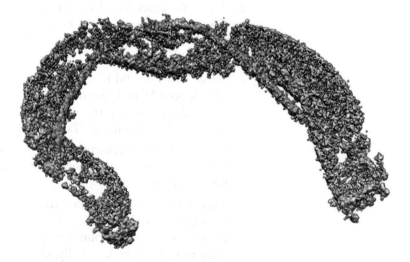

Fig. 3 Segmentation of cell body and periplasmic flagella. The cell body and periplasmic flagellum are indicated in gray and pink, respectively

3.5 Visualization of Periplasmic Flagella

1. Display all objects in the tomogram to visualize the periplasmic flagella in native cellular environment using UCSF Chimera program (Fig. 2).

2. Segment the cell body and periplasmic flagella using Segment Map option of UCSF Chimera program (Fig. 3).

4 Notes

1. $MgCl_2 \cdot 6H_2O$, $CaCl_2 \cdot 2H_2O$, glycerol, and Tween 80 solutions are sterilized by autoclaving at 121 °C for 20 m. $ZnSO_4 \cdot 7H_2O$, and sodium pyruvate solutions are sterilized by a filter with 0.22 μm pore size.

2. The 10 nm colloidal gold particle solution should be concentrated by 1.5 times before it is applied onto holey carbon EM grids.

3. EPU is generally used for data collection of single particle analysis, but Atlas mode of EPU is useful for scanning the whole grid.

4. It is preferable to avoid broken hole on the carbon film and near the edges of the grid for complete data collection to high-tilt angle.

5. Select the cell tip area, because the flagellar motor and periplasmic flagella are located at the cell tips in *Leptospira* spp.

6. Total dose is generally between 100 and 200 e/Å^2. The number of tilt images is dependent on tilt angles and the angular increment. For example, tilt series from −60° to 60° with tilt step of 2° or 3°, there will be 61 and 41 images, respectively.

7. To enhance contrast of the tomogram, use the "Use linear interpolation" with binning 2 or 4 and "2D Filter" options. The "Erase Gold" option is better for visualizing periplasmic flagella. The "Correct CTF" option is better for obtaining high-resolution subtomogram averaging structure.

8. SIRT reconstructions can generate high-contrast tomogram. It is better for visualizing periplasmic flagella and particle picking for subtomogram averaging.

Acknowledgments

The research study described in this chapter was supported by JSPS KAKENHI Grant Number 18K14639.

References

1. Berg HC (2003) The rotary motor of bacterial flagella. Annu Rev Biochem 72:19–54

2. Minamino T, Morimoto YV, Kawamoto A et al (2018) *Salmonella* flagellum. Salmonella: A Re-emerging Pathogen. 3–18

3. Izard J, Renken C, Hsieh CE et al (2009) Cryo-electron tomography elucidates the molecular architecture of *Treponema pallidum*, the Syphilis Spirochete. J Bacteriol 191:7566–7580

4. Zhao X, Zhang K, Boquoi T et al (2013) Cryoelectron tomography reveals the sequential assembly of bacterial flagellar in *Borrelia burgdorferi*. Proc Natl Acad Sci U S A 110:14390–14395

5. Raddi G, Morado DR, Tan J et al (2012) Three-dimensional structures of pathogenic and saprophytic *Leptospira* species revealed by cryo-electron tomography. J Bacteriol 194:1299–1306

6. Goldstein SF, Charon NW (1990) Multiple-exposure photographic analysis of a motile spirochete. Proc Natl Acad Sci U S A 87:4895–4899

7. Nakamura S, Leshansky A, Magariyama Y, Namba K, Kudo S (2014) Direct measurement of helical cell motion of the spirochete *Leptospira*. Boiphys J 106:47–54

8. Cox PJ, Twigg GI (1974) Leptospiral motility. Nature 250:260–261

9. Picardeau M, Brenot A, Saint Girons I (2001) First evidence for gene replacement in *Leptospira* spp. Inactivation of *L. biflexa flaB* results in non-motile mutants deficient in endoflagella. Mol Microbiol 40:189–199

10. Wunder EA Jr, Figueira CP, Benaroudj N et al (2016) A novel flagellar sheath protein, FcpA, determines filament coiling, translational motility and virulence for the *Leptospira* spirochete. Mol Microbiol 101:457–470

11. Sasaki Y, Kawamoto A, Tahara H et al (2018) Leptospiral flagellar sheath protein FcpA interacts with FlaA2 and FlaB1 in *Leptospira* biflexa. PLoS One 13:e0194923

12. Takabe K, Kawamoto A, Tahara H et al (2017) Implications of coordinated cell-body rotations for *Leptospira* motility. Biochem Biophys Res Commun 491:1040–1046

13. Pettersen EF, Goddard TD, Huang CC et al (2004) UCSF Chimera—a visualization system for exploratory research and analysis. J Comput Chem 25:1605–1612

14. Kremer JR, Mastronarde DN, McIntosh JR (1996) Computer visualization of three-dimensional image data using IMOD. J Struct Biol 116:71–76

Chapter 13

Measurement of the Cell-Body Rotation of *Leptospira*

Shuichi Nakamura

Abstract

Leptospira spp. swim in liquid and crawl on surfaces with two periplasmic flagella. The periplasmic flagella attach to the protoplasmic cylinder via basal rotary motors (flagellar motors) and transform the ends of the cell body into spiral or hook shape. The rotations of the periplasmic flagella are thought to gyrate the cell body and rotate the protoplasmic cylinder for propelling the cell; however, the motility mechanism has not been fully elucidated. Since the motility is a critical virulence factor for pathogenic leptospires, the kinematic insight is valuable to understand the mechanism of infection. This chapter describes microscopic methodologies to measure the motility of *Leptospira*, focusing on rotation of the helical cell body.

Key words *Leptospira*, Motility, Cell-body rotation, Rotation speed, Image analysis, Bead assay

1 Introduction

1.1 Background

Leptospira spp. have a thin (~0.14 μm in diameter), long (~10 μm in length), helical cell body with the helix pitch of ~0.7 μm and the helix diameter of ~0.2 μm. Two short flagella (~3 μm in length) reside within the periplasmic space, so-called periplasmic flagella (PFs), one PF per each cell end (Fig. 1). The PFs isolated from cells show coiled shape, and the existence of the curved and quasi-rigid PF within the periplasmic space results in bending of both ends of the cell body into spiral or hook shape. *Leptospira* spp. not only swim in liquids but also crawl on surfaces [1, 2]. Swimming involves gyrations of the spiral-shaped anterior (spiral-end) and hook-shaped posterior end (hook-end) and rolling of the short-pitch cell body (protoplasmic cylinder, PC) [3, 4]. When *Leptospira* spp. crawl, the PC rolling propels the spirochete attached to a solid surface, to which gyrations of spiral-end and hook-end hardly contribute [2]. The shapes of the cell end frequently change between spiral shape and hook shape (Fig. 2). When both ends show the same morphology ("spiral-spiral" and "hook-hook" in Fig. 2), the cell rotates but does not migrate.

Nobuo Koizumi and Mathieu Picardeau (eds.), *Leptospira spp.: Methods and Protocols*, Methods in Molecular Biology, vol. 2134, https://doi.org/10.1007/978-1-0716-0459-5_13, © Springer Science+Business Media, LLC, part of Springer Nature 2020

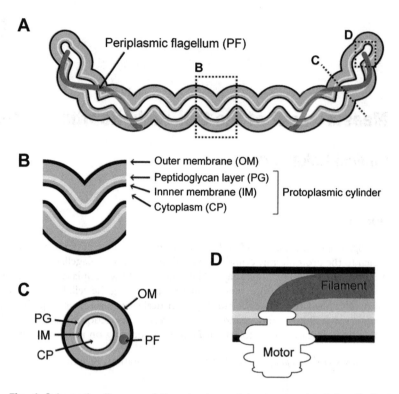

Fig. 1 Schematic diagram of the structure of *Leptospira*. (**a**) A longitudinal section. The area surrounded by a dashed square is the enlarged in (**b**), explaining the arrangement of membranes. (**c**) A cross section at the position indicated by a dashed line in **a**. (**d**) A basal motor and a filament of the periplasmic flagellum

Bacterial motility is a critical virulence factor of many pathogenic species [5, 6]. Likewise, motility in pathogenic leptospires is a key determinant of infectivity and host dissemination [7, 8]. The swimming motility is believed to be reliable strategy to invade host animals through lesions on the skin or membrane mucosa and reaches specific organs. Also, after attaching to surfaces of host cells, leptospires could use crawling to penetrate between cells of the tissues. To understand the mechanism of motility of *Leptospira*, many mutants lacking flagella or deficient in motility have been investigated [7–12]. "Motile fraction" and "swimming velocity" are measured primarily in motility assays of such mutants, though, in some cases, the cell-body rotation rate would be another kinematic parameter for more accurate characterization of the leptospiral motility.

1.2 Motility Parameters

To clarify significance of the measurement of the cell-body rotation, representative parameters related to the *Leptospira* motility are overviewed below.

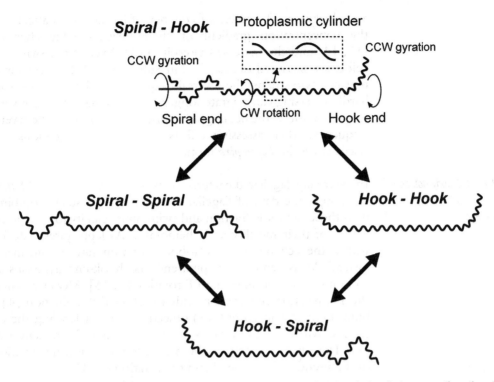

Fig. 2 Variation of the *Leptospira* morphology. Spiral-end is a left-handed spiral, generating thrust by counterclockwise (CCW) gyration when being viewed from hook-end to spiral-end, whereas the right-handed PC rotate clockwise (CW) for propelling a cell. Cells exhibiting spiral-hook or hook-spiral morphology translate with the anterior spiral and the posterior hook. Spiral-spiral and hook-hook cells rotate at one position without translation

1.2.1 Motile Fraction

When observing the wild-type *Leptospira* cells under a microscope, you will find motile cells and non-motile ones, no matter what species and serovars. The fraction of motile cells can be assessed easily just by counting the numbers of motile and non-motile cells, which is useful as the first quantitative evaluation of motility. Since directional movements (i.e., swimming and crawling) of leptospires involve coordination between two PFs, some cells show rotation without net migration (Fig. 2). Such rotation cells are "non-translation cells" but should be discriminated from non-motile cells. Thus, whether the cell rotate is a significant evaluation criterion.

1.2.2 Velocity of Translation

The velocities of swimming and crawling of individual cells are obtained by tracking the centroid of a cell body. When the cells are recorded at 30 frames per second (so-called video rate), the cell is assumed to move straight for 1/30 s.

1.2.3 Reversal Frequency

Leptospira cells frequently reverse the swimming and crawling direction. The reversal would be caused by switching of the rotational direction of the flagellar motor between counterclockwise and clockwise. The reversal frequency of leptospires is affected by the concentration gradient of chemotaxis-inducible chemicals [13, 14] and the increased viscosity [15]. Even if the swimming velocities measured as described in Subheading 1.2.2 were similar between two strains, the reversal frequency can be different. Cells hardly reversing can migrate longer distances, namely, the reversal frequency affects the net migration distance. Thus, note the reversal frequency, when assessing effects of mutations, chemicals, and viscosity on the *Leptospira* motility.

1.2.4 Cell-Body Rotation Rate (Spiral-End, PC, Hook-End)

In externally flagellated bacteria, such as *Escherichia coli* and *Salmonella* spp., the data of flagellar rotation rate is usually combined with those of motile fraction and swimming velocity for characterization of their motility and chemotaxis. As leptospires hide PFs within the cell body, the cell-body rotation rate is informative instead. Measurements of spiral-end and hook-end gyrations are an indirect evaluation of the PF rotation [2, 16]. Measurement of the PC rotation requires one-sided dark-field illumination [4], a high numerical aperture (NA) objective [16], or labeling the cell surface with probes such as microbeads and fluorescent dyes [2]. This chapter describes the methods for measuring gyrations of spiral-end and hook-end and the rotation of PC.

2 Materials

2.1 Bacteria and Media

1. *Leptospira biflexa* strain Patoc I.

2. EMJH basement: Dissolve 2.3 g of Difco™ Leptospira Medium Base EMJH (Becton Dickinson) in 900 mL of distilled water. Add 1 mL of glycerol (stock 10 g/100 mL glycerol) and 1 mL of sodium pyruvate (stock 10 g/100 mL sodium pyruvate).

3. EMJH supplement: Dissolve 10 g of bovine serum albumin in 50 mL of distilled water. Mix 1 mL of $ZnSO_4$ (stock 0.4 g/100 mL $ZnSO_4 \cdot 7H_2O$), 1 mL of $MgCl_2$ (stock1.5 g/100 mL $MgCl_2 \cdot 6H_2O$), 1 mL of $CaCl_2$ (1.5 g $CaCl_2 \cdot 2H_2O$), 12.5 mL of Tween 80 (stock 10 g/100 mL Tween 80), 1 mL of cyanocobalamin (stock 0.02 g/100 mL cyanocobalamin), and 10 mL of $FeSO_4$ (0.05 g/10 mL $FeSO_4 \cdot 7H_2O$, prepare just before making EMJH supplement), and adjust to 98 mL with sterilized water. Sterilize $MgCl_2$, $MgCl_2$, glycerol, and Tween 80 stock solutions at 121 °C for 20 min and $ZnSO_4$, cyanocobalamin, and sodium pyruvate stock solutions by using a filter with 0.22 μm pore size. Stock solutions can be stored at 4 °C.

4. Liquid EMJH medium: 902 mL of EMJH basement, 98 mL of EMJH supplement, pH 7.4. Store at 4 °C.

5. Motility medium: 20 mM sodium phosphate buffer, pH 7.4. Make 200 mM stock solution by mixing 81 mL of 200 mM Na_2HPO_4 and 19 mL of 200 mM NaH_2PO_4; measure pH using a pH meter; if necessary, adjust to pH 7.4 by the addition of Na_2HPO_4 or NaH_2PO_4 solution and then autoclave. Dilute the stock solution with distilled water to 20 mM.

6. Polymer solutions: Add polymers to motility medium to increase viscosity or viscoelasticity. Ficoll (5–30 w/v%) and methylcellulose (0.25–2 w/v%) are generally used in motility assays for microorganisms (*see* **Note 1**).

7. MES buffer: 50 mM 2-morpholinoethanesulfonic acid, pH 5.2.

8. 10 mM Tris–HCl buffer, pH 8.0.

2.2 Reagents for Preparation of Antibody-Coated Beads

1. Polystyrene beads with a diameter of 0.2 μm.

2. Antibody against *Leptospira* whole cell or lipopolysaccharide (*see* **Note 2**).

3. 1-(3-Dimethylaminopropyl)-3-ethylcarbodiimide (EDAC).

4. Centrifuge.

2.3 Microscopy

1. Dark-field microscope (*see* **Note 3**).

2. Video camera (*see* **Note 4**).

3. Computer.

4. Glass slide (0.8–1.0 mm in thickness).

5. 22 × 24 mm coverslip (0.13–0.17 mm in thickness).

6. Double-sided tape (5–10 μm in thickness).

2.4 Software for Data Analysis

The analyses described in this chapter are performed by ImageJ and Microsoft Excel.

3 Methods

3.1 Sample Preparation

1. Centrifuge 3–4 days culture at $1000 \times g$ for 10 min at room temperature.

2. Suspend the pellet into motility medium by tapping a tube without pipetting.

3. Make a flow chamber by attaching a coverslip with a glass slide via double-sided tape (Fig. 3).

4. Infuse the cell solution into the flow chamber (*see* **Note 5**).

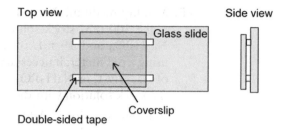

Fig. 3 A flow chamber for microscopic observation

3.2 Measurements of Spiral-End and Hook-End Gyrations

1. Place a flow chamber on a microscope stage.

2. Observe the cells with a 100× objective, and record the images on a computer using a high-speed camera (*see* **Note 4**).

3. Open the recorded movie on ImageJ.

4. Duplicate the area where the cell end gyration is displayed (Fig. 4a).

5. Set the threshold using *Image > Adjust > Threshold* to specify the bending cell ends to be analyzed.

6. Set parameters in *Analyze > Set Measurements >* check *Fit ellipse*.

7. Run analysis by *Analyze > Analyze Particle*, and save the result.

8. Open the result on Microsoft Excel, showing an angle versus time plot; *Fit ellipse* provides the long-axis angle of the ellipse fitted to spiral-end and hook-end (red dashed lines in Fig. 4b and c).

9. Determine the gyration speed by analyzing peak intervals (wavelengths) of angle-time plots or fast Fourier transform (FFT): The wave frequency is a reciprocal of the wavelength, and thus the gyration speed can be calculated from the average of several peak intervals.

3.3 Observation of PC Rotation Using Antibody-Coated Beads

1. Conjugate polystyrene beads with antibody by the following procedure:

   (a) Dilute 3 μL of carboxylated bead suspension (0.2 μm in diameter) into 300 μL of 50 mM MES buffer (pH 5.2).

   (b) Centrifugation at $17,000 \times g$ for 15 min at 23 °C.

   (c) Suspend the pellet in 200 μL of MES buffer and mix with 20 μL of antibody (*see* **Note 2**).

   (d) Dissolve 10 mg of 1-(3-dimethylaminopropyl)-3-ethyl-carbodiimide (EDAC) in 1 mL of MES buffer.

   (e) Add 20 μL of the EDAC solution to the bead suspension and incubate for 30 min at 23 °C.

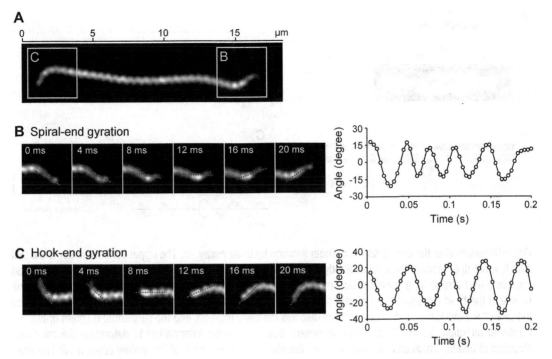

Fig. 4 Measurements of spiral-end and hook-end gyrations. (**a**) A still of a *Leptospira biflexa* cell swimming to the right, displaying spiral shape at the right (anterior side) (**b**) and hook shape at the left (posterior side) (**c**). The montages in **b** and **c** show about a half of one revolution of spiral-end and hook-end, respectively. The gyration speeds are determined from the periodic changes in the angles of the lines fitted to the cell ends (right panels in **b** and **c**). The angles of the fitted lines (red dashed lines in **b** and **c**) were given by *Fit ellipse* in ImageJ (see text). In this example data, the spiral-end speed and the hook-end speed are 30 Hz and 22 Hz, respectively

(f) Remove free antibodies and EDAC by centrifugation at 17,000 × *g* for 15 min at 23 °C, and suspend the pellet into 10 mM Tris–HCl buffer (pH 8.0).

2. Centrifuge 300 μL of *Leptospira* culture at 1000 × *g* for 10 min, and suspend the pellet into 500 μL of motility medium.

3. Mix 5 μL of the cell suspension with 15 μL of the antibody-coated bead.

4. Infuse the cell-bead mixture into a flow chamber, and observe motility of cells attached with beads by dark-field microscope.

5. Record the cell movements as described in Subheading 3.2.

6. Example data are shown in Fig. 5: The bead rotation rate is measured by the time trace of the bead position; information of the focal plane is required for determining the rotational direction (*see* the figure legend and **Note 6** for more details).

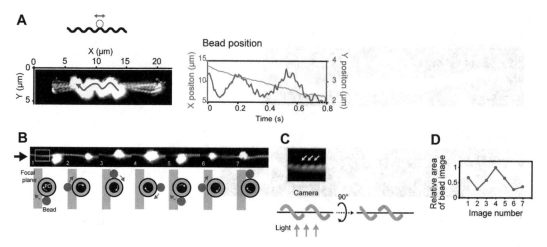

Fig. 5 Observation of the protoplasmic cylinder rotation by bead assay. (**a**) The upper schematic depicts a cell (black wavy line) labeled with an antibody-coated bead (red circle). The lower-left panel shows the bead trajectory (red arrow) by a superposition of sequential video images. The lower-right panel shows the time course of the bead position from the data shown in the left. In this data, the periodic change in the Y position indicates the bead rotation. (**b**) The montage displays the bead rotation, and the sequential diagram draws the rotational direction. (**c**) The focal plane of observation is essential information to determine the rotational direction of beads, which can be deduced from the visualized helix angle of PC (yellow arrows). (**d**) The bead movements in the Z-axis direction under dark-field microscope change the area of the bead images (halation). The X-axis indicates image numbers shown in **b**, which is useful to determine the rotational direction, too. The original data are shown in ref. 2

4   Notes

1. Since swimming abilities of spirochetes including *Leptospira* are known to be enhanced with an increase in viscosity, investigating viscosity dependencies of their motility parameters would lead to deeper understanding of the motility mechanism. When using polymers, note types of polymer. A major point when selecting a polymer in motility assays of *Leptospira* is whether the polymer solution is viscoelastic. According to verification by Berg and Turner [17], the addition of Ficoll, a branched polymer, to liquid makes homogeneous viscous solution, whereas methylcellulose forms a heterogeneous network in liquid, which is a gel-like, viscoelastic solution. Previous studies have shown that the swimming velocity of *Leptospira* is decreased monotonically with the increased Ficoll concentration but is increased with the concentration of methylcellulose [4, 18]. Numerical simulations based on a hydrodynamic theory predict that the observed differences between Ficoll and methylcellulose are attributed to differences in microscopic structure of viscous liquids, i.e., highly branched or gel-like structure [19, 20]. Viscosities of epithelium mucus layer in the cecum and colon of porcine are 7–12 mPa × s [21]. Since

mucin is a gel-like, viscous component of mucosa, 3% of commercial mucin solution (9.5 mPa × s) or 0.5% methylcellulose solution (7.5 mPa × s) is useful for examining *Leptospira* motility under conditions resembling environments near tissue surfaces [15]. Ficoll is better to investigate the effect of viscous drag on the bacterial movements, because mucin and methylcellulose change not only viscosity but also elasticity of polymer solutions.

2. According to the data provided by the company, the efficiency of antibody labeling depends on the amount of antibody and the bead size. It is necessary to examine appropriate conditions for each experiment.

3. *Leptospira* cells are too thin to be visualized by transmitted light through conventional optical microscopes, because the spatial resolution is about a half of wavelength of illumination light: When observed using 550 nm (green), the observation limit is about 280 nm. Largely refracted illumination through a high numerical aperture (NA) condenser of a dark-field microscope visualizes such thin cells as scattered objects in a dark field. High-intensity illumination system, such as combination of oil-immersion condenser and mercury lamp or laser, is required for high-speed recording.

4. Although either digital charge-coupled device (CCD) or complementary metal-oxide semiconductor (CMOS) video camera is available, a frame rate of recording enough to acquire target parameters should be set. Swimming speeds of *Leptospira* cells can be determined with normal video rate (30 frames per second, fps), whereas measurements of cell-body rotation rates (~100 revolutions per second) require more than 200 fps. We use IDP-Express R2000 (Photron, CMOS) for motility assays.

5. In general, bacterial suspension is placed between two glasses with different sizes, e.g., a glass slide and a coverslip. A possible event is unexpected flow of media, disturbing accurate measurements of motility. Especially, the effect of the flow on slow motile mutant is problematic, because the flow of media is usually faster than the actual movements of such mutant. To prevent the flow of media, use a flow chamber made by sticking a glass slide with a coverslip with double-sided tape (5–10 μm in thickness) (Fig. 3). Bacterial suspension placed on one side of the chamber spontaneously enters the space between two glasses by capillarity. Careful removal of bacterial suspension overflown from the sides stabilizes suspension within the chamber for several minutes. Sealing the opened sides of the flow chamber with silicone grease enables longer time observation.

6. Beads attached to the cell surface via anti-lipopolysaccharide (LPS) antibody move along the cell body in the bead-size-

dependent manner: Relatively large beads (0.5–1 μm in diameter) move fast relative to the cell body, whereas, small beads (0.1–0.2 μm) do not move on the cell body, i.e., staying at one position on the cell body. The phenomena suggest that LPS can freely move on the outer membrane and that the observed beads' movements are caused by drag force exerted on beads. In other words, beads are retarded from the swimming cell by viscous drag [2].

References

1. Cox PJ, Twigg GI (1974) Leptospiral motility. Nature 250:260–261
2. Tahara H, Takabe K, Sasaki Y et al (2018) The mechanism of two-phase motility in the spirochete *Leptospira*: swimming and crawling. Sci Adv 4:eaar7975
3. Goldstein SF, Charon NW (1990) Multiple-exposure photographic analysis of a motile spirochete. Proc Natl Acad Sci U S A 87:4895–4899
4. Nakamura S, Leshansky A, Magariyama Y, Namba K, Kudo S (2014) Direct measurement of helical cell motion of the spirochete *Leptospira*. Biophys J 106:47–54
5. Josenhans C, Suerbaum S (2002) The role of motility as a virulence factor in bacteria. Int J Med Microbiol 291:605–614
6. Lertsethtakarn P, Ottemann KM, Hendrixson DR (2011) Motility and chemotaxis in *Campylobacter* and *Helicobacter*. Ann Rev Microbiol 65:389–410
7. Lambert A, Picardeau M, Haake DA et al (2012) FlaA proteins in *Leptospira interrogans* are essential for motility and virulence but are not required for formation of the flagellum sheath. Infect Immun 80:2019–2025
8. Wunder EA, Figueira CP, Benaroudj N et al (2016) A novel flagellar sheath protein, FcpA, determines filament coiling, translational motility and virulence for the *Leptospira* spirochete. Mol Microbiol 101:457–470
9. Bromley DB, Charon NW (1979) Axial filament involvement in the motility of *Leptospira interrogans*. J Bacteriol 137:1406–1412
10. Picardeau M, Brenot A, Saint Girons I (2001) First evidence for gene replacement in *Leptospira* spp. inactivation of *L. biflexa flaB* results in non-motile mutants deficient in endoflagella. Mol Microbiol 40:189–199
11. Sasaki Y, Kawamoto A, Tahara H et al (2018) Leptospiral flagellar sheath protein FcpA interacts with FlaA2 and FlaB1 in *Leptospira biflexa*. PLoS One 13:e0194923
12. Wunder EA, Slamti L, Suwondo DN et al (2018) FcpB is a surface filament protein of the endoflagellum required for the motility of the spirochete *Leptospira*. Front Cell Infect Microbiol 8:130
13. Islam MS, Takabe K, Kudo S, Nakamura S (2014) Analysis of the chemotactic behaviour of *Leptospira* using microscopic agar-drop assay. FEMS Microbiol Lett 356:39–44
14. Affroze S, Islam MS, Takabe K, Kudo S, Nakamura S (2016) Characterization of leptospiral chemoreceptors using a microscopic agar drop assay. Curr Microbiol 73:202–205
15. Takabe K, Tahara H, Islam MS, Affroze S, Kudo S, Nakamura S (2017) Viscosity-dependent variations in the cell shape and swimming manner of *Leptospira*. Microbiology 163:153–160
16. Takabe K, Kawamoto A, Tahara H, Kudo S, Nakamura S (2017) Implications of coordinated cell-body rotations for *Leptospira* motility. Biochem Biophys Res Commun 491:1040–1046
17. Berg HC, Turner L (1979) Movement of microorganisms in viscous environments. Nature 278:349–351
18. Takabe K, Nakamura S, Ashihara M, Kudo S (2013) Effect of osmolarity and viscosity on the motility of pathogenic and saprophytic *Leptospira*. Microbiol Immunol 57:236–239
19. Magariyama Y, Kudo S (2002) A mathematical explanation of an increase in bacterial swimming speed with viscosity in linear-polymer solutions. Biophys J 83:733–739
20. Nakamura S, Adachi Y, Goto T, Magariyama Y (2006) Improvement in motion efficiency of the spirochete *Brachyspira pilosicoli* in viscous environments. Biophys J 90:3019–3026
21. Naresh R, Hampson DJ (2010) Attraction of *Brachyspira pilosicoli* to mucin. Microbiology 156:191–197

Chapter 14

In Vivo Imaging of Bioluminescent Leptospires

Frédérique Vernel-Pauillac and Catherine Werts

Abstract

The study of pathological processes is often limited to in vitro or ex vivo assays, while understanding pathogenesis of an infectious disease requires in vivo analysis. The use of pathogens, genetically modified to express with luminescent enzymes, combined to charge-coupled device (CCD) cameras, constitutes a major technological advance for assessing the course of infection in an intact, living host in real time and in a noninvasive way. This technology, also called bioluminescence imaging, detects the photons emitted from biological sources of light through animal tissues. Here, we describe the method we developed to monitor leptospirosis in a mouse model, by following in a spatiotemporal scale, the dissemination and spread of leptospires. These bacteria have been genetically modified to express the firefly luciferase, which produces light in the presence of the substrate D-luciferin. This useful and accessible technology facilitates the study of the kinetics of blood and tissue dissemination of live leptospires, and the pharmacological impact of treatments and host directed therapeutics.

Key words Bioluminescence, Luciferase, Leptospires, In vivo live imaging, Bacterial dissemination, Spatiotemporal kinetics, In vivo therapeutic monitoring

1 Introduction

In vivo assays are an essential step to advance our understanding of pathological processes involved in an infectious disease. To assess features leading to the spread of an infection and the preferential localization of the pathogen, or tissue niche, conventional techniques have relied on end point assays or repeated sacrifices of animals at designated time points.

Nowadays, each experiment conducted with live animals is ethically evaluated and animal welfare must be ensured. Therefore, in vivo live imaging constitutes an alternative and useful way to reduce and refine experiments; it allows multi-animal, high-throughput studies at relatively low cost, useful to assess infection evolution with a high signal sensitivity [1–5]. Bioluminescence imaging is based on the activity of luminescent reporter genes stably inserted into the genome of the microorganism of interest [6–8]. The expression of the reporter and its bioluminescent

Nobuo Koizumi and Mathieu Picardeau (eds.), *Leptospira spp.: Methods and Protocols*, Methods in Molecular Biology, vol. 2134, https://doi.org/10.1007/978-1-0716-0459-5_14, © Springer Science+Business Media, LLC, part of Springer Nature 2020

activity can be measured and quantified in real time providing some spatiotemporal features about the disease progression in an intact living host [9, 10].

In the lab, bioluminescent pathogenic and saprophytic strains of *Leptospira* have been constructed by transposon insertion of a cassette expressing the firefly luciferase gene under the control of a leptospiral promoter, into the genome [9]. The firefly luciferase is a useful reporter system that can generate visible light through the oxidation of D-luciferin, its substrate, in the presence of oxygen and ATP as a source of energy. Of note, the reaction is catalyzed only in metabolically active bacteria producing both ATP and the luciferase. Therefore, we can evaluate the spread of live bacteria by tracking bioluminescence. The light generated is measured in a light-tight box using new generation optical cameras that enable a highly sensitive detection of the photons transmitted. Moreover, bioluminescence imaging based on luciferase doesn't require excitation, avoiding high background signal. The absolute luciferase activity can be analyzed and quantified using the In Vivo Imaging Software (IVIS) and compared with a baseline or control conditions [11].

With this method, we were able to observe, in live mice, the spatiotemporal evolution and intensity of an experimental infection with leptospires [9]. Moreover, this method proves to be reliable for the determination of critical experimental end points.

2 Materials

All solutions are reconstituted, prepared, or diluted in sterile Dulbecco's phosphate buffered saline (DPBS) or guaranteed endotoxin-free reagents. Prepare and store aliquots of working solutions at −20 °C (unless indicated otherwise). Thaw and warm all solutions at room temperature before administration to animals. Diligently follow all waste disposal regulations when disposing soiled sharp objects or other waste materials.

1. *Leptospira biflexa* sevorar Patoc strain PFLum7 (saprophytic bioluminescent leptospires).

2. *Leptospira interrogans* serovar Manilae strain MFLum1 (virulent *L. interrogans* bioluminescent strain).

3. Bacterial culture medium Ellinghausen-McCullough-Johnson-Harris (EMJH): 1.0 g/L sodium phosphate dibasic, 0.3 g/L monopotassium phosphate, 1.0 g/L sodium chloride, 0.25 g/L ammonium chloride, 5 mg/L thiamine. EMJH medium is supplemented with 10 g/L bovine serum albumin, 4 mg/L zinc sulfate heptahydrate, 50 mg/L iron sulfate heptahydrate, 15 mg/L hexahydrate magnesium chloride, 15 mg/L sodium chloride dihydrate, 0.1 g/L sodium pyruvate, 0.12 mg/L vitamin B12, 0.4 g/L glycerol, and 1.25 g/L Tween 80.

4. Dulbecco's phosphate buffered saline (DPBS): 8 g/L sodium chloride, 200 mg/L potassium chloride, 1.44 g/L sodium phosphate dibasic, 240 mg/L monopotassium phosphate, pH 7.4.

5. 1 mL single-use syringe.

6. Ventilated 25 cm^2 flask.

7. 50 mL tube.

8. Petroff-Hausser chamber.

9. 24-well black plate.

10. 96-well black plate.

11. Adult mice, preferably white, with no age restriction or sex.

12. D-Luciferin: 30 mg/mL solution in DPBS. Aliquot 600 μL of D-luciferin solution per tube.

13. Luminometer.

14. IVIS Spectrum imaging system. Combines 2D optical and 3D optical tomography in one platform (Fig. 1).

15. XGI-8 Gaseous Anesthesia System (Fig. 2).

3 Methods

3.1 Leptospire Culture

Leptospira strains are grown in liquid EMJH medium at 30 °C with no agitation. The generation time is 18–20 h for pathogenic *L. interrogans* and 4–7 h for saprophytic *L. biflexa* strains.

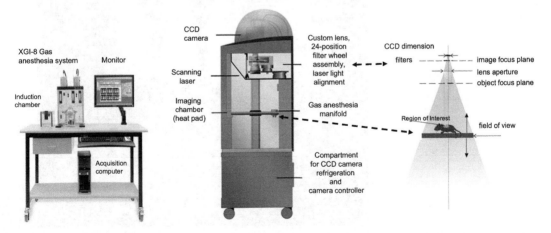

Fig. 1 Schematic picture of the IVIS Spectrum system. Representation of the in vivo imaging system combining 2D optical and 3D tomography in one platform. Mice are anesthetized before imaging in the induction chamber and during imaging by using the XGI-8 Gas Anesthesia System, designed to work with the IVIS Spectrum [11]. On the right is a schematic representation of the major components for bioluminescence acquisition and analysis using the in vivo imaging software [3]

1. Strains are diluted weekly from a 7-day culture, to 5×10^6 leptospires/mL in a 25 cm^2 flask with ventilated cap, into a 20 mL culture volume.

2. Incubate for 7 days at 28 °C.

3.2 Preparation of Leptospira Inoculum

To ensure an optimal infection, take care to perform the following steps carefully and without delay.

1. Prior to the animal infection, check the fitness of the strain by measuring the light emitted with a known quantity of leptospires in the presence of D-luciferin, using a luminometer (*see* **Note 1**).

2. Transfer the culture in a 50 mL tube and centrifuge 25 min at $3345 \times g$, at room temperature (RT).

3. Remove the supernatant (in respect to regulations for disposal of infectious waste), and resuspend the pellet with 5 mL of sterile DPBS.

4. Take a 1 mL aliquot of the homogenized suspension and prepare a 1:100 diluted solution in DPBS. Count the bacteria

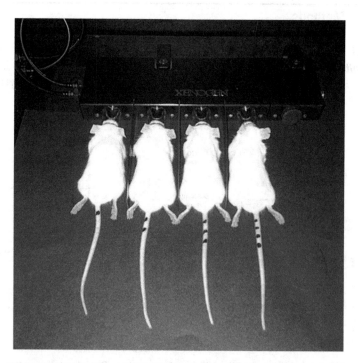

Fig. 2 Five-port anesthesia manifold. This system including light baffle dividers and transparent nose cones is inserted within the IVIS imaging chamber to maintain mice under anesthesia during the imaging

with a dark-field microscope according to the recommendations for Petroff-Hausser chamber use.

5. Dilute the initial homogenized suspension to the appropriate infectious dose in DPBS, such that the desired quantity of leptospires is injected in a 200 μL volume, usually 2×10^7 for MFLum1 to get a sublethal infection.

6. Transfer the inoculum from the lab to the animal care facility using a box certified for the transport of infectious substances.

3.3 Mice Conditioning

The sensitivity of bioluminescence detection is impacted by the color of the animal's fur. It is necessary to shave the areas to be imaged when black mice are used.

1. At least 1 week prior to the experiment, mice are divided into five individuals per cage, corresponding to the number of homogeneous experimental groups to evaluate.

2. One day prior to the infection, mice within each group are individually ear-tagged (*see* **Note 2** and Fig. 3) and, if necessary, shaved ventrally (*see* **Note 3**). Additional marks on the tail facilitate the manipulation and proper positioning of mice in the induction box.

3.4 Animal Inoculation

1. Homogenize the inoculum by inverting the tube and carefully fill a syringe with 1 mL of the suspension. This allows five mice to be injected per syringe.

2. Take a cage from the ventilated rack, and deliver 200 μL of bacterial suspension (or DPBS for control) into the peritoneal cavity of each identified mouse.

3. Put mice back in a clean cage, and note the experimental parameters on the identification sheet attached to the cage.

4. Continue the infection with the subsequent groups.

3.5 IVIS Imaging Preparation

1. Switch on the IVIS imaging station with a personal access code, and initialize the system using the IVIS Acquisition control screen as required when connection occurs.

2. Wait for the system to self-calibrate. The heating pad (allowing for the proper retention of body temperature of the anesthetized mouse) within the dark box is warm as soon as the temperature indicator becomes green (Fig. 4).

3. Select basic acquisition parameters according to the manufacturer's instructions:

   (a) Imaging mode luminescence with exposure time auto or 5 min, binning medium, F/Stop 1, excitation filter in block position, emission filter open.

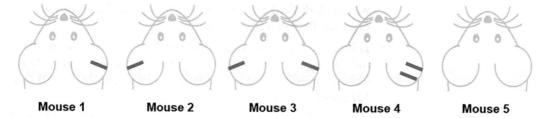

Mouse 1   **Mouse 2**   **Mouse 3**   **Mouse 4**   **Mouse 5**

Fig. 3 Ear tag identification. Make one short cut on the right ear to start with the mouse identified as No. 1, one cut on the left ear for No. 2. Mouse No. 3 will have a cut on each ear, No. 4 two cuts on the right ear, while mouse No. 5 will have no cut

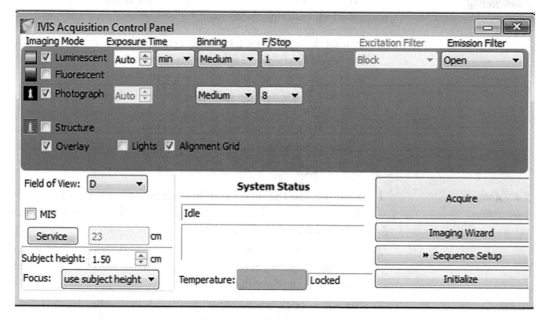

Fig. 4 IVIS imaging settings. Screenshot showing all imaging parameters selected after the IVIS imaging system has been initialized and ready to acquire image

  (b) Imaging mode photograph with binning medium, F/Stop 8.

  (c) Field of view = D (*see* **Note 4**).

  (d) Overlay and alignment grid.

4. Prepare the imaging platform within the IVIS spectrum. Place a thick black paper sheet on the imaging tray. Clean transparent nose cones and dividers with 70% ethanol, dry, and insert the nose cones into the anesthesia manifold (*see* **Note 5**).

5. Start the anesthesia system by opening the oxygen (O_2) supply and set a 1.5% flow rate. Check the level of isoflurane available, open and set the system to deliver a constant 2.5% flow per minute to the induction chamber.

6. Turn on anesthesia to the IVIS spectrum.

3.6 In Vivo Mouse Imaging

For each experiment, imaging sessions should be scheduled as appropriate. The first one is realized 30 min post-infection (p.i.) to ensure that the inoculum has been correctly injected. During the acute phase of infection (from day 1 to day 5 p.i.), imaging sessions are usually performed every day in ventral position to follow disease progression in vivo (*see* **Note 6**). From day 14 p.i., the level of renal colonization can be monitored in the dorsal position, with imaging sessions at determined time points. For any experiment, an image should be acquired with control mice to establish luminescence background.

1. Thaw aliquots of D-luciferin and warm to room temperature, as many aliquots as experimental groups. Luciferin injection and anesthesia are performed sequentially.

2. Ten to twelve minutes before imaging, inject 100 μL of D-luciferin in the peritoneal cavity of the first experimental group of mice to be imaged.

3. Place mice in the induction chamber and maintain them under anesthesia for 10 min (*see* **Note 7**).

4. Transfer each anesthetized mouse (be careful to handle infected mice on a tissue to prevent the urine from contaminating the environment) immediately to the dark box imaging system on a disposable black sheet (to avoid cross contamination), and position the head gently in the transparent cones inserted in the anesthesia manifold to maintain the anesthesia. Depending on the type of imaging, place animals in ventral or dorsal position (*see* **Note 8**). Center the animals between each divider and align the legs and tail of each mouse. Pay attention that the order of individual mice should be respected for individual follow-up over time.

5. Close the door of the dark box and click Acquire on the control screen. First, acquire in automatic exposure time to get the absolute light value and then with a 5 min exposure time for a picture (*see* **Note 9**). When the first image capture is activated, autosave function is proposed to the user. Select Yes to save all images in a folder specifically created for the experiment.

6. While the first group of mice is imaged, inject the luciferin to the second group of mice, and put them under anesthesia for 10 min before imaging.

7. After imaging, remove mice from the dark box system, and return them to their respective cage, in the security position, on their right side. Monitor for complete recovery from anesthesia (approximately 2 min for healthy mice) before putting the cage back on the ventilated rack.

8. Perform imaging of subsequent groups of mice according to the same temporal sequence of luciferin administration and anesthesia procedure.

9. Once the imaging of different experimental groups is complete, remove all equipment (nose cones, dividers, black sheet) to decontaminate with a disinfectant. Decontaminate the imaging tray, check for any residue, close, and disinfect the door.

3.7 Ex Vivo Tissue Imaging

Blood, urine, and specific tissues such as kidneys, liver, lungs, and brain from infected mice can be harvested after mice are euthanatized and imaged. This allows to correlate a bioluminescent signal to the number of bacteria determined by quantitative PCR in a targeted organ (*see* **Note 10**).

1. Euthanize animals in compliance with ethics regulations.

2. Aseptically remove the tissue(s) from one mouse, cut longitudinally (e.g., a kidney), and place each piece of tissue into individual wells of a 24-well black plate. Keep one well empty for the blank control of bioluminescence. For urine or blood, use a 96-well black plate.

3. Immerse tissues in 200 µL of 3 mg/mL D-luciferin per well including the blank control well. Do not cover the plate. Transfer the plate to the imaging tray of the IVIS spectrum. Close the door of the dark box.

4. Select imaging parameters as in **step 3** in Subheading 3.4 with exposure time = 5 min, field of view = C.

5. Click on Acquire on the control screen.

6. After imaging, remove the plate and clean the imaging chamber and door. Put the plate in a sealed bag, close it, and transfer to −80 °C until the DNA extraction and PCR quantification can be performed.

3.8 In Vivo Image Analysis

Bioluminescence imaging is a powerful technology providing valuable information about the spatiotemporal course of the disease in each animal. The progression of the infection can be monitored by analyzing and quantifying the light emitted in a selected anatomical region, at different time points, in the same mouse. By comparing the level of luminescence in a chosen region, the efficacy of a potential therapeutic treatment can be evaluated.

1. Select the image acquired in auto-mode acquisition. First, determine the limits of a region of interest (ROI) that can be applied to all individuals in the same experiment, using the mouse with the highest light level.

2. Add ROI for the mouse from the ROI Tools menu. Select the desired ROI form (*see* **Note 11**), adjust the size, and the position to the targeted region.

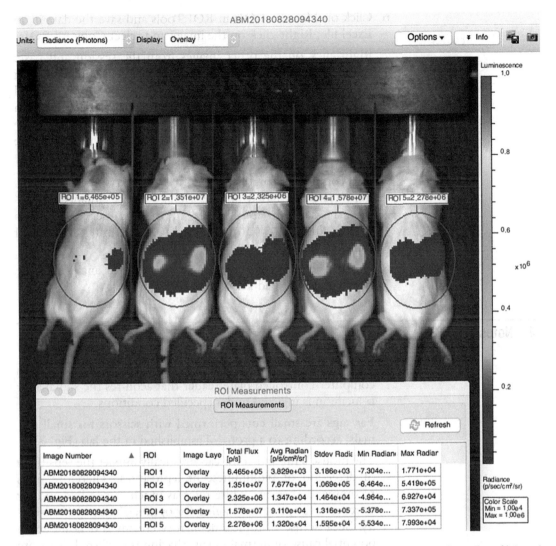

Fig. 5 ROI analysis in infected mice. To quantify the level of light emitted in each mouse, the region of interest (ROI, shown in red) is defined and applied to all mice within the same experimental group. For each mouse, total flux (p/s) is indicated in the respective white tag, and Avg radian (p/s/cm²/sr) is reported in the measurement window

3. Adjust the color scale in Image Adjust menu from a minimum of 1.0×10^4 to 1.0×10^6; this allows background signal to be removed for presentations without changing the detected signal value, saved for quantification analysis.

4. When this ROI is drawn, copy and paste ROI, selected by right-clicking on ROI (maintaining the same ROI size), to the other individuals imaged in the same picture (Fig. 5).

5. If needed, reposition the ROI to the selected region but without modifying the initial ROI size.

6. Click on Measure ROIs in ROI Tools and save the data in an Excel file. Right-click on the image and select Copy All ROIs.

7. Right-click on the other images from the session and paste ROIs. Position the ROIs if necessary but without changing the size of the initial ROI. This strategy allows optimal and consistent comparison of luminescent signal across data from different groups of mice.

8. Save all data in the Excel file and images without ROIs in TIFF format, preferable for publications.

9. The comparison of ROIs, according to experimental conditions tested, is based on the average (Avg) radiance ($p/s^2/cm^2/sr$) data. It corresponds to the number of photons (p) per flux (s^2), in a determined surface during an optimal exposure time (sr).

10. Perform graph representations and statistical analysis of the data with GraphPad Prism software.

4   Notes

1. Measurement of bioluminescence is performed using a computer-controlled plate reader that acquires the signal after D-luciferin is injected under specified conditions.

2. Ear tags are small cuts performed with scissors for small animals, according to a protocol established in the lab (Fig. 3).

3. The mouse's fur and color can drastically impact the sensitivity of detection of bioluminescent leptospires. Therefore, it is recommended to use white mice as much as possible and/or to remove fur prior to imaging for optimal/increased sensitivity. Because use of topical depilatory creams is delicate and a potential cause of dermal burns, shaving is preferred. Mice are shaved on the ventral part to monitor luminescence in the first week after the infection and shaved on the back to follow renal colonization from the second week post-infection. Prior to imaging for longitudinal studies, shaving has to be performed to avoid decreased sensitivity due to hair regrowth.

4. Field of view (FOV) is a parameter that depends on the number of mice to be imaged in the experiment: D = 4 to 5 mice, C = 3 mice, B = 1 complete mouse, and A = a part of a mouse.

5. Working with bacterial pathogens such as leptospires, which can be shed at high level in urine, requires a risk assessment and rigorous cleaning of materials before and after use. Moreover, and in contrast to most disinfectants, 70% ethanol leaves negligible background luminescence in the nose cones.

6. In compliance with ethical aspects, end points have been established to avoid unnecessary suffering. According to our expertise, if an infected mouse shows an Avg radiance $>10^6$ at day 3 p.i., it will not survive the infection and must be euthanized.

7. After injection of the mouse, D-luciferin needs 10 min to be optimally bio-distributed, meaning D-luciferin should be injected before the anesthesia is carried out. An optimal anesthesia occurs after 7 min exposition with isoflurane, but may need more time as mice habituate to the anesthesia.

8. To check peritoneal and systemic dissemination of the infection, mice are positioned on the back for a ventral view. To measure the light emitted from the kidney, mice are positioned on the abdominal side for a dorsal view. Kidney colonization is detectable, according to our experience [9], from day 8 postinfection and better observed 2 weeks after infection.

9. Automatic time exposure ensures a time of exposure adapted to the light emitted, avoiding any saturated signal.

10. Although we established a linear correlation of bioluminescence in vitro with quantitative PCR of *L. interrogans* [9], the bioluminescence in organs or urine may depend on the oxygen availability or decreased fitness of the luciferase in acidic conditions.

11. IVIS software is dedicated for defining ROIs and measuring photon flux. Select the most appropriate ROI form according to the region/tissue, for example, a square to analyze dissemination in the peritoneal cavity or an ellipse to check for renal colonization (Fig. 5). Multiple regions can also be analyzed if several ROIs are defined.

Acknowledgments

This work was supported by PTR 2017-66 grant to CW. We thank Richard Wheeler for English editing. This work was performed at the UtechS Photonic BioImaging (PBI) platform, member of France Life Imaging network (grant ANR-11-INBS-0006).

References

1. Contag CH, Contag PR, Mullins JI, Spilman SD, Stevenson DK, Benaron DA (1995) Photonic detection of bacterial pathogens in living hosts. Mol Microbiol 18(4):593–603. https://doi.org/10.1111/j.1365-2958.1995.mmi_18040593.x

2. Contag CH, Bachmann MH (2002) Advances in in vivo bioluminescence imaging of gene expression. Annu Rev Biomed Eng 4:235–260. https://doi.org/10.1146/annurev.bioeng.4.111901.093336

3. Sjollema J, Sharma PK, Dijkstra RJ, van Dam GM, van der Mei HC, Engelsman AF et al (2010) The potential for bio-optical imaging of biomaterial-associated infection in vivo. Biomaterials 31(8):1984–1995. https://doi.org/10.1016/j.biomaterials.2009.11.068

4. Badr CE (2014) Bioluminescence imaging: basics and practical limitations. Methods Mol Biol 1098:1–18. https://doi.org/10.1007/978-1-62703-718-1_1

5. Avci P, Karimi M, Sadasivam M, Antunes-Melo WC, Carrasco E, Hamblin MR (2018) In-vivo monitoring of infectious diseases in living animals using bioluminescence imaging. Virulence 9(1):28–63. https://doi.org/10.1080/21505594.2017.1371897

6. Zhang Y, Pullambhatla M, Laterra J, Pomper MG (2012) Influence of bioluminescence imaging dynamics by D-luciferin uptake and efflux mechanisms. Mol Imaging 11(6):499–506

7. Keyaerts M, Caveliers V, Lahoutte T (2012) Bioluminescence imaging: looking beyond the light. Trends Mol Med 18(3):164–172. https://doi.org/10.1016/j.molmed.2012.01.005

8. Mezzanotte L, van't Root M, Karatas H, Goun EA, Lowik C (2017) In vivo molecular bioluminescence imaging: new tools and applications. Trends Biotechnol 35(7):640–652. https://doi.org/10.1016/j.tibtech.2017.03.012

9. Ratet G, Veyrier FJ, Fanton d'Andon M, Kammerscheit X, Nicola MA, Picardeau M et al (2014) Live imaging of bioluminescent *leptospira interrogans* in mice reveals renal colonization as a stealth escape from the blood defenses and antibiotics. PLoS Negl Trop Dis 8(12):e3359. https://doi.org/10.1371/journal.pntd.0003359

10. Acuff NV, Li X, Latha K, Nagy T, Watford WT (2017) Tpl2 promotes innate cell recruitment and effector T cell differentiation to limit *Citrobacter rodentium* burden and dissemination. Infect Immun 85(10). https://doi.org/10.1128/IAI.00193-17

11. IVIS-guide. https://www.yumpu.com/en/document/view/40278852/xenogen-ivis-200-user-manual. 2006

Chapter 15

Cell Monolayer Translocation Assay

Elsio A. Wunder Jr.

Abstract

An essential property associated with leptospiral virulence is the pathogen's ability to translocate across host cells, enabling *Leptospira* to evade the host immune response, disseminate, and establish infection. Cell monolayer translocation assay allows for the quantification of *Leptospira* strain's competence to cross cell barriers while measuring the integrity of the polarized eukaryotic cell monolayer during this process.

Key words *Leptospira*, Polarized monolayers, Transwell®, Translocation, Eukaryotic cells, Electrical resistance, Leptospiral pathogenesis

1 Introduction

One of the most important features of leptospiral pathogenesis is the aptitude of *Leptospira* to expeditiously penetrate mammalian cell barriers and disseminate to multiple organs throughout the host [1, 2]. Previous studies have emphasized the ability of pathogenic *Leptospira* to translocate across polarized mammalian cell monolayers [2, 3]. Together with adhesion to extracellular surfaces (*see* Chapter 16), rapid translocation of host cell barriers is used by pathogenic *Leptospira* as means of evading the bodily immune response and establishing infection, essential properties associated with bacterial virulence [4, 5].

Polarized cell monolayer translocation assay with transepithelial/transendothelial electrical resistance (TEER) measurement was first developed to study endothelial macromolecular transport [6]. Since then, this method has been widely used to assess the ability of a diverse range of pathogens to translocate across a cell barrier [7–10], including *Leptospira* [11, 12] and other spirochetes [13, 14]. This in vitro assay to assess bacterial translocation is based on the culture of eukaryotic cells on a permeable membrane contained within a plastic insert and placed on a plastic well, establishing an upper and lower compartment (Fig. 1). Culture medium is added to each compartment, promoting cell growth and

Nobuo Koizumi and Mathieu Picardeau (eds.), *Leptospira spp.: Methods and Protocols*, Methods in Molecular Biology, vol. 2134, https://doi.org/10.1007/978-1-0716-0459-5_15, © Springer Science+Business Media, LLC, part of Springer Nature 2020

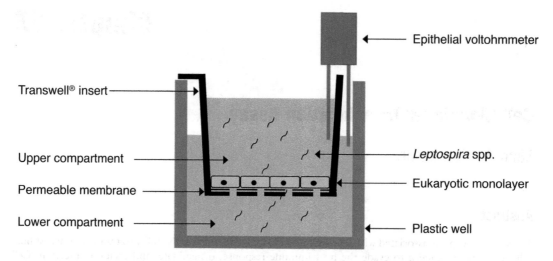

Fig. 1 Schematic representation showing the polarized eukaryotic cell monolayer and the components of the system to assess leptospiral translocation, including the epithelia voltohmmeter used to measure the epithelial/endothelial electrical resistance (TEER)

establishment of polarized monolayer. Pathogen is added to the upper compartment, and the ability of translocation is measured by assessing the number of pathogens present in the lower compartment after determined periods of time post-infection. The integrity of the polarized monolayers is monitored using an epithelial voltohmmeter apparatus (Fig. 1).

Here we describe an in vitro cell monolayer translocation assay that enables the investigation of a key feature of *Leptospira* pathogenesis, providing assessment of the phenotype of new isolates or potential mutants of interest. In vitro surrogates to investigate key steps of pathogenesis contribute to the principle of replacing, reducing, and refining the use of animals in experimentation.

2   Materials

Use sterile ultrapure water to prepare media and solutions. All media and solutions should be pre-warmed at 37 °C. Use 10% NaOH or 1 N HCl solutions to adjust pH.

2.1 Growth and Seeding of Eukaryotic Cells (MDCK)

1. Complete growth medium: Dulbecco's Modified Eagle Medium (DMEM) + 10% fetal bovine sera (FBS) (*see* **Note 1**).

2. Phosphate buffered saline (PBS): 137 mM NaCl, 2.68 mM KCl, 10.1 mM Na_2HPO_4, 1.47 mM KH_2PO_4, pH 7.4.

3. Trypsin/EDTA solution: PBS, 0.05% trypsin, 0.02% EDTA.

4. Neubauer counting chamber.

5. 12 mm diameter Transwell® filter units with 3 μm pores.

2.2 Cell Monolayer Resistance

1. Electrode and transepithelial/transendothelial electrical resistance (TEER) measurement device (*see* **Note 2**).

2. Complete growth medium.

3. Sterile PBS.

4. 10% trypan blue solution: 100 µL trypan blue, 900 µL water.

2.3 Infection

1. EMJH base: 2.3 g of Difco Leptospira Medium Base EMJH (Becton Dickinson) in 890 mL of water + 0.01% glycerol, pH 7.4.

2. EMJH supplement: 1% bovine serum albumin (BSA), 0.005% Thiamine hydrochloride, 0.01% $CaCl_2$, 0.01% $MgCl_2$, 0.004% $ZnSO_4$, 0.0002% vitamin B12, 0.05% $FeSO_4$, 1.25% Tween 80, 0.0003% of $MgSO_4$ in 100 mL of water, pH 7.4.

3. Enriched EMJH supplement: 1% of lactalbumin hydrolysate, 0.04% of sodium pyruvate, 1KU superoxide dismutase, and 10 mL of heat-inactivated rabbit serum in 100 mL of EMJH supplement.

4. Leptospira EMJH Medium: 890 mL of EMJH base + 110 mL of Enriched EMJH supplement.

5. Sterile PBS.

6. *Leptospira* sp. cultures in mid-log growth.

7. Petroff-Hausser counting chamber.

8. Trypsin/EDTA solution.

9. Complete growth medium.

10. EMJH/DMEM medium: 1:2 v/v ratio of DMEM and Leptospira EMJH media.

3 Methods

Follow strict aseptic technique to perform every procedure. Eukaryotic cells and *Leptospira* strains must be handled in a Class II biosafety cabinet to prevent foreign contamination and cross contamination. All surfaces should be cleaned with 70% ethanol before and after use. Class II biosafety cabinet should be treated with UV light for 20 min before and after use.

3.1 Preparation of Leptospira EMJH Medium [15, 16]

1. Prepare stock solutions that can be kept at −20 °C for up to 1 month: 10% glycerol, 10KU superoxide dismutase (SOD), 0.4% $ZnSO_4$, 1% $CaCl_2$, 1% $MgCl_2$, 0.5% thiamine hydrochloride, 0.02% vitamin B12, and 0.3% $MgSO_4$.

2. Prepare fresh stock solutions: 0.5% $FeSO_4$ and 10% Tween 80.

3. To prepare 100 mL of EMJH supplement, dissolve 10 g of bovine serum albumin (BSA) into 50 mL of distilled water, and stir slowly overnight to avoid foam. Once the BSA is fully

dissolved, add 1 mL of the following stock solutions: thiamine, $CaCl_2$, $MgCl_2$, $ZnSO_4$, and vitamin B12. Also add 10 mL of $FeSO_4$, 12.5 mL of Tween 80, and 0.1 mL of $MgSO_4$ stock solutions. Adjust pH to 7.4 with NaOH and adjust volume to 100 mL with water. The EMJH supplement can be stored at −20 °C indefinitely until use.

4. For 1 L of EMJH medium, dissolve 2.3 g of Difco Leptospira Medium Base EMJH in 890 mL of water. Add 0.9 mL of 10% glycerol stock solution and adjust pH to 7.4 with HCl.

5. Thaw 100 mL of EMJH supplement and add 1 g of lactalbumin hydrolysate, 0.04 g of sodium pyruvate, 100 μL of SOD stock solution, and 10 mL of heat-inactivated rabbit serum.

6. Add the enriched EMJH supplement to the Medium Base EMJH. Sterilize by filtration through a 0.22 μm filter system, and create aliquots as needed for experiments. The final EMJH medium can be stored at 4 °C.

3.2 Establishing Polarized Cell Monolayers

1. Thaw a working aliquot of your eukaryotic cells into a T25 flask with 5–7 mL of complete growth medium (*see* **Note 3**).

2. Incubate at 37 °C with 5% CO_2 for growth until cells reach approximately 80–100% confluence.

3. Wash cells twice with 3 mL of PBS.

4. Add 1 mL of trypsin/EDTA solution for trypsinization.

5. Incubate at 37 °C with 5% CO_2 until cells are completely disassociated.

6. Add 2 mL of complete growth medium and passage ~1.5 mL of trypsinized cells into 10–15 mL of complete growth medium in a T75 flask.

7. Incubate at 37 °C with 5% CO_2 for growth.

8. Once the cells have reached an approximate 80–100% confluence, wash them twice with 9 mL of PBS.

9. Add 2 mL of trypsin/EDTA solution for trypsinization.

10. Incubate at 37 °C with 5% CO_2 until cells are completely disassociated.

11. Add 4–6 mL of complete growth medium and transfer the cells into a 15 mL tube.

12. Enumerate the eukaryotic cells with a Neubauer counting chamber (*see* **Note 4**).

13. Dilute the cells in complete growth medium to reach a concentration of 4×10^5 cells/mL.

14. Add 1.5 mL of complete growth media to the lower compartment of each well on the plate (*see* **Note 5**).

15. Put the plate at 37 °C with 5% CO_2 for 20 m to equilibrate.

16. Add 500 μL of the medium containing 4×10^5 cells/mL to the upper compartment of each well on the plate (*see* **Note 6**).

17. Incubate at 37 °C with 5% CO_2 (*see* **Note 7**).

18. Wash cells daily with PBS, adding fresh medium to the compartments (*see* **Note 5**).

3.3 Measuring Cell Monolayer Resistance

1. Evaluate the transepithelial/transendothelial electrical resistance (TEER) daily to see if the monolayer is continuous (*see* **Note 8**).

2. Sterilize the electrode (*see* **Note 9**).

3. Connect the electrode to the meter.

4. Precondition the electrode, by placing it in your culture medium for a few minutes (*see* **Note 10**).

5. Set the *Function* switch to *Ohms*.

6. Measure the blank resistance by placing the electrode in a cup or a cell culture insert filled with the growth medium without cells. Record the value (*see* **Note 11**).

7. Perform your experimental resistance measurements, cleaning the electrode between each reading (*see* **Note 9**).

8. To obtain the true resistance value of your cultured cell monolayers, subtract the blank resistance value (*see* **Note 12**).

9. When you finish making measurements, clean and store your electrodes dry.

3.4 Confirmation of Polarized Cell Monolayer

1. When cells reach the electrical resistance considered to be compatible with established polarized cell monolayers, an extra assay can be performed to confirm this status (*see* **Note 13**).

2. Empty the compartments from one well and wash the cells twice with PBS (*see* **Note 14**).

3. Add 1.5 mL of PBS to the lower compartment.

4. Add 500 μL of the trypan blue solution to the upper compartment.

5. Wait for 5 min, and visually check to see if any trypan blue has seeped into the lower compartment (*see* **Note 15**).

6. If no trypan blue goes through, the monolayer is established, and the cells are considered to be ready for the infection experiment.

7. If the trypan blue goes through, the cells are not in an established monolayer and should be kept in culture until the expected results are achieved.

3.5 Determining the Concentration of Eukaryotic Cell

1. Trypsinize one well of your eukaryotic cell Transwell® plate to check the concentration of cells (*see* **Note 16**).

2. Wash one individual well of the Transwell® plate with PBS.

3. Add 600 μL of trypsin/EDTA solution to the lower compartment and 250 μL to the upper compartment.

4. Incubate the plate at 37 °C incubator with 5% CO_2 for 10 m.

5. Scratch membrane to collect all cells into 1.5 mL tube.

6. Determine the number of eukaryotic cells with a Neubauer chamber.

3.6 Infection of Polarized Cell Monolayer

1. Prepare mid-log phase *Leptospira* sp. cultures.

2. *Leptospira biflexa* serovar Patoc strain Patoc1 should be used as a negative control for the experiment (*see* **Note 17**).

3. *Leptospira interrogans* serovar Copenhageni strain Fiocruz L1-130 should be used as a positive control for the experiment (*see* **Note 18**).

4. Perform the infection of wells in duplicate or triplicate for each strain to be tested (*see* **Note 19**).

5. For a *Leptospira* strain that reaches ~10^8 cells/mL, one 5 mL tube of culture should be enough for one Transwell® plate.

6. Centrifuge the *Leptospira* sp. culture at 12,000 × *g* for 10 m.

7. Remove supernatant and resuspend pelleted cells in 5 mL of PBS.

8. Centrifuge the cells once more at 12,000 × *g* for 10 m.

9. Resuspend the pellet in 200 μL of PBS.

10. Determine the number of *Leptospira* spp. (leptospires per milliliter) using a Petroff-Hausser counting chamber.

11. Determine the number of leptospires that need to be infected per well using a multiplicity of infection (MOI) of 100 leptospires (1:100) (*see* **Note 20**).

12. Wash the remaining wells of the Transwell® plate with PBS.

13. Add 1.5 mL of complete growth medium into the lower compartment of each experimental well.

14. Add 500 μL of leptospires diluted in EMJH/DMEM medium into the upper compartment of the Transwell® plate (*see* **Note 20**).

15. Immediately measure the TEER.

16. Incubate the plate at 37 °C incubator with 5% CO_2.

17. Determine the time points that will be used to assess the translocation and TEER (*see* **Note 21**).

3.7 Determining Translocation Through Polarized Cell Monolayer

1. At each time point selected for your analysis after the period of interaction/infection, measure the TEER and collect an aliquot of 50 µL from the bottom compartment of each well.

2. Immediately after the collection of the aliquot, proceed to assess how many leptospires have translocated through the cell monolayer using a Petroff-Hausser counting chamber.

3. Count in triplicate the number of leptospires that translocate in each well, and calculate the mean and standard deviation.

4. The ability of leptospires to translocate MDCK polarized monolayers is determined by calculating the proportion of leptospires in the lower compartment in comparison to the initial inoculums (*see* **Note 22**).

5. Calculate the mean and standard deviation of the percentage of translocation of the monolayer for each leptospiral strain tested, considering the duplicate or triplicate wells (*see* **Note 19**).

6. Plot the results in a graph, showing the percentage of translocation in a time line series, including the variation of the average measured TEER of the wells for each strain.

4 Notes

1. The procedure described here reflects the use of Madin-Darby Canine Kidney (MDCK) cells, and DMEM medium is ideal for those cells. Other eukaryotic cell lines can be used for this experiment, and the media should reflect the growth needs of the cell line in use.

2. We used the EVOM2 epithelial voltohmmeter (World Precision Instruments, Sarasota, FL) in our experiments, and the instruction manual can be found at https://www.wpiinc.com/clientuploads/pdf/EVOM2_IM.pdf.

3. Cell passage number can affect the electrical resistance of eukaryotic cell lines. Thus, it is ideal to keep a working seed of your cells and a uniform number of passages (<30) when performing your experiments.

4. MDCK cells are counted in a Neubauer counting chamber. Trypan blue is used to selectively stain dead cells, as the dye is only able to cross the membrane of a dead cell. By counting live cells (clear) and dead cells (blue), you can determine the percentage viability of your cells, taking that in consideration when determining the number of cells.

5. To avoid disturbing flow between the compartments, which will perturb the cell monolayers, it is very important to always follow this sequence: when taking out medium from the

compartments, start with the upper compartment and then from the lower compartment; when adding medium to the compartments, start always with the lower compartment and then add medium to the upper compartment.

6. When you seed 500 μL of cells into 500 μL of medium in the upper compartment of the Transwell® plate, this will allow you to have 2×10^5 cells/cm^2.

7. The time of incubation to achieve polarized monolayers is dependent on cell lineage, number of cells seeded, medium composition, and other factors. For the protocol described here, MDCK cells took 41–43 h to achieve monolayer establishment.

8. A high electrical resistance correlates with the presence of well-developed tight junctions, while a decreased in the electrical resistance suggests a disruption in the integrity of the cell monolayer. TEER measurements can be repeated in duplicate or triplicate for each measurement during the experimentation.

9. The electrode should be placed into the Class II biosafety cabinet and sterilized with UV light for 20 min before and after use. In between measurements, the electrode can be sterilized with 70% ethanol.

10. For the equipment EVOM2, when used for resistance measurements only, the electrode does not need to be equilibrated or preconditioned before use.

11. Blank resistance should result in a steady ohm reading of the solution, with an average measurement of $100 \, \Omega/\text{cm}^2$ depending on the medium. The value of the blank always adds to the total resistance measured across a tissue culture membrane.

12. The blank resistance should be measured and then subtracted from the resistance reading across tissue in order to obtain the true tissue resistance. However, since the value is constant, this procedure is not essential.

13. For the MDCK cells used, which this protocol is based, the TEER compatible with monolayer reached a range of 200 and $300 \, \Omega/\text{cm}^2$, which was confirmed by trypan blue assays. Those numbers can vary accordingly to the cell lines being used, as much as medium and other factors [17]. MDCK cells take 41–43 h to reach polarized monolayers after seeding in Transwell® plate at a concentration of 2×10^5 cells/cm^2.

14. When using several plates in one experiment, it is ideal to perform the trypan blue assay in one well from EACH experimental plate. However, considering that the cells, medium, and other procedures are standardized, one well measurement, together with the individual TEER measurement, should be sufficient to determine monolayer establishment.

15. For more accurate results, the trypan blue in the lower compartment of the well can also be quantified by absorbance measurements with spectrophotometer.

16. When using several plates in one experiment, the average of eukaryotic cells per well should ideally be determined from one well of EACH experimental plate. However, considering that the cells, medium, and other procedures are standardized, one well measurement should be sufficient to determine this number for the entire experiment.

17. Saprophyte species of *Leptospira* spp., e.g. *Leptospira biflexa* serovar Patoc strain Patoc1, are not able to rapidly translocate through polarized monolayers (REF) and are good controls to make sure that your polarized monolayer is well established and the translocation system is working.

18. The strain Fiocruz L1-130 has a well-known ability to translocate [3, 18] and for that reason is recommended. However, the positive control should be any leptospiral strain in which ability to translocate is well established and can be used to determine the efficiency and accuracy of the assay. An extra positive control can be used to determine the accuracy of the TEER measurements, like a pathogenic *Salmonella* sp. strain [3].

19. Duplicate or triplicate infection for each strain allows for evaluation of reproducibility and variations within the experiment. Negative and positive controls should also be tested in duplicate or triplicate. Furthermore, each plate should have the following: one well without any cells for measurement of the blank resistance (*see* **Note 12**), one well for the trypan blue assay, and one well for determination of eukaryotic cell average.

20. The number of leptospires that need to be inoculated with the eukaryotic cells is calculated based on the average number of eukaryotic cells per well and the number of leptospires per milliliter. The final number of leptospires should be diluted in 500 μL of the EMJH/DMEM medium.

21. Previous experiments showed that pathogenic *Leptospira* spp. are able to translocate after 15 min [3]. When establishing the times for assessment, remember to take in consideration the amount of time that takes to measure the TEER of all the wells and to count all the leptospires that were able to translocate in each well. Time assessment over 8–12 h should take in consideration the generation time for leptospires.

22. Another method to calculate the ability of leptospires to translocate MDCK polarized monolayers is by calculating the proportion of leptospires measured in the lower compartment in a well with eukaryotic monolayer in comparison to the number of leptospires that are present in the lower compartment of a well without eukaryotic cells.

Acknowledgments

This protocol was established and validated with the essential technical support from Claudio P. Figueira, Weinan Zhu, Joshua F. Ackerman, Mary Catherine (Cate) Muenker, and Haritha Adhikarla.

References

1. Wunder EA Jr, Figueira CP, Santos GR et al (2016) Real-time PCR reveals rapid dissemination of *Leptospira interrogans* after Intraperitoneal and Conjunctival inoculation of hamsters. Infect Immun 84:2105–2115

2. Martinez-Lopez DG, Fahey M, Coburn J (2010) Responses of human endothelial cells to pathogenic and non-pathogenic Leptospira species. PLoS Negl Trop Dis 4:e918

3. Barocchi MA, Ko AI, Reis MG et al (2002) Rapid translocation of polarized MDCK cell monolayers by *Leptospira interrogans*, an invasive but nonintracellular pathogen. Infect Immun 70:6926–6932

4. Picardeau M (2017) Virulence of the zoonotic agent of leptospirosis: still terra incognita? Nat Rev Microbiol 15:297–307

5. Ko AI, Goarant C, Picardeau M (2009) Leptospira: the dawn of the molecular genetics era for an emerging zoonotic pathogen. Nat Rev Microbiol 7:736–747

6. Navab M, Hough GP, Berliner JA et al (1986) Rabbit beta-migrating very low density lipoprotein increases endothelial macromolecular transport without altering electrical resistance. J Clin Invest 78:389–397

7. Clyne M, Duggan G, Dunne C et al (2017) Assays to study the interaction of campylobacter jejuni with the mucosal surface. Methods Mol Biol 1512:129–147

8. Cruz N, Lu Q, Alvarez X et al (1994) Bacterial translocation is bacterial species dependent: results using the human Caco-2 intestinal cell line. J Trauma 36:612–616

9. Finlay BB, Falkow S (1990) Salmonella interactions with polarized human intestinal Caco-2 epithelial cells. J Infect Dis 162:1096–1106

10. Nataro JP, Hicks S, Phillips AD et al (1996) T84 cells in culture as a model for enteroaggregative *Escherichia coli* pathogenesis. Infect Immun 64:4761–4768

11. Merien F, Baranton G, Perolat P (1997) Invasion of Vero cells and induction of apoptosis in macrophages by pathogenic *Leptospira interrogans* are correlated with virulence. Infect Immun 65:729–738

12. Thomas DD, Higbie LM (1990) In vitro association of leptospires with host cells. Infect Immun 58:581–585

13. Thomas DD, Navab M, Haake DA et al (1988) Treponema pallidum invades intercellular junctions of endothelial cell monolayers. Proc Natl Acad Sci U S A 85:3608–3612

14. Comstock LE, Thomas DD (1989) Penetration of endothelial cell monolayers by Borrelia burgdorferi. Infect Immun 57:1626–1628

15. Slamti L, Picardeau M (2012) Construction of a library of random mutants in the spirochete *Leptospira biflexa* using a mariner transposon. Methods Mol Biol 859:169–176

16. Johnson RC, Harris VG (1967) Differentiation of pathogenic and saprophytic letospires. J Bacteriol 94:27–31

17. Srinivasan B, Kolli AR, Esch MB et al (2015) TEER measurement techniques for in vitro barrier model systems. J Lab Autom 20:107–126

18. Wunder EA Jr, Figueira CP, Benaroudj N et al (2016) A novel flagellar sheath protein, FcpA, determines filament coiling, translational motility and virulence for the Leptospira spirochete. Mol Microbiol 101:457–470

Chapter 16

Cell Adhesion Assay to Study Leptospiral Proteins: An Approach to Investigate Host-Pathogen Interaction

Aline F. Teixeira and Ana L. T. O. Nascimento

Abstract

The adhesion of pathogenic bacteria to host cells and the extracellular matrix (ECM) is considered an important step in the pathogenesis of microorganisms. It has been described that *Leptospira* spp. bind to multiple receptors on host cells and to the ECM to initiate infection. Most studies of *Leptospira* adherence described until now have focused on the in vitro attachment of recombinant *L. interrogans* proteins to ECM components. These putative adhesins may be involved in the colonization of the host, contributing to the bacterial invasion process. Certainly, in vitro cell adhesion studies have contributed to the elucidation of leptospiral pathogenesis mechanisms. Here, we describe a cell adhesion assay that can be used for studying the interactions between putative leptospiral adhesins and host components.

Key words *Leptospira*, Cell adhesion, Extracellular matrix, Host-pathogen interaction, Leptospirosis

1 Introduction

Studies of cell adhesion have been widely explored for many important purposes in both cell biology and biomedical research. The interaction between pathogens and cells or the extracellular matrix (ECM) can influence cell control, leading to subversion of its functions and establishment of disease [1–3]. ECM plays an important role in the host, such as regulating eukaryotic cell adhesion, differentiation, migration, proliferation, shape, and function, besides supporting and connecting cells and tissues [4]. Two main classes of macromolecules compose the mammalian ECM, the glycosaminoglycans (GAGs) and fibrous proteins, such as collagens, elastin, fibronectin, and laminin [5, 6]. This network of extracellular macromolecules under normal conditions is not exposed to bacteria. However, pathogens may gain access to these components after tissue injury. It has been well documented that *Leptospira* interacts with ECM and plays a role in the initial colonization of host tissues. This interaction is likely to occur via leptospiral surface-exposed molecules known as adhesins, which act as a

Nobuo Koizumi and Mathieu Picardeau (eds.), *Leptospira spp.: Methods and Protocols*, Methods in Molecular Biology, vol. 2134, https://doi.org/10.1007/978-1-0716-0459-5_16, © Springer Science+Business Media, LLC, part of Springer Nature 2020

bridge between the bacteria and the host. Several leptospiral adhesins have been identified and characterized as ECM-binding molecules [7–20]. The adhesion process is considered as a critical step in promoting host infections. Therefore, it is a key event that deserves to be investigated in leptospiral pathogenesis.

The process of cell adhesion is complex and includes receptor-ligand binding, changes in intracellular signaling pathways, and modulation of cytoskeletal assembly [21]. Most of the methods used measure the ability of a bacterial cell to interact with a specific adhesion molecule or evaluate this interaction in the presence of inhibitors [2]. The first interaction assay performed with leptospires used radiolabeled leptospiral suspensions and host cell monolayers [22]. Despite the fact that assays involving radioactive-labeled bacteria have been used successfully, hazards involved in these manipulations have decreased their use. Moreover, assays that involve host cells are considered laborious and expensive. The development of recombinant DNA technology has made an impressive impact in the number of recombinant proteins that can be produced, and consequently, new methods to evaluate the adhesion process have been established. Here, we describe a method based on a simple, fast, and sensitive colorimetric assay to investigate the initial interaction between recombinant leptospiral proteins and host components.

2   Materials

2.1   Extracellular Matrix Components

All host macromolecules and the control proteins can be purchased from Sigma-Aldrich, and solutions prepared and stored as described by the manufacturer. Dilute all macromolecules to a working concentration of 1 μg/well in PBS.

1. Phosphate-buffered saline (PBS): 137 mM NaCl, 2.7 mM KCl, 10 mM Na_2HPO_4, 2 mM KH_2PO_4, pH 7.4. Autoclave and store at room temperature.

2. Tris buffer: 0.2 M Tris–HCl, pH 8.8.

3. Laminin: The solution of laminin from Engelbreth-Holm-Swarm mouse sarcoma basement membrane is kept at −20 °C (*see* **Note 1**).

4. Collagen type 1: The solution of collagen type 1 of rat tail tendon is stored at 2–8 °C (*see* **Note 2**).

5. Collagen type 4: Collagen type 4 from Engelbreth-Holm-Swarm mouse sarcoma basement membrane is reconstituted in PBS containing 0.25% acetic acid at a concentration of 0.5–2.0 mg/mL, which is kept for several hours at 4 °C with occasional shaking (*see* **Note 3**).

6. Elastin: Elastin from human aorta is dissolved in Tris buffer to a concentration of 1 mg/mL and stored at 2–8 °C.

7. Fibronectin: The solution of fibronectin from human fibroblasts is kept at −20 °C (*see* **Note 4**).

2.2 Cloning, Expression, and Purification of Recombinant Proteins

1. Genomic DNA of *L. interrogans* serovar Copenhageni strain FIOCRUZ L1-130 (*see* **Note 5**).

2. Specific primers for sequence amplification.

3. Thermocycler.

4. Microtubes, 0.2 mL volume.

5. DNA and Gel Band Purification Kits.

6. Vectors for cloning.

7. Restriction enzymes.

8. Tris-acetate-EDTA (TAE) buffer: 40 mM Tris, 20 mM acetic acid, 1 mM EDTA. Store at room temperature.

9. 1% agarose gel: Dissolve 1 g of agarose in 100 mL TAE buffer.

10. Vectors for expression.

11. *E. coli* BL21.

12. Luria-Bertani (LB) medium: 1% NaCl, 0.5% yeast extract, 1% tryptone. For solid medium, use 1.5% bacteriological agar. Adjust the pH to 7.0 and sterilize by autoclaving for 20 min at 121 °C.

13. Isopropyl-β-D-thiogalactopyranoside (IPTG): 1 M solution in water. Sterilize with a 0.2 μm filter and store in 1 mL aliquots at −20 °C.

14. Ampicillin: 100 mg/mL solution in water. Sterilize by filtration, aliquot, and store at −20 °C.

15. 5-Bromo-4-chloro-3-indolyl-β-D-galactopyranoside (X-Gal): 20 mg/mL solution in dimethylformamide. Sterilize with a 0.2 μm filter, and store in the dark in aliquots of 0.5 mL at −20 °C.

16. Chloramphenicol: 30 mg/mL solution in ethanol. Sterilize by filtration, aliquot, and store at −20 °C.

17. Phenylmethylsulfonyl fluoride (PMSF): 0.1 M solution in isopropanol (*see* **Note 6**). PMSF is unstable in aqueous solution. Store at 20–25 °C.

18. Lysozyme: 10 mg/mL solution in water. Aliquot and store at −20 °C.

19. Lysis buffer: 10 mM Tris–HCl, pH 8.0, 150 mM NaCl, 100 μg/mL lysozyme, 1% Triton X-100, 2 mM PMSF.

20. Cell disruptor.

21. Soluble fraction buffer: 10 mM Tris–HCl, pH 8.0, 150 mM NaCl.

22. Insoluble fraction buffer: 10 mM Tris–HCl, pH 8.0, 150 mM NaCl, 8 M urea.

23. Urea: 9 M solution in water. Dissolve 540.54 g urea in 600 mL of warm water (~30 °C), and mix vigorously with a stirring bar. Afterward, add water to a final volume of 1 L.

24. Imidazole: 2 M solution in water.

25. Binding buffer: 100 mM Tris–HCl, pH 8.0, 500 mM NaCl, 5 mM imidazole.

26. Wash buffer: 100 mM Tris–HCl, pH 8.0, 500 mM NaCl, 20–100 mM imidazole.

27. Elution buffer: 100 mM Tris–HCl, pH 8.0, 500 mM NaCl, 1 M imidazole.

28. 1.5 mL microtubes.

29. Chromatographic column.

30. Chelating Sepharose.

31. Nickel sulfate (Ni^{2+}): 0.3 M solution in water.

32. Thirty percent acrylamide/BIS solution (29:1) acrylamide: BIS: Weigh out 29 g acrylamide and 1 g bisacrylamide and add water to a final volume of 100 mL, with shaking until totally dissolved. Filter the solution through a 0.22 μm filter, and store at 4 °C, in a bottle wrapped with aluminum foil.

33. Ammonium persulfate: 10% solution in water.

34. N,N,N',N'-Tetramethylethylenediamine (TEMED): store at 4 °C.

35. Resolving gel buffer: 1.5 M Tris–HCl, pH 8.8. Autoclave and store at room temperature.

36. Stacking gel buffer: 1 M Tris–HCl, pH 6.8. Autoclave and store at room temperature.

37. SDS-PAGE sample buffer (5×): 250 mM Tris–HCl, pH 6.8, 0.1% SDS, 50% glycerol, 6.7% β-mercaptoethanol, 0.5 mg/mL bromophenol blue. Mix 7.5 mL of 1 M Tris–HCl, pH 6.8, 0. 3 mL of 10% SDS, 15 mL of glycerol, and 15 mg bromophenol blue in water a final volume of 28 mL, and heat the solution at 65 °C for total dissolution. At the end, add 2 mL of β-mercaptoethanol.

38. Sodium dodecyl sulfate (SDS): 10% solution in water.

39. Tris-glycine buffer (5×): 25 mM Tris–HCl, pH 8.3, 192 mM glycine, 0.5% SDS. Store at room temperature.

40. Glass plates.

41. Gel caster.

42. Combs.

43. Coomassie Brilliant Blue: 2.5 mg/mL solution. Weigh out 0.25 g Coomassie Brilliant Blue, dissolve in 10 mL of glacial acetic acid, and add 1/1 (vol/vol) methanol and water to a final volume of 100 mL. Stir the solution for 3–4 h and then filter through filter paper. Store at room temperature.

44. Destaining solution: 10% acetic acid, 45% methanol. Mix 10 mL of glacial acetic acid and 45 mL of methanol and add water to a final volume of 100 mL.

45. Dialysis membrane with a cutoff of 10 kDa.

2.3 Interaction Assay

1. ELISA microplates (96-well) for high binding.

2. Absorbance microplate reader.

3. 37 °C incubator.

4. Microplate washer.

5. 50 mL tubes.

6. 1.5 mL vial.

7. Sealing tape for 96-well plates.

8. Wash buffer: 1× PBS, 0.05% Tween 20 (PBS-T).

9. Blocking buffer: PBS-T, 10% dry milk or 1% BSA. Weigh out 10 g skimmed dry milk, and dilute the powder in 100 mL PBS-T. For preparing BSA, weigh out 1 g BSA, and dissolve in 100 mL of PBS-T.

10. Mouse antiserum against protein of interest.

11. HRP-conjugated anti-His monoclonal antibodies.

12. HRP-conjugated anti-mouse IgG.

13. Bovine serum albumin: 1 mg/mL in water. Aliquot and store at −20 °C.

14. Fetuin from fetal bovine serum: 1 mg/mL in water. Aliquot and store at −20 °C.

15. *o*-Phenylenediamine (OPD).

16. Hydrogen peroxide (H_2O_2).

17. Citrate-phosphate buffer: 0.1 M citrate, 0.2 M phosphate, pH 5.0.

18. Sulfuric acid: 2 M solution in water.

3 Methods

3.1 Cloning and Expression of Recombinant Protein in E. coli

1. Genes of interest are amplified without the signal peptide (*see* **Note** 7) from the genomic DNA of *L. interrogans* by PCR with specific primers. Sample preparation is performed following the protocol described in Table 1.

Table 1
Reagents used for PCR preparation

Recipe for PCR amplification		
Reagents	Use concentration	Stock concentration
PCR buffer	1×	10×
$MgCl_2$	3 mM	50 mM
Primer	1 μM	10 μM
dNTPs	200 μM	10 mM
Taq polymerase	0.5 μL	
DNA	100 ng	
Pure water	Complete to a final volume 50 μL	

Table 2
Recipe for preparation of a ligation reaction

Ligation reaction for insertion of a fragment into a cloning vector	
Reagents	Concentration
Buffer T4 DNA ligase	1×
T vector	50 ng
PCR product	[a]
T4 DNA ligase	1 μL
Pure water	Complete to a final volume 10 μL

[a]Use three times more DNA insert in relation to vector

Use the following conditions for the thermocycler:

(a) Initial denaturation step: 94 °C—4 min.

(b) Denaturation: 94 °C—30 s.

(c) Annealing: 62 °C—30 s.

(d) Extension: 72 °C—2 min (generally, 1 min for 1000 bp).

2. PCR-amplified products are visualized in an agarose gel. DNA bands are cut out and purified. Purified fragments are cloned into a cloning vector, according to Table 2. Incubate the reaction mixture for 1 h at room temperature.

3. After the ligation reaction, plasmids are used to transform competent *E. coli* cells, which are then seeded on LB agar plates containing 100 μg/mL ampicillin, 0.5 mM IPTG, and 80 μg/mL X-gal for blue/white screening of recombinant bacterial colonies with the lac[+] genotype. Plates are incubated overnight at 37 °C.

Table 3
Recipe for a ligation reaction on pAE vector

Ligation reaction for cloning into an expression vector	
Reagents	Concentration
Buffer T4 DNA ligase	1×
pAE vector	100 ng
DNA fragment	a
T4 DNA ligase	0.5 µL
Pure water	Complete to a final volume 20 µL

aUse 3–5 times more DNA insert in relation to vector

4. White colonies are selected, plasmids are extracted, and positive clones are confirmed by plasmid DNA restriction; mix 3 µL of plasmid, 1 µL of buffer 10×, 0.25 µL of each enzyme, and 5.5 µL of water, and incubate mixture for 1 h at 37 °C.

5. DNA fragments are purified and ligated into the *E. coli* expression vector as described in Table 3. Incubate the reaction mixture for 50 min at 22 °C.

6. Plasmids are used to transform competent *E. coli* BL21, which are then plated on LB agar containing 50 µg/mL ampicillin and 34 µg/mL chloramphenicol. Plates are incubated overnight at 37 °C. Positive clones are confirmed by DNA restriction analysis, as described above.

7. Select one positive colony and prepare a pre-inoculum in LB medium containing the antibiotics of interest. Allow it to grow until saturation, generally, overnight at 37 °C.

8. Dilute 20-fold the saturated inoculum in a larger volume of LB medium containing antibiotics, and place the flask in a 37 °C incubator with continuous shaking until an absorbance of 0.6 at 600 nm is reached. For recombinant protein synthesis, add 0.1–1 mM IPTG, and keep under continuous shaking at 37 °C for 3 h.

9. Centrifuge the bacterial suspensions at 3,075 × *g* for 15 min at 4 °C. Resuspend the bacterial cell pellets in ten volumes of lysis buffer in relation to the culture volume used, place on ice for 15 min, and lyse cells with the aid of a cell disruptor.

10. Centrifuge the suspension at 12,000 × *g* for 15 m at 4 °C, and separate the soluble and insoluble fractions. If the protein is expressed in the insoluble form, resuspend the pellet in the resuspension buffer containing 8 M urea. Incubate overnight with shaking. Cloning and expression steps are shown in Fig. 1.

3.2 Purification of Recombinant Protein

The pAE vector used allows the expression of a recombinant protein with a 6×His fused to its N-terminal. Thus, proteins can be purified by immobilized metal affinity chromatography (IMAC). Usually, we use divalent nickel as the metal.

1. Fill a column with chelating Sepharose and wait for resin sedimentation. Afterward, wash the column with 3–5 volumes of water to remove ethanol used as preservative.

2. Charge the column with 3–5 volumes of nickel solution and incubate for 30 min. Wash with 3–5 volumes of water.

3. Equilibrate the column with binding buffer.

4. Add protein extract to column.

5. For soluble proteins, wash the column with ten volumes of wash buffer containing 20 mM imidazole.

6. Wash the column with five volumes of wash buffer containing 40–100 mM imidazole.

7. Elute the bound protein with 3–5 column volumes of elution buffer. Collect 1 mL fractions in microtubes.

8. In case of insoluble proteins, do on-column refolding by gradually removing the urea. Pass on-column five volumes of binding buffer containing from 8 to 0 M urea. After urea elimination, proceed starting at **step** 7.

9. Evaluate the purified proteins using 12% SDS-PAGE. This percentage is typically used for proteins from 20 to 60 kDa. Mix 3.3 mL of water, 4 mL of 30% acrylamide mix, and 2.5 mL of resolving gel buffer. Add 100 μL of SDS, 100 μL of 10% ammonium persulfate, and 10 μL of TEMED; homogenize and pour the mixture into a gel cassette. Overlay the gel with isopropanol until complete polymerization. Prepare 5 mL stacking gel by mixing 3.4 mL of water, 830 μL of acrylamide mixture, and 630 μL of stacking gel buffer. Add 50 μL of SDS, 50 μL of 10% ammonium persulfate, and 5 μL of TEMED. Insert the gel comb carefully. Add 4 μL of 5× SDS-PAGE buffer to 16 μL of protein aliquots, and heat the samples for 5 min at 96 °C. When running the gel, include protein molecular weight standards. Set the amperage to 25 mA per gel, and run the gel until the dye reaches the bottom of the gel.

10. Remove the polyacrylamide gel from the glass, and place it in a suitable container with a lid; stain the gel with Coomassie blue solution for 30 min or overnight with shaking. Remove the excess dye using the destaining solution with shaking. Change the solution until visualization of the protein bands. Figure 1 shows the purification process steps.

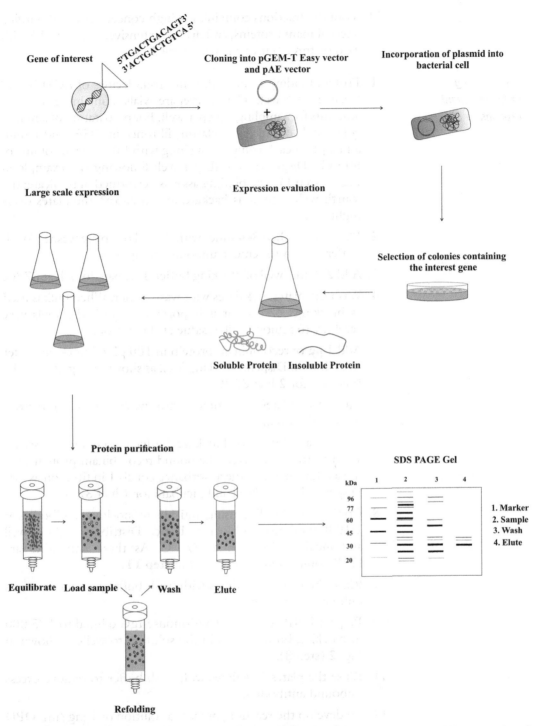

Fig. 1 Flowchart of cloning, expression, and purification of recombinant protein. The gene of interest is cloned into a cloning vector and subcloned into a bacterial expression vector. The recombinant constructs are used to transform *E. coli* bacterial strains. Colonies containing the gene of interest are selected, and induction protein expression is performed. After checking if the proteins are expressed in their soluble or insoluble fraction, a large-scale production is completed for protein purification. Using a Ni^{2+}-charged chromatography column, the recombinant proteins are purified at their native or denatured condition. Purified samples are assessed by SDS-PAGE

11. Pool the fractions containing a high concentration of purified recombinant proteins, and dialyze extensively against PBS with two or three buffer changes every 2 h.

3.3 Binding of Recombinant Proteins to ECM

1. Protein binding to individual macromolecules of ECM is analyzed in 96-well plates. Prepare vials containing enough amounts for sampling 1 μg per well. For preparation of laminin, collagen I, collagen IV, elastin, fibronectin, BSA, and fetuin, add 5 μL of each component along with PBS to final volume of 500 μL. Dispense 100 μL per well following the example in Fig. 2 (step 1). The binding assay is performed in triplicate; the fourth well is used as background. Incubate the plates overnight at 4 °C.

2. Plates are washed 3–4 times with the aid of a plate washer. Wash buffer is used to remove unbound component.

3. Add 200 μL/well of blocking buffer. Incubate for 2 h at 37 °C.

4. Wash the plates 3–4 times with wash buffer. When milk is used as blocking solution, it is important to wash the wells very carefully to remove milk residue at the bottom.

5. Add 1 μg of recombinant protein in 100 μL of blocking buffer to each well. Dispense 100 μL/well as shown in Fig. 2 (step 2). Incubate for 2 h at 37 °C.

6. Rinse the plates 3–4 times with wash buffer to remove unbound protein.

7. Prepare a solution of blocking buffer containing antiserum against protein to detect the bound recombinant proteins. We use a dilution where the absorbance equals 1 in titration assays. Dispense 100 μL/well and incubate for 1 h at 37 °C.

8. Confirm the binding using anti-His monoclonal antibodies at 1:10,000 dilution in blocking buffer. Distribute 100 μL/well and incubate for 1 h at 37 °C. As these antibodies are HRP-conjugated, continue from **step 11**.

9. Rinse the plates 3–4 times with wash buffer to remove excess unbound antibodies.

10. Prepare HRP-conjugated anti-mouse IgG diluted to 1/5,000 in blocking buffer, and add the solution to wells as shown in Fig. 2 (step 3).

11. Rinse the plates 3–4 times with wash buffer to remove excess unbound antibodies.

12. To develop the reaction, prepare a solution of 1 mg/mL OPD in 10 mL of citrate phosphate buffer, and add 10 μL of H_2O_2; dispense 100 μL/well of this reagent. Incubate for 15 min at room temperature for color development (step 4 in Fig. 2).

13. Add 50 μL/well of 2 M H_2SO_4 to stop the reaction.

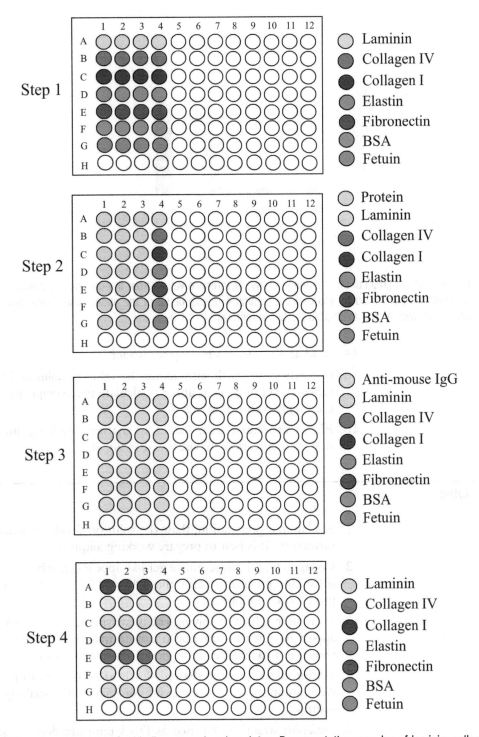

Fig. 2 Experimental design for binding assay in microplates. Representative samples of laminin, collagen I, collagen IV, elastin, fibronectin, BSA, and fetuin are shown in quadruplicate in the coating step (step 1). Assays are performed in triplicate; the fourth well is used as background. Thus, after blocking, 1 μg of recombinant protein is added in the first, second, and third columns of the plate (step 2). Anti-protein and HRP-conjugated anti-mouse IgG are plated in all wells (step 3). A solution containing the substrate for HRP is added and incubated for color development (step 4)

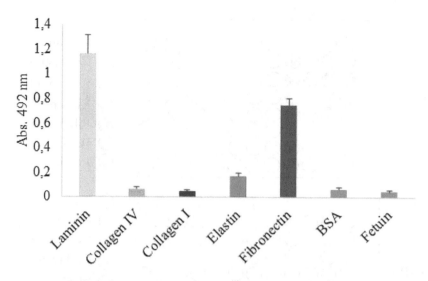

Fig. 3 Demonstrative graph of an interaction assay. Absorbance values from the columns 1, 2, and 3 of the reaction plate are subtracted by values from column 4 (background) of the plate. Absorbance average is calculated, and then, a column chart is plotted to represent the data

14. Read at 492 nm in a microplate reader.

15. Construct the graph subtracting the value of column 4 from values of columns 1, 2, and 3. Calculate the average and plot the graph as shown in Fig. 3.

16. ELISA procedure steps including coating, blocking, interaction, and detection are visualized in Fig. 4.

4   Notes

1. For laminin preparation, thaw the solution slowly to avoid gel formation. It is best to prepare working aliquots.

2. Collagen type 1 can acquire a gel-like appearance when cold; to solve this problem, warm the solution 37 °C. Note if the product is solubilized.

3. Even after 4 h of dissolving collagen type 4, the solution can appear hazy with some non-dissolved material. This has been shown not to affect its biological performance of experiments.

4. Fibronectin should be thawed slowly at 4 °C until forming a clear solution. Avoid vortexing or vigorously shaking the solution.

5. Different strain can be used as DNA template depending on the bacterium model used.

6. PMSF has a short half-life; therefore, a freshly prepared solution is used in the purification steps.

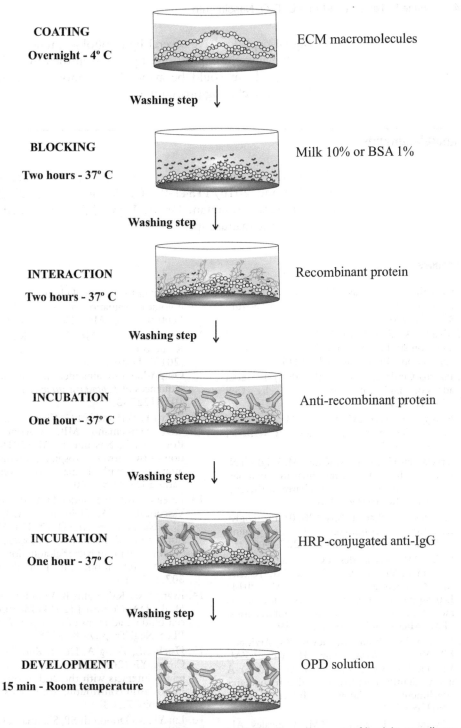

Fig. 4 Schematic representation of an interaction assay by ELISA. One microgram of laminin, or collagen I, or collagen IV, or elastin, or fibronectin is adsorbed into ELISA plate in PBS buffer. BSA and fetuin are used as control proteins. On the following day, remove the buffer, wash the plate, and block the non-specific sites of the components using blocking buffer. After incubation for 2 h at 37 °C, repeat the washing steps, and add 1 μg recombinant per well to allow interaction to the immobilized components. Bound proteins are detected by addition of an antiserum against the recombinant protein, followed by addition of an enzyme conjugated IgG antibody. The reaction is developed using OPD as substrate for horseradish peroxidase (HRP) that produces a yellow-orange product detectable at 492 nm

7. Signal peptides have several hydrophobic amino acids, which could cause the expression of proteins in their insoluble form. This problem could be avoided by removing this sequence before the cloning procedures.

Acknowledgments

This work was financially supported by FAPESP (grant 2014/50981-0 and 2016/11541-0), CNPq (grant 301229/2017-1), and Fundação Butantan. Dr. A. Leyva (USA) provided English editing of the manuscript.

References

1. Gruenheid S, Finlay BB (2003) Microbial pathogenesis and cytoskeletal function. Nature 422:775–781

2. Khalili AA, Ahmad MR (2015) A review of cell adhesion studies for biomedical and biological applications. Int J Mol Sci 16:18149–18184

3. Pizarro-Cerdá J, Cossart P (2006) Bacterial adhesion and entry into host cells. Cell 124:715–727

4. Frantz C, Stewart KM, Weaver VM (2010) The extracellular matrix at a glance. J Cell Sci 123:4195–4200

5. Järveläinen H, Sainio A, Koulu M, Wight TN, Penttinen R (2009) Extracellular matrix molecules: potential targets in pharmacotherapy. Pharmacol Rev 61:198–223

6. Schaefer L, Schaefer RM (2010) Proteoglycans: from structural compounds to signaling molecules. Cell Tissue Res 339:237–246

7. Vieira ML, Fernandes LG, Domingos RF, Oliveira R, Siqueira GH, Souza NM, Teixeira AR, Atzingen MV, Nascimento AL (2014) Leptospiral extracellular matrix adhesins as mediators of pathogen-host interactions. FEMS Microbiol Lett 352:129–139

8. Barbosa AS, Abreu PA, Neves FO, Atzingen MV, Watanabe MM, Vieira ML, Morais ZM, Vasconcellos SA, Nascimento AL (2006) A newly identified leptospiral adhesin mediates attachment to laminin. Infect Immun 74:6356–6364

9. Oliveira R, de Morais ZM, Gonçales AP, Romero EC, Vasconcellos SA, Nascimento AL (2011) Characterization of novel OmpA-like protein of Leptospira interrogans that binds extracellular matrix molecules and plasminogen. PLoS One 6:e21962

10. Fernandes LG, Vieira ML, Alves IJ, de Morais ZM, Vasconcellos SA, Romero EC, Nascimento AL (2014) Functional and immunological evaluation of two novel proteins of Leptospira spp. Microbiology 160:149–164

11. Teixeira AF, de Morais ZM, Kirchgatter K, Romero EC, Vasconcellos SA, Nascimento AL (2015) Features of two new proteins with OmpA-like domains identified in the genome sequences of Leptospira interrogans. PLoS One 10:e0122762

12. Silva LP, Fernandes LG, Vieira ML, de Souza GO, Heinemann MB, Vasconcellos SA, Romero EC, Nascimento AL (2016) Evaluation of two novel leptospiral proteins for their interaction with human host components. Pathog Dis 74:ftw040

13. Pereira PRM, Fernandes LGV, de Souza GO, Vasconcellos SA, Heinemann MB, Romero EC, Nascimento ALTO (2017) Multifunctional and redundant roles of Leptospira interrogans proteins in bacterial-adhesion and fibrin clotting inhibition. Int J Med Microbiol 307:297–310

14. Evangelista KV, Hahn B, Wunder EA, Ko AI, Haake DA, Coburn J (2014) Identification of cell-binding adhesins of Leptospira interrogans. PLoS Negl Trop Dis 8:e3215

15. Hsieh CL, Tseng A, He H, Kuo CJ, Wang X, Chang YF (2017) Immunoglobulin-like protein B interacts with the 20th exon of human tropoelastin contributing to. Front Cell Infect Microbiol 7:163

16. Lin YP, McDonough SP, Sharma Y, Chang YF (2010) The terminal immunoglobulin-like repeats of LigA and LigB of Leptospira enhance their binding to gelatin binding domain of fibronectin and host cells. PLoS One 5:e11301

17. Stevenson B, Choy HA, Pinne M, Rotondi ML, Miller MC, Demoll E, Kraiczy P, Cooley

AE, Creamer TP, Suchard MA, Brissette CA, Verma A, Haake DA (2007) *Leptospira interrogans* endostatin-like outer membrane proteins bind host fibronectin, laminin and regulators of complement. PLoS One 2:e1188

18. Rossini AD, Teixeira AF, Souza Filho A, Souza GO, Vasconcellos SA, Heinemann MB, Romero EC, Nascimento ALTO (2020) Identification of a novel protein in the genome sequences of *Leptospira interrogans* with the ability to interact with host's components. J Microbiol Immunol Infect 53(1):163–175

19. Cosate MR, Siqueira GH, de Souza GO, Vasconcellos SA, Nascimento AL (2016) Mammalian cell entry (Mce) protein of *Leptospira interrogans* binds extracellular matrix components, plasminogen and β2 integrin. Microbiol Immunol 60:586–598

20. Figueredo JM, Siqueira GH, de Souza GO, Heinemann MB, Vasconcellos SA, Chapola EG, Nascimento AL (2017) Characterization of two new putative adhesins of *Leptospira interrogans*. Microbiology 163:37–51

21. Humphries MJ (2009) Cell adhesion assays. Methods Mol Biol 522:203–210

22. Thomas DD, Higbie LM (1990) In vitro association of leptospires with host cells. Infect Immun 58:581–585

Chapter 17

Complement Resistance Assays

Lourdes Isaac and Angela Silva Barbosa

Abstract

Like many other pathogens of medical importance, pathogenic *Leptospira* employ diverse strategies to circumvent Complement System activation. Under physiological conditions, this central humoral arm of innate immunity is tightly controlled by negative Complement regulatory proteins. However, upon infection, pathogenic microorganisms interfere with normal Complement host defense mechanisms by recruiting or mimicking Complement regulators and by secreting endogenous proteases or acquiring host's proteases that inactivate key Complement components. In this chapter, we describe in detail some of the most frequently used assays to evaluate *Leptospira* Complement resistance.

Key words Complement, *Leptospira*, Complement regulators, Serum susceptibility, Cleavage of Complement proteins, Proteases

1 Introduction

The Complement System is important for the innate and acquired immune responses against several pathogens. Once activated, this System generates several important biological functions that may help to control infection and bacteria dissemination such as (a) generation of opsonins (fragments C3b, iC3b, and C3d) which contribute to enhance microorganism internalization and killing by macrophages and other phagocytic cells; (b) formation of a pore on the microorganism surface called membrane attack complex ($C5b6789_n$) which may lead to pathogen lysis; (c) production of chemoattractant factors (fragments C3a and C5a) attracting more inflammatory cells toward the infected tissue; and (d) stimulating the production of specific antibodies against pathogens (fragment C3d/C3dg and receptor CR2) (reviewed in [1]).

To avoid excessive activation of this System and consequently host cell damage, several regulatory proteins control the Alternative Pathway [e.g., Factor H (FH), FH-like (FHL)-1, Factor I] as well as the Classical and the Lectin Pathways [e.g., C1 inhibitor,

Nobuo Koizumi and Mathieu Picardeau (eds.), *Leptospira spp.: Methods and Protocols*, Methods in Molecular Biology, vol. 2134, https://doi.org/10.1007/978-1-0716-0459-5_17, © Springer Science+Business Media, LLC, part of Springer Nature 2020

Membrane Cofactor Protein (MCP), C4b binding protein (C4BP), Factor I] and the common Terminal Pathway (e.g., CD59, vitronectin, clusterin). Some pathogens are able to evade the Complement System because they acquire Complement regulatory proteins from the host.

As early as in the mid-1960s, a bactericidal activity of normal human serum (NHS) against saprophytic but not pathogenic *Leptospira* strains was reported by Johnson and Muschel [2]. Since then, it became clear that the Complement System had a role in eliminating nonpathogenic *Leptospira* strains. More recently, a number of studies aimed at unraveling the mechanisms by which virulent *Leptospira* circumvent Complement activation have been conducted (reviewed in [3]). Some of the general strategies used by pathogenic *Leptospira* to counteract Complement include: (1) binding of negative Complement regulators including FH, FHL-1, C4BP, and vitronectin [4–7]; (2) cleavage of Complement proteins of the Alternative, Lectin, and Classical Pathways by secreted proteases [8, 9], and (3) the acquisition of host proteases such as plasminogen, known to degrade Complement proteins once converted to its active form, plasmin [10, 11].

In the following sections, we describe in detail some of the most frequently used assays to evaluate *Leptospira* Complement resistance:

1. *Serum susceptibility assay*—Leptospires susceptible to Complement attack fragment into fine particles, losing their typical morphology. They acquire a finely granulated aspect, as observed under the microscope. Serum susceptibility can be assessed by incubating freshly harvested leptospires with normal human serum (NHS) followed by counting viable bacteria under a dark-field microscope.

2. *Serum adsorption assays using intact leptospires to evaluate binding of negative Complement regulators*—Acquisition of negative Complement regulators such as FH, FHL-1, C4BP, or vitronectin by intact leptospires can be assessed by incubating bacteria in NHS-EDTA. Surface-bound proteins are subjected to Western blotting, and Complement regulators are detected by specific antibodies.

3. Cleavage of Complement proteins by secreted proteases—Pathogenic *Leptospira* such as *L. interrogans* serovar Pomona strain Pomona, *L. interrogans* serovar Kennewicki strain Fromm, *L. interrogans* serovar Copenhageni strain 10A, *L. interrogans* serovar Icterohaemorrhagiae strain RGA, *L. interrogans* serovar Pyrogenes strain Salinem, *L. kirshneri* serovar Cynopteri strain 3522C, and *L. noguchii* serovar Panama strain CZ 214 secrete proteases that cleave several Complement proteins like C3 and Factor B (Alternative Pathway),

C4 and C2 (Classical and Lectin Pathways, [8]), and C5-C9 proteins (Terminal Pathway, [9]). Proteases present in the supernatants of nonpathogenic *L. biflexa* serovar Andamana strain CH11 or *L. biflexa* serovar Patoc strain Patoc I do not have a significant proteolytic activity on these substrates.

4. *Inhibition of Complement activation by proteases secreted by pathogenic Leptospira*—The proteolytic activity of secreted proteases from pathogenic *Leptospira* cultures cleaves several Complement proteins either purified or present in NHS, affecting global activation of the Alternative, the Classical, and the Lectin Pathways. Proteases present in the supernatants of nonpathogenic *Leptospira* cultures have no significant effect on the cleavage of Complement proteins [8], and they do not interfere with Complement activation. The activation of each Complement Pathway is evaluated independently, according to Roos et al. [12] and Fraga et al. [8].

2 Materials

2.1 Serum Susceptibility Assay

1. *Leptospira* cultures. Cultivate in modified Ellinghausen-McCullough-Johnson-Harris (EMJH) medium. Alternatively, leptospires may also be cultivated in EMJH medium supplemented with Difco™ Leptospira Enrichment EMJH (Becton Dickinson) (*see* **Note 1**). Use *Leptospira* cultures in mid to late logarithmic phase.

2. Modified EMJH medium: Difco™ Leptospira Medium Base EMJH (Becton Dickinson) supplemented with 10% normal rabbit serum previously inactivated for 30 min at 56 °C, containing 0.015% L-asparagine, 0.001% sodium pyruvate, 0.001% calcium chloride, 0.001% magnesium chloride, 0.03% peptone, and 0.03% meat extract.

3. Phosphate-buffered saline (PBS): 0.01 M Na_2HPO_4, 0.0018 M KH_2PO_4, pH 7.4, 0.137 M NaCl, 0.0027 M KCl.

4. A pool of normal human serum (NHS): Serum samples can be used either fresh or thawed from aliquots stored at −80 °C (*see* **Note 2**). Heat-inactivated NHS (HI-NHS) is obtained by incubating NHS in a water bath or in a dry block incubator for 30 min at 56 °C.

5. Refrigerated centrifuge with appropriate adaptors for 50 mL tubes or for microfuge tubes.

6. Microscope with a dark-field condenser.

7. Bacterial counting chamber [cell depth: 0.02 mm (1/50 mm)].

8. Water bath/dry block incubator.

2.2 Acquisition of Complement Regulatory Proteins by Leptospira

1. *Leptospira* cultures (as described in Subheading 2.1).
2. Filtered Complement fixation diluent buffer (CFD): 4 mM sodium barbitone, 0.145 M NaCl, 0.83 mM $MgCl_2$, 0.25 mM $CaCl_2$, pH 7.3.
3. NHS-EDTA: NHS containing 10 mM EDTA pH 8.0 (*see* **Note 2**).
4. Glycine buffer: 0.1 M glycine-HCl, pH 2.0.
5. Primary antibodies: anti-FH, anti-C4BP, or anti-vitronectin produced in mouse, rabbit, goat, or sheep.
6. Secondary antibodies: horseradish peroxidase (HRP) conjugated antibodies with specificity for the antibody species and isotype of the primary antibody being used (e.g., anti-mouse IgG-HRP, anti-rabbit IgG-HRP, anti-goat IgG-HRP, or anti-sheep IgG-HRP).
7. Centrifuge with appropriate adaptors for 50 mL tubes/a microcentrifuge for 1.5–2.0 mL tubes.
8. Bacterial counting chamber [cell depth: 0.02 mm (1/50 mm)].
9. Water bath/dry block incubator.

2.3 Direct Cleavage of Complement Proteins by Leptospira-Secreted Proteases

1. *Leptospira* cultures (as described in Subheading 2.1).
2. A pool of NHS: aliquot and maintain at −80 °C until use (*see* **Note 2**).
3. Purified human Complement proteins: keep at −80 °C until use (*see* **Note 3**).
4. Primary antibodies: anti-C3, anti-C4, anti-C5, anti-C6, anti-C7, anti-C8, or anti-C9 produced in rabbit, goat, or sheep.
5. Secondary antibodies: HRP-conjugated antibodies with specificity for the antibody species and isotype of the primary antibody being used (e.g., anti-rabbit IgG-HRP, anti-goat IgG-HRP, or anti-sheep IgG-HRP).
6. Refrigerated centrifuge with appropriate adaptors for 50 mL tubes or for microcentrifuge tubes.
7. Bacterial counting chamber [cell depth: 0.02 mm (1/50 mm)].
8. Water bath/dry block incubator.
9. 0.22 μm sterile syringe filters.
10. A kit to determine the total protein concentration.

2.4 Inhibition of Complement Activation by Proteases Secreted by Pathogenic Leptospira

1. *Leptospira* cultures (as described in Subheading 2.1).
2. A pool of NHS: aliquot and maintain at −80 °C until use (*see* **Note 2**).
3. *Escherichia coli* lipopolysaccharide.
4. Human IgG or human IgM.

5. Mannan from *Saccharomyces cerevisiae*.

6. Bovine serum albumin (BSA).

7. Primary antibodies: anti-human C3 or anti-human C4 produced in rabbit, goat, or sheep.

8. Secondary antibodies: HRP-conjugated antibodies with specificity for the antibody species and isotype of the primary antibody being used (e.g., anti-rabbit IgG-HRP, anti-goat IgG-HRP, or anti-sheep IgG-HRP).

9. Refrigerated centrifuge with appropriate adaptors for 50 mL tubes.

10. Bacterial counting chamber [cell depth: 0.02 mm (1/50 mm)].

11. Water bath/dry block incubator.

12. 0.22 μm sterile syringe filters.

13. Spectrophotometer with a filter for 492 nm wavelength.

14. A kit to determine total protein concentration.

15. ELISA plates.

16. Citrate buffer: 0.1 M $Na_3C_6H_5O_7$, 0.2 M NaH_2PO_4, pH 5.0.

17. *o*-Phenylenediamine dihydrochloride (OPD): 0.05% solution in citrate buffer.

18. H_2O_2: 0.015% solution in citrate buffer.

19. Wash buffer: PBS, 0.05% Tween (PBS-T).

20. Blocking buffer: PBS-T, 3% BSA.

21. Carbonate buffer: 100 mM $NaHCO_3$, 100 mM Na_2CO_3, pH 9.6.

22. Alternative Pathway (AP) buffer: 0.96 mM sodium barbital, 144 mM NaCl, 2.48 mM barbituric acid, 1.4 mM $MgCl_2$, 10 mM EGTA, pH 7.3–7.4.

23. Gelatin Veronal Buffer (GVB^{++}): 0.96 mM sodium barbital, 144 mM NaCl, 2.48 mM barbituric acid, 0.83 mM $MgCl_2$, 0.25 mM $CaCl_2$, pH 7.3–7.4.

2.5 SDS Polyacrylamide Gel

1. Apparatus for SDS-PAGE electrophoresis (migration cell and power supply).

2. Resolving gel buffer: Tris–HCl, pH 8.8. 1.5 M Tris Base (187 g); add circa 900 mL ddH_2O, stir the solution at least 10 m, adjust pH to 8.8 with 1 M HCl, fill up to 1 L with ddH_2O, and store at 4 °C.

3. Stacking gel buffer: Tris–HCl, pH 6.8. 0.5 M Tris Base (60.5 g); add circa 900 mL ddH_2O, stir the solution at least 10 m, adjust pH to 6.8 with 1 M HCl, fill up to 1 L with ddH_2O, and store at 4 °C.

4. Sodium dodecyl sulfate (SDS): 10% solution in double distilled water. 100 g SDS, add circa 700 mL ddH$_2$O, stir the solution at least 10 m, adjust pH to 6.8 with 1 M HCl, fill up to 1 L with ddH$_2$O, and store at 4 °C.

5. Ammonium persulfate (APS): 10% solution in double distilled water. Store for at most a month.

6. 30% acrylamide/Bis solution: 290 g acrylamide and 10 g bis-acrylamide; fill up to 1 L with ddH$_2$O, filter through a 0.2 μm filter, and store at 4 °C.

7. Sample buffer (Laemmli buffer): 250 mM Tris–HCl, pH 6.8, 10% SDS, 50% glycerol, 0.5% bromophenol blue; fill up to 30 mL with ddH$_2$O, and store at −20 °C. Add 5% β-mercaptoethanol before use.

8. 5× running buffer: 1.92 M glycine, 0.25 M Tris Base; add circa 800 mL ddH$_2$O and 50 mL 10% SDS, adjust pH to 8.3 with acetic acid, and fill up to 1 L.

9. Tetramethylethylenediamine (TEMED).

2.6 Immunoblotting

1. A Western blot transfer system and a chemiluminescence imaging system.

2. Transfer buffer: 100 mL of 10× Transfer buffer (stock 0.25 M Tris Base, 1.92 M glycine, 1% SDS), 200 mL of ethanol, 700 mL ddH$_2$O.

3. PBS.

4. Blocking buffer: PBS, 5–10% non-fat dry milk, 0.05% Tween 20.

5. Wash buffer: PBS, 0.05% Tween 20.

6. Nitrocellulose membrane.

7. Sponges.

8. Filter paper.

9. Chemiluminescent reagent kit for Western blot.

3 Methods

3.1 Serum Susceptibility Assay

1. Harvest leptospires by centrifugation at 5500 × g for 25 min. Discard supernatant, and wash the pelleted bacteria in 20 mL of PBS. Collect the cells by centrifugation as described above.

2. Resuspend the sedimented cells in 5 mL of PBS, and count the number of bacteria by dark-field microscopy using a bacterial counting chamber as follows:

The number of leptospires per mL = number of bacteria counted* × dilution (if used) × 50,000**

*Count all bacteria within the center square millimeter (1 mm × 1 mm area), which is ruled into 25 groups of 16 small squares.

**50,000 = 50 (cell depth is 1/50) × 1000 (1000 mm^3 = 1 mL)

3. Incubate bacteria (5×10^7 to 5×10^8) in NHS (final volume of 1 mL) for 60 min or 120 min at 37 °C. A serum concentration between 20 and 60% is generally adopted in the majority of protocols assessing leptospiral serum resistance [4, 5, 13]. As a control, incubate the same number of spirochetes in 20–60% HI-NHS for 60 or 120 min at 37 °C.

4. Assess leptospiral survival after each treatment by counting viable bacteria as described above. Viable leptospires keep their movement and helicoidal shape, while dead leptospires lose their characteristic form as a result of fragmentation, as mentioned above. Serum sensitivity is estimated by comparing the counts in NHS-incubated samples to those incubated in HI-NHS (100% survival). At least three independent experiments must be performed.

3.2 Acquisition of Complement Regulatory Proteins by Leptospira

1. Prepare SDS polyacrylamide gel (1 mm thickness) prior to the assay. Glass plates (10 cm × 8 cm) must be cleaned with ethanol before assembling casting stand. For running two 12% SDS-PAGE, prepare 10 mL of the resolving (running) gel by mixing 3.4 mL of ddH$_2$O, 4 mL of 30% Acrylamide/Bis, 2.5 mL of resolving gel buffer, and 0.1 mL of 10% SDS. Add 0.1 mL of 10% APS and 0.004 mL of TEMED shortly before pouring the gels. Pour 5 mL of the running gel into each set of glass plates, add 0.1 mL of isopropanol to prevent dehydration, and allow polymerizing for 40–60 min. Remove isopropanol and place a filter paper between the glass plates to dry. For two stacking gels, mix 2.7 mL of ddH$_2$O, 0.67 mL of 30% Acrylamide/Bis, 0.5 mL of stacking gel buffer, and 0.04 mL of 10% SDS. Add 0.04 mL of 10% APS and 0.004 mL of TEMED shortly before pouring the gels. For each gel, pour approximately 1–2 mL over the running gel, and insert the comb. Once polymerized, assemble the running unit, fill both the upper and lower compartments with the running buffer, and remove the comb.

2. Harvest leptospires by centrifugation at $5500 \times g$ for 25 min. Discard supernatant, and wash the pelleted bacteria three times in 20 mL of CFD.

3. Incubate 10^8 to 10^9 leptospires in NHS-EDTA for 60 min at 37 °C with gentle agitation. For practical reasons, total volume must not exceed 1 mL.

4. Collect the cells by centrifugation (9000 × *g* for 10 min), and wash five times with 0.5–1.0 mL of CFD. The last wash fraction must be collected.

5. Elute proteins bound to the surface of leptospires by adding 100 μL of 0.1 M glycine-HCl to the sedimented bacteria. After resuspension, incubate for 5 min.

6. Collect the supernatant by centrifugation (9000 × *g* for 10 min).

7. Add 5 μL of sample buffer (5× concentrated) to 20 μL of the eluate fractions as well as 20 μL of the last wash fraction, and incubate at 95 °C for 5 min for complete disruption of molecular interactions. Apply the samples, and run the gel at a constant current of 25 mA until the bromophenol blue dye reaches the bottom of the gel. Turn off the power supply, and keep the gel in running buffer until ready to transfer.

8. Soak nitrocellulose membrane and wet the sponges and filter papers in Transfer buffer. Assemble sandwich as follows: black side transfer—sponge—filter paper—gel—membrane—filter paper—sponge—red side transfer. Remove air bubbles between layers by rolling a pipette over the "sandwich." Fill apparatus with Transfer buffer. For a Mini-Transblot, transfer for 1 h at 350 mA at 4 °C.

9. Block nonspecific binding sites by incubating the membranes with Blocking buffer overnight at 4 °C.

10. Incubate the membranes with primary antibodies (anti-FH, anti-C4BP, or anti-vitronectin) diluted according to manufactures' recommendations in Blocking buffer for 1 h at room temperature on a rocking platform. Wash the membranes three times with Wash buffer for 5 min each.

11. Add the secondary peroxidase-conjugated antibodies diluted according to manufactures' recommendations in Blocking buffer for 1 h at room temperature. Wash the membranes three times with Wash buffer for 5 min each.

12. Incubate the membranes with a chemiluminescent reagent according to manufactures' recommendations for 1 min to detect proteins bound to peroxidase-conjugated antibodies. Positive signals can be detected by either a chemiluminescence imaging system or using a X-ray film in a darkroom.

3.3 Direct Cleavage of Complement Proteins by Leptospira-Secreted Proteases

1. Prepare SDS polyacrylamide gel prior to the assay as described in Subheading 3.2.

2. Spin down *Leptospira* cultures at 3200–5400 × *g* for 25 min at 4 °C, and wash the pellet three times with PBS. Resuspend in modified EMJH medium, and determine the number of leptospires.

3. Incubate 1×10^9 leptospires in fresh modified EMJH medium for 4 h at 37 °C. Harvest supernatants after centrifugation at 3200–5400 × g for 25 min at 4 °C, and pass through a 0.22 μm sterile syringe filter.

4. Determine total protein concentration in the supernatants.

5. Incubate 0.25 μg of each purified Complement protein (e.g., C3, fragment C3b, C6) or NHS (use a volume corresponding to 0.25 μg of the selected Complement protein) with 3 μg of proteins from the leptospiral supernatants for 1 h at 37 °C (*see* **Note 4**).

6. Analyze the cleavage products (25 μL of reaction mixture) by SDS-PAGE and Western blot as described in Subheading 3.2. After electrophoresis in 12–15% SDS-polyacrylamide gels, transfer proteins to nitrocellulose membranes for 1 h at 350 mA.

7. Block nonspecific binding sites by incubating the membranes with Blocking buffer overnight at 4 °C.

8. Incubate the membranes with primary antibodies (anti-C3, anti-C4, anti-C5, anti-C6, anti-C7, anti-C8, or anti-C9) diluted according to manufactures' recommendations in Blocking buffer for 1 h at room temperature on a rocking platform. Wash the membranes three times with Wash buffer for 5 min each.

9. Add the secondary peroxidase-conjugated antibodies diluted according to manufactures' recommendations in Blocking buffer for 1 h at room temperature. Wash the membranes three times with Wash buffer for 5 min each.

10. Incubate the membranes with a chemiluminescent reagent according to manufactures' recommendations for 1 min to detect proteins bound to peroxidase-conjugated antibodies. Positive signals can be detected by either a chemiluminescence imaging system or using a X-ray film in a darkroom.

3.4 Inhibition of Complement Activation by Proteases Secreted by Pathogenic Leptospira

1. Spin down the cultures at 3200–5400 × g for 25 min at 4 °C, and wash the pellet three times using PBS. Resuspend in modified EMJH medium, and determine the number of leptospires.

2. Incubate 1×10^9 leptospires in fresh modified EMJH medium for 4 h at 37 °C. Harvest supernatants after centrifugation at 3200–5400 × g for 25 min at 4 °C, and pass through a 0.22 μm sterile syringe filter.

3. Determine total protein concentration in the supernatants (*see* **Note 5**).

4. Coat 96-well plates with 100 μL (1 μg/well) of *Escherichia coli* LPS (for the Alternative Pathway), human IgG or human IgM (0.4 μg/well; for the Classical Pathway), or mannan from

Saccharomyces cerevisiae (1 μg/well; for the Lectin Pathway) in carbonate buffer overnight at 4 °C.

5. Before each of the following steps, wash the plates three times with Wash buffer.

6. Add 200 μL of Blocking buffer to each well and incubate for 2 h at 37 °C.

7. Use NHS as a source of Complement proteins. Use NHS diluted (1:7–1:16) in cold AP buffer to measure Alternative Pathway activation or NHS diluted (1:50 to 1:120) in cold GVB^{++} to assess Classical and Lectin Pathways activation.

8. Incubate diluted NHS samples with leptospiral culture supernatants (3 μg of total secreted proteins) in 1.5 mL microcentrifuge tubes for 1 h at 37 °C. As a negative control, incubate samples of diluted NHS with PBS under similar conditions.

9. Transfer the reactions to the coated plates (with LPS, IgG/IgM, or mannan), and incubate for 1 h at 37 °C.

10. Evaluate Complement activation by measuring C3b deposition (Alternative Pathway) or C4b deposition (Classical and Lectin Pathways) by using polyclonal antibodies against human C3 or human C4, respectively, both diluted 1:5000 (or according to the manufactures' recommendations) for 1 h at 37 °C.

11. Wash the plates three times with Wash buffer, and incubate with peroxidase-conjugated anti-IgG (diluted according to the manufactures' recommendations) for 1 h at 37 °C. Prepare OPD solution (4 mg in 10 mL citrate buffer: 0.1 M Na$_3$C$_6$H$_5$O$_7$, 0.2 M NaH$_2$PO$_4$, pH 5.0.) containing 5 μL of 30% H$_2$O$_2$. H$_2$O$_2$ should be added to solution immediately before use. Add 100 μL/well of OPD containing H$_2$O$_2$.

12. Stop the reaction by adding 50 μL/well of 4 M H$_2$SO$_4$.

13. Determine the absorbance using a spectrophotometer at 492 nm. Results are expressed as O.D.

14. Repeat the experiments at least three times to compare C3b or C4b deposition in the presence or absence of leptospiral secreted proteases (NHS treated only with PBS).

4   Notes

1. Leptospires may also be cultivated in EMJH medium supplemented with Difco™ Leptospira Enrichment EMJH (Becton Dickinson), which contains albumin, polysorbate 80, and additional growth factors.

2. For obtention of NHS, harvest blood in dry tubes without additives or with a separator gel. Keep the blood at room

temperature for a maximum of 30 min, and then centrifuge at 4 °C for 600–700 × g for 15 min. After withdrawn the serum, split it in aliquots and immediately transfer to −70 °C. Store serum aliquots at this temperature until use, and avoid repeated freeze/thaw cycles.

3. To avoid ongoing in vitro Complement activation, maintain buffers on ice until use. Thaw serum samples and purified Complement proteins/fragments immediately before use keeping the vials on ice as well.

4. It is recommended to evaluate the specific proteolytic activity in the *Leptospira* supernatants by using different concentrations of proteins from supernatants of *Leptospira* cultures (0.6–0.06 µg) and higher concentrations (3 µg) of Complement proteins.

5. If necessary, check for the presence of LPS by using Limulus Amebocyte Lysate kit.

Acknowledgments

This work was supported by the São Paulo Research Foundation (FAPESP) grants # 2017/12924-3 (LI and ASB) and 2018/12896-2 (ASB) and the National Council for Scientific and Technological Development (CNPq) grants # 307780/2017-1 (LI) and 305114/2017-4 (ASB).

References

1. Hajishengallis G, Reis ES, Mastellos DC et al (2017) Novel mechanisms and functions of complement. Nat Immunol 18(12):1288–1298. https://doi.org/10.1038/ni.3858

2. Johnson RC, Muschel LH (1965) Antileptospiral activity of normal serum. J Bacteriol 89:1625–1626

3. Barbosa AS, Isaac L (2018) Complement immune evasion by spirochetes. Curr Top Microbiol Immunol 415:215–238. https://doi.org/10.1007/82_2017_47

4. Meri T, Murgia R, Stefanel P et al (2005) Regulation of complement activation at the C3-level by serum resistant leptospires. Microb Pathog 39:139–147

5. Barbosa AS, Abreu PA, Vasconcellos SA et al (2009) Immune evasion of *Leptospira* species by acquisition of human complement regulator C4BP. Infect Immun 77:1137–1143. https://doi.org/10.1128/IAI.01310-08

6. Castiblanco-Valencia MM, Fraga TR, Silva LB et al (2012) Leptospiral immunoglobulin-like proteins interact with human complement regulators factor H, FHL-1, FHR-1, and C4BP. J Infect Dis 205:995–1004. https://doi.org/10.1093/infdis/jir875

7. da Silva LB, Miragaia LS, Breda LC et al (2015) Pathogenic *Leptospira* species acquire factor H and vitronectin via the surface protein LcpA. Infect Immun 83:888–897. https://doi.org/10.1128/IAI.02844-14

8. Fraga TR, Courrol DS, Castiblanco-Valencia MM et al (2014) Immune evasion by pathogenic *Leptospira* strains: the secretion of proteases that directly cleave complement proteins. J Infect Dis 209:876–886. https://doi.org/10.1093/infdis/jit569

9. Amamura TA, Fraga TR, Vasconcellos SA et al (2017) Pathogenic *Leptospira* secreted proteases target the membrane attack complex: a potential role for thermolysin in complement inhibition. Front Microbiol 8:958. https://doi.org/10.3389/fmicb.2017.00958

10. Vieira ML, de Morais ZM, Vasconcellos SA et al (2011) *In vitro* evidence for immune evasion activity by human plasmin associated to pathogenic *Leptospira interrogans*. Microb Pathog 51:360–365. https://doi.org/10.1016/j.micpath.2011.06.008

11. Castiblanco-Valencia MM, Fraga TR, Pagotto AH et al (2016) Plasmin cleaves fibrinogen and the human complement proteins C3b and C5 in the presence of *Leptospira interrogans* proteins: a new role of LigA and LigB in invasion and complement immune evasion. Immunobiology 221:679–689. https://doi.org/10.1016/j.imbio.2016.01.001

12. Roos A, Bouwman LH, Munoz J et al (2003) Functional characterization of the lectin pathway of complement in human serum. Mol Immunol 39:655–668. https://doi.org/10.1016/S0161-5890(02)00254-7

13. Castiblanco-Valencia MM, Fraga TR, Breda LC et al (2016) Acquisition of negative complement regulators by the saprophyte *Leptospira biflexa* expressing LigA or LigB confers enhanced survival in human serum. Immunol Lett 173:61–68. https://doi.org/10.1016/j.imlet.2016.03.005

Chapter 18

Evaluation of Intracellular Trafficking in Macrophages

Claudia Toma and Toshihiko Suzuki

Abstract

Macrophages are phagocytic cells that constitute the primary barrier against pathogens. After phagocytosis a single-membraned vesicle that contains the pathogen is formed. This phagosome undergoes a maturation process to acquire an increasingly antimicrobial environment. Leptospiral uptake by macrophages induces the formation of a *Leptospira*-containing phagosome (LCP). The kinetics of lysosomal marker recruitment by the LCP is correlated with virulence. This chapter presents a protocol to study the intracellular trafficking of *Leptospira* spp. within macrophages by fluorescent labeling bacteria and different markers of the phagocytic pathway. We also describe a method to evaluate the bacterial survival within macrophages.

Key words Macrophage, *Leptospira*, Intracellular trafficking, Phagosome maturation, Immunofluorescence, Survival

1 Introduction

Pathogens internalized by macrophages are sequestered in phagosomes, which sequentially acquire their microbicidal properties through a complex maturation process leading to the formation of phagolysosomes. Thus, the intracellular trafficking of phagocytosed bacteria can be characterized by the acquisition of different markers such as the early endosomal antigen (EEA1), the late endosomal antigen (LAMP1), or cathepsin D [1].

Macrophage uptake of pathogenic leptospires is enhanced by outer membrane proteins [2, 3]. After phagocytosis, *L. interrogans* can evade host defense mechanisms by delaying phagosome maturation [4] and resisting to reactive oxygen species [5]. These strategies allow the successful dissemination of the pathogen to the target organs [6]. Although *L. interrogans* serovar Manilae does not induce macrophage cell death, several studies have reported that macrophages undergo cell death by *L. interrogans* infection when using serovar Lai or serovar Icterohaemorrhagiae, suggesting that strain- and/or serovar-specific differences exist during *Leptospira*-macrophage interactions [7, 8].

Nobuo Koizumi and Mathieu Picardeau (eds.), *Leptospira spp.: Methods and Protocols*, Methods in Molecular Biology, vol. 2134, https://doi.org/10.1007/978-1-0716-0459-5_18, © Springer Science+Business Media, LLC, part of Springer Nature 2020

Pathogenic *L. interrogans* can adhere very strongly to biotic and abiotic surfaces; thus, it is crucial to extensively wash extracellular leptospires to accurately evaluate the fate of intracellular bacteria. In this chapter, we describe a protocol of macrophage infection, in which extracellular bacteria are completely removed in order to study the intracellular trafficking of *Leptospira* by fluorescent labeling bacteria and different markers of the phagocytic pathway. We also describe a method to evaluate the bacterial survival within macrophages.

2 Materials

2.1 Seeding Macrophages on Culture Plates

1. Bone marrow-derived macrophages (BMDMs) (*see* **Note 1**).
2. Sterile coverslips (18 × 24 cm).
3. Tissue culture 6-well plates.
4. Macrophage seeding medium: RPMI medium supplemented with 10% fetal bovine serum (FBS).
5. Conical sterile polypropylene centrifuge tubes (50 mL).
6. Cell counting chamber.

2.2 Leptospira Culture

1. Ellinghausen–McCullough–Johnson–Harris (EMJH) broth: Difco Leptospira Medium Base EMJH (Becton Dickinson), Difco Leptospira Enrichment EMJH (Becton Dickinson), 100 µg/mL of 5-fluorouracil. For 1 L of EMJH liquid medium, dissolve 2.3 g of Leptospira Medium Base EMJH into 900 mL distilled water, and sterilize by autoclaving. Add 100 mL of Leptospira Enrichment EMJH and 10 mL of 0.22 µm-filtrated 10 mg/mL 5-fluorouracil, and then aliquot as needed for experiments.
2. Dark-field microscope.
3. Polypropylene microcentrifuge tubes (1.5 mL).
4. Phosphate-buffered saline (PBS): 2.68 mM KCl, 1.47 mM KH_2PO_4, 136.89 mM NaCl, 8.10 mM Na_2HPO_4, pH 7.4.
5. Cell counting chamber.

2.3 Macrophage Infection

1. Starvation and washing medium: RPMI without FBS.
2. Plate centrifuge.
3. Extracellular *Leptospira*-killing medium: RPMI with 25 µg/mL gentamicin (*see* **Note 2**).
4. Tissue culture 6-well plates.

2.4 Immuno-fluorescence Staining of Leptospira, LAMP1, and Cathepsin D

1. Tissue culture 6-well plates.

2. 2% paraformaldehyde in PBS (2% PFA).

3. Cold methanol.

4. Tris-buffered saline (TBS): 25 mM Tris–HCl, pH 7.4, 137 mM NaCl, 2.68 mM KCl.

5. Permeabilization/blocking buffer: 0.2% saponin, 10% Blocking One (Nacalai Tesque) in TBS.

6. Goat polyclonal anti-cathepsin D (G-19, Santa Cruz Biotechnologies).

7. Anti-*Leptospira* antibody (*see* **Note 3**).

8. Rat monoclonal anti-mouse LAMP1 (1D4B, eBioscience).

9. Secondary antibodies: antibodies conjugated with a fluorochrome of choice (e.g., TRITC-conjugated donkey anti-goat, Cy5-conjugated donkey anti-rabbit, FITC-conjugated donkey anti-rat) (*see* **Note 4**).

10. Mounting medium (e.g., Vectashield, Vector Laboratories).

11. Immunofluorescence microscope slides.

12. Confocal laser microscope.

2.5 Recovery of Intracellular Surviving Bacteria

1. Tissue culture 6-well plates.

2. Sterile H_2O.

3. Cell scrappers.

4. EMJH broth.

5. 10× Tris-EDTA buffer: 100 mM Tris-HCl, pH 8.3, 10 mM EDTA.

6. 2× proteinase K buffer: 20 mM Tris–HCl, pH 8.3100 mM KCl, 5 mM $MgCl_2$, 1% Tween 20, 800 µg/mL of proteinase K.

7. Heat Block.

3 Methods

3.1 Seeding Macrophages on Culture Plates

1. Place a coverslip into each well of a 6-well plate.

2. Prepare a suspension of macrophages (around 5×10^5 cells per mL) in RPMI/10% FBS medium (e.g., for one 6-well plate, prepare around 6×10^6 cells in 12 mL).

3. Transfer 2 mL of the cell suspension (around 1×10^6 cells) into each well of a 6-well plate, and briefly shake plates to distribute the cells evenly.

4. Push the coverslip gently with a tip to ensure that it is place at the bottom of the well (*see* **Note 5**).

5. Incubate cells at 37 °C in a 5% CO_2 atmosphere overnight to allow cell attachment.

6. Proceed as indicated in **step 1** in Subheading 3.3.

3.2 *Leptospira* Culture

1. Grow bacterial culture for 5–7 days at 30 °C in EMHJ broth until it reached exponential phase ($\sim 1 \times 10^8$/mL) (*see* **Note 6**).

2. The day of macrophage infection, count bacteria in a Neubauer chamber (*see* **Note 7**).

3. Centrifuge bacteria $10,000 \times g$ for 10 min and resuspend in the proper amount of PBS (*see* **Note 8**).

4. Proceed as indicated in **step 2** in Subheading 3.3.

3.3 Macrophage Infection

1. Aspirate the RPMI medium containing 10% FBS, and replace with 4 mL of starvation and washing medium for 1 h at 37 °C in a 5% CO_2 atmosphere.

2. Add 100 μL of the *Leptospira* suspension to each well, and centrifuge the plate at $500 \times g$ for 10 min to synchronize the phagocytosis process.

3. Place cells at 37 °C in a 5% CO_2 atmosphere.

4. At 1 h post-infection (p.i.), wash the coverslips extensively with pre-warmed RPMI medium.

5. Transfer the coverslips to a new 6-well plate containing RPMI medium with gentamicin (2 mL/well) to kill extracellular bacteria.

6. After 1 h of gentamicin treatment (2 h p.i.), wash extensively the coverslips again with pre-warmed RPMI medium to remove remaining loosely attached bacteria.

7. Transfer the coverslip to a new 6-well plate containing RPMI medium (*see* **Note 9**).

8. Place cells at 37 °C in a 5% CO_2 atmosphere.

9. At each desire time point, proceed as indicated in **step 1** in Subheading 3.4 or **step 1** in Subheading 3.5.

3.4 Immuno-fluorescence Staining of Leptospira, LAMP1, and Cathepsin D

1. At each time point, transfer carefully one coverslip into a well of a 6-well plate, containing 2 mL of 2% PFA to fix.

2. Keep the coverslips at 4 °C, overnight.

3. Transfer each coverslip into a well of a 6-well plate containing 2 mL of cold-methanol for postfixation.

4. Incubate the plate for 10 min on ice or at −20 °C (*see* **Note 10**).

5. Wash the plate three times with TBS (*see* **Note 11**).

6. Aspirate TBS from the last wash, add 2 mL of permeabilization/blocking buffer, and incubate 2 h at room temperature (RT) (*see* **Note 12**).

7. Incubate the plate for 1 h at 37 °C with anti-cathepsin D antibody (*see* **Note 13**).

8. Wash the plate three times with permeabilization/blocking buffer (about 10 min each wash).

9. Incubate the plate for 1 h at 37 °C with the desired secondary antibody conjugated to fluorochrome (e.g., TRITC-conjugated donkey anti-goat).

10. Wash the plate three times with permeabilization/blocking buffer (about 10 min each wash) (*see* **Note 14**).

11. Incubate the plate for 1 h at 37 °C with anti-*Leptospira* antibody.

12. Wash the plate three times with permeabilization/blocking buffer (about 10 min each wash).

13. Incubate the plate for 1 h at 37 °C with the desired secondary antibody conjugated to fluorochrome (e.g., Cy5-conjugated donkey anti-rabbit).

14. Wash the plate three times with permeabilization/blocking buffer (about 10 min each wash).

15. Incubate the plate for 2 h at RT with anti-LAMP 1 (1:100 in permeabilization/blocking buffer).

16. Wash the plate three times with permeabilization/blocking buffer (about 10 min each wash).

17. Incubate the plate for 1 h at 37 °C with the desired secondary antibody conjugated to fluorochrome (e.g., FITC-conjugated donkey anti-rat).

18. Wash the plate three times with permeabilization/blocking buffer (about 10 min each wash).

19. Wash the plate twice with DW (*see* **Note 15**).

20. Mount the coverslips on a slide glass with mounting medium.

21. Analyze the samples using a confocal laser scanning microscope (*see* **Note 16**).

3.5 Evaluation of Intracellular Surviving Bacteria

1. At each time point, transfer carefully each coverslip into a well of a 6-well plate containing 1 mL of distilled water, and detach the macrophages using a cell scrapper.

2. Forcefully pipette up and down to lyse and disaggregate the cells.

3. *To evaluate surviving bacteria by culturing*, transfer 0.1 mL of the suspended cells to 2 mL of EMJH medium, and incubate at 30 °C for more than 4 days. Check daily by dark-field microscope for the presence of motile leptospires.

4. *To evaluate intracellular bacteria by qPCR*, transfer 0.1 mL of the suspended cells to a microcentrifuge tube, and centrifuge at $20,000 \times g$ for 10 min.

5. Aspirate carefully the supernatant and resuspend the cells in 36 μL of distilled water.

6. Add 4 μL of 10 × Tris-EDTA buffer and 60 μL of 2 × proteinase K buffer.

7. Incubate the tube for 90 min at 56 °C, followed by 10 min incubation at 95 °C.

8. Centrifuge the lysate at 9200 × g for 3 min.

9. Transfer the supernatant, which contain DNA, to a new microcentrifuge tube.

10. The extracted DNA can be used as template DNA to quantify intracellular *Leptospira* by qPCR.

4   Notes

1. Instead of primary cell cultures, macrophage cell lines such as RAW264.7 cells could be used; however, they are less phagocytic.

2. Antibiotic susceptibility is strain dependent; thus, concentration of gentamicin should be adjusted accordingly.

3. Any species- or serovar-specific *Leptospira* antibody can be used; however, an antibody raised in a different host from the other primary antibodies should be selected in order to avoid cross-reactivity when adding the fluorochrome-conjugated secondary antibodies.

4. Secondary antibodies that exhibit minimal cross-reaction with heterologous host serum proteins are recommended.

5. It is important to avoid bubble formation under the coverslips to prevent macrophage growth at the bottom of the well.

6. It is important to work with pathogenic strains that have been in vitro cultivated for similar number of passages to obtain comparable and reproducible results (e.g., downregulation of many outer membrane proteins during long-term in vitro cultivation is reported).

7. *Leptospira* growth can be monitored by the optical density at 420 nm (OD_{420}).

8. At this step, centrifuge the volume of *Leptospira* culture enough for a multiplicity of infection (MOI) of 50 (5×10^7 *Leptospira* per well). Calculate the amount of PBS to resuspend the bacteria after centrifugation as 100 μL per well. Prepare two wells for each time point if you are going to perform immunofluorescence staining and the evaluation of intracellular surviving bacteria in the same experiment.

9. Extensive washing is an essential step to decrease the background of any remaining leptospiral antigen and to ensure

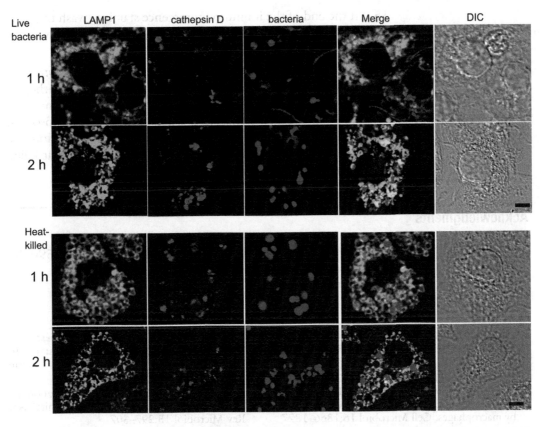

Fig. 1 Acquisition of LAMP-1 and cathepsin D by the *Leptospira*-containing phagosomes. BMDMs were infected with either live or heat-killed *L. interrogans* at a MOI of 50 as described in Subheading 3.3. At the indicated time points, cells were fixed and processed for immunofluorescence as described in Subheading 3.4. DIC: differential interference contrast. Scale bars, 5 μm

that the recovery of *Leptospira* in Subheading 3.5 is from intracellular.

10. Cold methanol postfixation can be skipped if cathepsin D is not stained.

11. After this step, the procedure can be stopped, and the coverslips can be stored in TBS at 4 °C.

12. Exchange the solutions quickly, and avoid drying of the coverslips during all the immunostaining procedure.

13. The dilution ratio and the incubation time may vary depending on the antibody (we used 1:50 dilution in TBS for cathepsin D).

14. After the addition of the first fluorochrome-conjugated secondary antibody, perform all incubations and washing steps in the dark to prevent fluorochromes from fading.

15. At the end of the immunofluorescence staining, wash the plate with DW to remove any salt crystals that can interfere with the microscopic observation.

16. Analyze the samples by quantifying the number of *Leptospira*-containing phagosomes (LCPs) with LAMP-1 and cathepsin D. LCPs recruit LAMP1 during phagosome maturation; LCPs will further recruit cathepsin D when fused with the lysosomes. As a control of normal phagolysosome fusion, a non-pathogenic strain (e.g., *L. biflexa* Patoc I) or a heat-killed pathogenic *Leptospira* can be used (Fig. 1).

Acknowledgments

This work was supported by JSPS KAKENHI grant JP21590484 (C.T.).

References

1. Stuart LM, Ezekowitz RA (2005) Phagocytosis: elegant complexity. Immunity 22:539–550

2. Toma C, Murray GL, Nohara T, Mizuyama M, Koizumi N, Adler B, Suzuki T (2014) Leptospiral outer membrane protein LMB216 is involved in enhancement of phagocytic uptake by macrophages. Cell Microbiol 16:1366–1377

3. Zhang L, Zhang C, Ojcius DM, Sun D, Zhao J, Lin X, Li L, Li L, Yan J (2012) The mammalian cell entry (Mce) protein of pathogenic *Leptospira* species is responsible for RGD motif-dependent infection of cells and animals. Mol Microbiol 83:1006–1023

4. Toma C, Okura N, Takayama C, Suzuki T (2011) Characteristic features of intracellular pathogenic Leptospira in infected murine macrophages. Cell Microbiol 13:1783–1792

5. Eshghi A, Lourdault K, Murray GL, Bartpho T, Sermswan RW, Picardeau M, Adler B, Snarr B, Zuerner RL, Cameron CE (2012) *Leptospira interrogans* catalase is required for resistance to H_2O_2 and for virulence. Infect Immun 80:3892–3899

6. Picardeau M (2017) Virulence of the zoonotic agent of leptospirosis: still terra incognita? Nat Rev Microbiol 15:297–307

7. Hu W-L, Dong H-Y, Li Y, Ojcius DM, Li S-J, Yan J (2017) Bid-induced release of AIF/EndoG from mitochondria causes apoptosis of macrophages during infection with *Leptospira interrogans*. Front Cell Infect Microbiol 7:471

8. Du P, Li SJ, Ojcius DM, Li KX, Hu WL, Lin X, Sun AH, Yan J (2018) A novel Fas-binding outer membrane protein and lipopolysaccharide of *Leptospira interrogans* induce macrophage apoptosis through the Fas/FasL-caspase-8/−3 pathway. Emerg Microbes Infect 7:135

Chapter 19

Biofilm Formation and Quantification Using the 96-Microtiter Plate

Roman Thibeaux, Malia Kainiu, and Cyrille Goarant

Abstract

Biofilm formation in microtiter plates is certainly the most commonly used method to grow and study biofilm. This simple design is very popular due to its high-throughput screening capacities, low cost, and easy handling. In the protocol described here, we focus on the use of 96-well optically clear, polystyrene flat-bottom plate to study biofilm formation by *Leptospira* spp. and quantify the biofilm formation by crystal violet (CV) staining. We also describe an alternative method, based on phase contrast image analysis that we believe is more suitable for accurately quantifying biofilm growth by reducing handling of this fragile structure.

Key words Biofilm, 96-well plate, Static culture, Crystal violet, Phase contrast images, Quantification

1 Introduction

Microbial biofilms are commonly defined as sessile microbial consortia established in a three-dimensional structure and consist of multicellular communities composed of prokaryotic cells embedded in a matrix synthesized by the microbial community. Biofilm formation is a multistage process that starts with microbial adhesion with a subsequent production and accumulation of an extracellular matrix mainly composed of polymeric substances such as polysaccharides, extracellular DNA, or proteins [1]. Microbial biofilm constitutes a serious problem for public health because of the increased resistance to antimicrobial action and their potential to cause infections in patients with indwelling medical devices [2].

Over the last couple of decades, a number of model systems have been tested for in vitro study of biofilm [3]; among them microtiter plates are being widely used. Originally developed to investigate bacteria attachment [4], it further proved to be compatible with the study of sessile development [5]. The advantages of the microtiter plate assay are its simplicity, the use of basic laboratory materials, the adaptability to small or large numbers of

Nobuo Koizumi and Mathieu Picardeau (eds.), *Leptospira spp.: Methods and Protocols*, Methods in Molecular Biology, vol. 2134, https://doi.org/10.1007/978-1-0716-0459-5_19, © Springer Science+Business Media, LLC, part of Springer Nature 2020

samples, the ease of replication, and the variety of samples that can be tested in a single assay.

In the classical procedure, bacterial cells are grown in the wells of a polystyrene microtiter plate [6].

At different time points, the wells are washed to remove planktonic cells before staining the biofilm attached to the surface of the wells. The most widely used method to follow biofilm formation in microtiter plates is the crystal violet staining method [7], which measures biofilm biomass at the bottom of the well. Crystal violet (CV) belongs to the family of triphenylmethane dyes, which bind to the bacterial cellular components by ionic interactions. CV is a basic dye that stains both live and dead cells by binding to negatively charged surface molecules including DNA [8], proteins [9], and polysaccharides [10]. Quantitation of biofilm growth can be obtained by solubilization of the CV and measurement of the absorbance using a spectrophotometer. This approach has been successfully used to characterize in vitro the initial step of *Leptospira* biofilm formation [11].

In this chapter, we also detail an alternative method based on phase contrast image quantification instead of CV quantification. We found this method as robust as the CV method to quantify biofilm formation, but most importantly, the latter has the advantage of limiting sample handling and thereby reducing erroneous quantification resulting from biofilm disruption during washing steps. Nevertheless, the need for a good microscope and a CCD camera may appear to be an obstacle for some laboratories.

2 Materials

2.1 Biofilm Growing

1. Actively motile exponential-phase *Leptospira interrogans* serovar Manilae strain L495 (*see* **Note 1**).

2. Ellinghausen-McCullough-Johnson-Harris (EMJH) culture medium [12, 13]: Dissolve 2.3 g of Difco™ Leptospira Medium Base EMJH (Becton Dickinson) in 900 mL of purified water. Add 100 mL of Difco™ Leptospira Enrichment EMJH (Becton Dickinson) (*see* Chapter 1).

3. 96-well optically clear, sterile polystyrene flat-bottom plate with low evaporation lid (*see* **Note 2**).

4. Flat-bottom, sterile glass culture tubes with screw cap.

5. 5 mL Combitips advanced.

6. Multipette plus.

7. Petroff-Hausser cell-counting chamber (*see* **Note 3**).

8. Upright dark-field microscope with a 20× objective.

9. 30 °C incubator with controlled humidity (*see* **Notes 7** and **8**).

10. Lab coat and nitrile gloves.

2.2 Biofilm Quantification

2.2.1 Crystal Violet Quantification (See **Note 4***)*

1. 0.1% Crystal Violet (CV) solution: 0.1% CV in water.

2. Phosphate-buffered saline (PBS): 10 mM Na_2PO_4, 1.8 mM KH_2PO_4, 137 mM NaCl, 2.7 mM KCl, pH 7.4.

3. Fixative buffer: 4% paraformaldehyde (PFA) in PBS.

4. Pipetter.

5. Multichannel pipette.

6. 200 µL tips.

7. Multichannel pipette reservoir.

8. Container for liquid waste.

9. 30% glacial acetic acid solution: 30% glacial acetic acid in water.

10. 96-well optically clear, flat-bottom plate (can be non-sterile).

11. Spectrophotometer equipped with plate reading capability (550–600 nm).

12. Timer.

13. Deionized H_2O.

2.2.2 Phase Contrast Image Quantification

1. Upright phase contrast microscope with a 5× objective and equipped with a CCD digital camera.

2. Open-source Image Software analysis Fiji [14] (https://fiji.sc/).

3 Methods

3.1 Growing Leptospira *Biofilm*

1. Inoculate bacteria into 5 mL EMJH growth medium in sterile culture tubes and grow to exponential phase. Typically, bacteria are grown at 30 °C for 4 days without shaking (*see* **Note 5**).

2. Count bacteria under a dark-field microscope using a Petroff-Hausser cell-counting chamber (*see* **Note 6**), and dilute the initial culture with fresh EMJH to a final concentration of 1×10^6 bacteria per mL.

3. Under sterile conditions, aliquot 150 µL of diluted cultures into the wells of a 96-well microtiter plate. Be sure to include blank wells containing sterile EMJH medium as a control. It is recommended not to use the peripheral wells of the plate, as border effect is known to occur (*see* Fig. 1 and **Note 7**). For quantitative assays, we typically use 6–18 replicate wells for each treatment.

4. Cover the plate (*see* **Note 8**), and incubate at 30 °C for several weeks in static conditions without changing the medium [11] (*see* **Note 9**). When working with pathogenic strains, an incubation of 3 weeks is encouraged to obtain an adherent, mature biofilm.

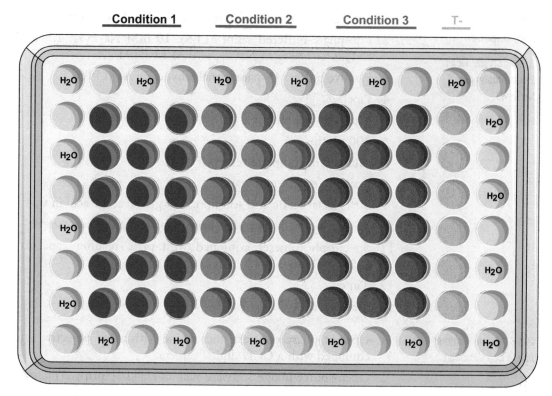

Fig. 1 Classical working plate layout including 8 blank wells containing sterile EMJH medium and 18 technical replicates for each conditions tested (three conditions maximum per plate in this configuration). Also, note the presence of sterile distilled water in the peripheral wells of the plate to ensure correct moisture level and reduce border effect

3.2 Crystal Violet Biofilm Staining

All these steps are completed at room temperature.

1. Following desired incubation time, remove planktonic cells by gently pipetting the supernatant and add 150 µL of fixative buffer (4% PFA) to increase biofilm stability and incubate 30 min.

2. Gently remove the fixative buffer, and rinse two times with 200 µL of PBS by gentle pipetting.

3. Tilt the plate and remove as much PBS as possible.

4. Pipette 175 µL of 0.1% crystal violet solution into wells. This volume ensures that the stain will cover the biofilm (*see* **Note 4**).

5. Incubate for 10 min.

6. Gently remove the crystal violet solution and rinse two times with 200 µL PBS.

7. Tilt the plate and remove as much PBS as possible and dry for a few hours or overnight.

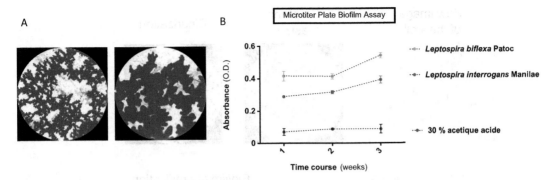

Fig. 2 (a) Phase contrast images illustrating crystal violet staining of *Leptospira* biofilm after 3 weeks of incubation. **(b)** Example of data that may be obtained by using spectrophotometry to quantitate biofilm formation

8. At this point, you should be able to see dot-like or reticulated structures at the bottom of the wells (*see* **Note 10**). The wells can be photographed for qualitative illustration (*see* Fig. 2a).

3.3 CV-Stained Biofilm Quantitation

1. Pipette 200 μL of 30% acetic acid solution into each well. This will solubilize the CV.

2. Incubate for 10–15 min.

3. Pipette up and down to assure that the stained biofilm is well solubilized, and then transfer 100 μL of each sample to a new 96-well optically clear, flat-bottom plate.

4. Read optical density of all samples in plate at a wavelength of 550–600 nm [15].

5. Plot and analyze the data (example given in Fig. 2b).

3.4 Biofilm Phase Contrast Image Quantitation

Alternative to CV quantitation, we also found useful to directly quantify the biofilm formation over time through phase contrast image analysis. This method significantly reduces sample manipulation and allows to work with native, unfixed, unstained biofilm.

1. Every week, take phase contrast images of the growing biofilm (*see* **Note 11**). Make sure that the center of the well is into focus. Typically, we take pictures of six replicate wells per independent experiment.

2. Using an image analysis software, crop the outer part of the image where spherical aberrations and field curvature artifacts are visible (*see* **Note 12**).

3. Create a binary image from your 2D grayscale image and using the particle analysis tool quantify the surface of the biofilm as a function of time (*see* Fig. 3).

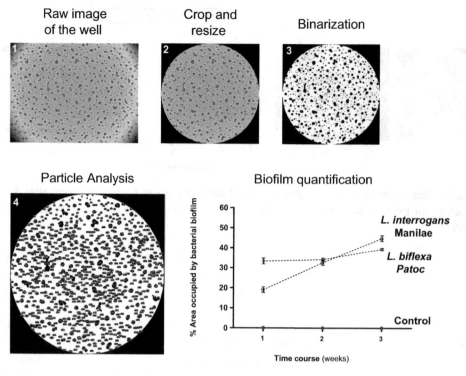

Fig. 3 Workflow of image analysis performed from unstained, unfixed biofilm images (1). The raw image is cropped to removed artifacts (2), and a binary image is created (3) prior to biofilm particle quantification (4). The graphic represents an example of quantitative biofilm data that may be obtained using this particular image quantification method

4   Notes

1. Biofilm formation has been observed in several strains including pathogenic, intermediate, and saprophyte strains. The timeline of biofilm formation, the amount, as well as its morphology might differ according to the strain used. We recommend to calibrate the kinetic of biofilm production with regard to the strain of interest.

2. Several types of polystyrene microtiter plates are available. U-bottom plates or V-bottom plates are often used. We prefer to use flat-bottom plates since they are more suitable for taking images of the growing biofilm.

3. Petroff-Hausser cell-counting chamber 0.02 mm deep in combination with a 1.5 mm optically planed is recommended when counting *Leptospira* cells. The shallow thickness of the area to be counted limits the number of cells out of focus, improving the counting, while the #1.5H coverslip allows a good visualization by dark-field microscopy.

4. When working with crystal violet, wear gloves and a lab coat, and be careful to avoid spilling the powder, which will readily stain clothes and surfaces. Stains can be removed to some degree with 70 or 95% ethanol, but it is difficult to remove all crystal violet once it has stained a surface. It is best to dedicate a bottle and stir bar exclusively to preparation of crystal violet solution. Crystal violet is classified as a hazardous substance (H226, flammable liquid and vapor; H319, causes serious eye irritation; H351, suspected of causing cancer; H411, toxic to aquatic life with long-lasting effects). The use of a chemical hood while working with crystal violet is strongly advised.

5. *Leptospira* 'doubling time can vary from 6 to 24 h. We recommend to precisely determine the growth kinetic of the strain of interest in order to identify the exponential growth phase. With our strain *L. interrogans* Manilae L495, the exponential growth phase is observed between the second and fifth days.

6. We recommend performing the counting from a 1/100 dilution of the initial culture. It is also useful to wait a few minutes before starting counting the bacteria to ensure that the liquid flow within the counting chamber is equilibrated and that no movement of liquid will disturb the counting.

7. Border effects are very common when working with a 96-well plate, especially if the incubator does not have a humidity control system. Therefore we recommend to add 200 μL of sterile distilled water in the peripheral wells of the plate.

8. Placing the covered plate in a disposable plastic storage container can reduce evaporation of media in the incubation chamber.

9. We observed that *Leptospira* biofilm is weakly adherent and can rapidly be lost when pipetting. Therefore we recommend to let the biofilm grow under strict static conditions for at least 3 weeks when it eventually starts to be adherent (although still fragile). Limiting the numbers of pipetting event is crucial to avoid the loss of biofilm which might give erroneous quantitative results.

10. If there is no visible biofilm formation in any of your samples, there are several things you may need to adjust (the length of incubation) or check (assay medium evaporation). We also recommend to take extra care during the pipetting steps as they are mainly responsible for biofilm detachment.

11. During long-term incubation, condensation may form on the lid and can impede well observation or be responsible for artifact if pictures are taken. We recommend drying the lid under a sterile hood for 5 min prior to sample examination.

12. We recommend you to crop the images using the same defined template to be sure that the further analysis area will be similar for all images.

Acknowledgments

This work was supported by an AXA Research Funds grant "AXA Postdoctoral Fellowship," 15-AXA-PDOC-037 granted to RT.

References

1. Flemming H-C, Wingender J (2010) The biofilm matrix. Nat Rev Microbiol 8(9):623

2. Costerton JW, Stewart PS, Greenberg EP (1999) Bacterial biofilms: a common cause of persistent infections. Science 284 (5418):1318–1322

3. Coenye T, Nelis HJ (2010) In vitro and in vivo model systems to study microbial biofilm formation. J Microbiol Methods 83(2):89–105. https://doi.org/10.1016/j.mimet.2010.08. 018

4. Fletcher M (1977) The effects of culture concentration and age, time, and temperature on bacterial attachment to polystyrene. Can J Microbiol 23(1):1–6

5. O'toole GA, Kolter R (1998) Flagellar and twitching motility are necessary for *Pseudomonas aeruginosa* biofilm development. Mol Microbiol 30(2):295–304

6. Djordjevic D, Wiedmann M, McLandsborough L (2002) Microtiter plate assay for assessment of *Listeria monocytogenes* biofilm formation. Appl Environ Microbiol 68 (6):2950–2958

7. Christensen GD, Simpson WA, Younger J, Baddour L, Barrett F, Melton D, Beachey E (1985) Adherence of coagulase-negative staphylococci to plastic tissue culture plates: a quantitative model for the adherence of staphylococci to medical devices. J Clin Microbiol 22 (6):996–1006

8. Yang Y, Jung DW, Bai DG, Yoo GS, Choi JK (2001) Counterion-dye staining method for DNA in agarose gels using crystal violet and methyl orange. Electrophoresis 22 (5):855–859. https://doi.org/10.1002/ 1522-2683()22:5<855::aid-elps855>3.0. co;2-y

9. Bonnekoh B, Wevers A, Jugert F, Merk H, Mahrle G (1989) Colorimetric growth assay for epidermal cell cultures by their crystal violet binding capacity. Arch Dermatol Res 281 (7):487–490

10. Colvin KM, Gordon VD, Murakami K, Borlee BR, Wozniak DJ, Wong GC, Parsek MR (2011) The pel polysaccharide can serve a structural and protective role in the biofilm matrix of *Pseudomonas aeruginosa*. PLoS Pathog 7(1):e1001264. https://doi.org/10. 1371/journal.ppat.1001264

11. Ristow P, Bourhy P, Kerneis S, Schmitt C, Prevost MC, Lilenbaum W, Picardeau M (2008) Biofilm formation by saprophytic and pathogenic leptospires. Microbiology 154 (Pt 5):1309–1317

12. Ellinghausen HC Jr, McCullough WG (1965) Nutrition of Leptospira Pomona and growth of 13 other serotypes: fractionation of oleic albumin complex and a medium of bovine albumin and polysorbate 80. Am J Vet Res 26:45–51

13. Johnson RC, Harris VG (1967) Differentiation of pathogenic and saprophytic leptospires. I. Growth at low temperatures. J Bacteriol 94(1):27–31

14. Schindelin J, Arganda-Carreras I, Frise E, Kaynig V, Longair M, Pietzsch T, Preibisch S, Rueden C, Saalfeld S, Schmid B, Tinevez JY, White DJ, Hartenstein V, Eliceiri K, Tomancak P, Cardona A (2012) Fiji: an open-source platform for biological-image analysis. Nat Methods 9(7):676–682. https://doi.org/ 10.1038/nmeth.2019

15. Merritt JH, Kadouri DE, O'Toole GA (2005) Growing and analyzing static biofilms. Curr Protoc Microbiol . Chapter 1:Unit 1B 1. https://doi.org/10.1002/ 9780471729259.mc01b01s00

Chapter 20

Survival Tests for *Leptospira* spp.

Clémence Mouville and Nadia Benaroudj

Abstract

Measuring viability is an important and necessary assessment in studying microorganisms. Several methods can be applied to *Leptospira* spp., each with advantages and inconveniencies. Here, we describe the traditional colony-forming unit method, together with two other methods based, respectively, on the reducing capacity of live cells (Alamar Blue® Assay) and differential staining of live and dead cells (LIVE/DEAD *Bac*Light®). The Alamar Blue® Assay uses the blue reagent resazurin, which can be reduced into the pink reagent resorufin by live cell oxidoreductases. Production of resorufin can be quantified by absorbance or fluorescence reading. The LIVE/DEAD *Bac*Light® assay uses a mixture of two nucleic acid dyes (Syto9 and propidium iodide) that differentially penetrate and stain nucleic acid of cells with decreased membrane integrity. The colony-forming unit method is labor-intensive but the most sensitive and linear method. The two other methods are not laborious and well-adapted to high-throughput studies, but the range of detection and linearity are limited.

Key words Spirochetes, *Leptospira*, Survival, Plating, Colony-forming unit, Resazurin, Resorufin, Syto9, Propidium iodide, Fluorescence, Absorbance

1 Introduction

Assessing and quantifying the viability of a microorganism is a basic but fundamental procedure in microbiology. This is particularly important when the efficiency of an antibiotic or other antimicrobial compound needs to be determined or to characterize bacterial species and their ability to survive under different environments.

Evaluating whether a bacterium is alive or dead is rather complex as finding the best criteria to define bacterial viability is still matter of debate. Several different qualitative and quantitative techniques are available to measure bacterial survival, each having their advantages and limitations. Here, we will describe three methods that can be applied to *Leptospira* spp.

The gold standard is the colony-forming unit (CFU) method based on the ability of a bacterium to form visible colonies on agar plates [1]. In this method, bacterial viability is defined by their cultivability and ability to divide on solid medium. This technique

Nobuo Koizumi and Mathieu Picardeau (eds.), *Leptospira spp.: Methods and Protocols*, Methods in Molecular Biology, vol. 2134, https://doi.org/10.1007/978-1-0716-0459-5_20, © Springer Science+Business Media, LLC, part of Springer Nature 2020

allows the determination of absolute number of bacteria, and the range of detection is unlimited, although this method does not take into consideration and cannot be applied to viable but non-cultivable bacteria. After applying a treatment to a bacterial population, the bacterial suspension is diluted to obtain isolated, readily countable colonies on agar plate. This method is easy to implement and does not require sophisticated equipment and is thus feasible with *Leptospira* spp. However, optimization to find the right bacterial suspension dilution for obtaining isolated colonies on agar plates might be time-consuming [2]. Another limitation lies in the generation time of *Leptospira* spp., and, as a consequence, the results are obtained after 1 week or 3–4 weeks of incubation for saprophytes and pathogenic species, respectively. An additional potential issue is that *Leptospira* spp. grow below the surface of the solid agar medium and appear as faint colonies, rendering their macroscopic visualization difficult. Nevertheless, this method was applied to *Leptospira* spp. and has allowed enumerating and quantifying their viability under different growth conditions [3–6].

Alternative methods using other viability criteria, such as metabolic activities, were developed and are now preferentially used due to their adaptability to high-throughput screenings. One method is based on the fact that live and actively respiring cells maintain a reducing environment. This method uses resazurin (7-hydroxy-10-oxidophenoxasin-10-ium-3-one), a chromogen electron acceptor that is reduced into resorufin by cell oxidoreductases using NADH as cofactor (Fig. 1) [7]. Resazurin is permeable through cell membranes, and its reduction is therefore an indicator of the oxidoreduction state of the cells. Resazurin and resorufin have different spectrophotometric and fluorometric properties [8, 9]. Resazurin is a blue non-fluorescent reagent, and resorufin is a pink highly fluorescent reagent. When mixed with live cells, upon accepting electron, the blue-colored non-fluorescent resazurin is transformed into the pink fluorescent resorufin. Reduction of resazurin can be qualitatively determined by observing the visible change of the blue color into pink, indicative of the presence of live cells (Fig. 2a). Resazurin reduction can be also quantified by

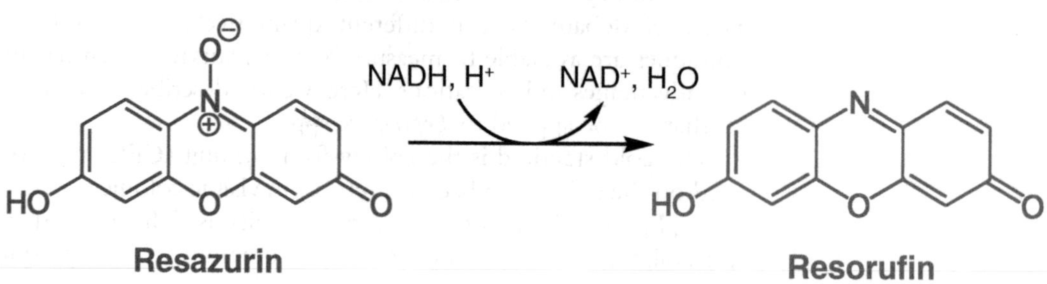

Fig. 1 Reaction of resazurin reduction into resorufin. Resazurin (a blue reagent) is an electron receptor that is reduced into resorufin (a pink reagent) in the presence of cellular oxidoreductases and NADH

absorbance (Fig. 2b) or fluorescence measurement (Fig. 2c). This method is easy to perform and not labor-intensive, as it does not require a high number of steps, and is therefore compatible with processing multiple samples. Reduction of resazurin can be visible after a few hours, and the signal of reduced resazurin is stable for several days. In addition, it allows flexibility in the measurement method used (qualitative or quantitative). This method, commercially known as the Alamar Blue® Assay, has been extensively used with bacteria including the slow-growing *Mycobacterium tuberculosis* [10]. Since the EMJH medium used to cultivate *Leptospira* does not interfere with the reduction of resazurin, Alamar Blue® Assay was also successfully used to measure *Leptospira* viability [11–14].

Another criterion that can be used to rapidly assess bacterial viability is the membrane integrity. In this method, bacterial viability is correlated with the level of membrane damage and is assessed using a mixture of two fluorescent nucleic acid stains, the Syto9 and the propidium iodide (PI). These two dyes differ in their spectroscopic properties and in their capacities to penetrate bacteria. The green fluorescent Syto9 penetrates bacteria regardless their membrane integrity and stains DNA of all bacteria in a population. On the contrary, the red fluorescent PI penetrates only permeable bacteria and stains DNA of only bacteria with damaged membranes. Importantly, PI has a higher affinity for nucleic acids than Syto9 [15]. When a bacterial population containing intact and membrane-damaged bacteria is exposed to a mixture of Syto9 and PI, intact bacteria will be stained by green fluorescence, whereas membrane-damaged bacteria will be stained by red fluorescence. Because the two dyes have different fluorescent properties, respective nucleic acid staining can be quantified by measuring fluorescence of the bacterial suspension at different wavelengths. The green/red fluorescence ratio is proportional to the relative number of live bacteria, and a standard curve will determine the proportionality factor. This method is known as the LIVE/DEAD *Bac*Light stain. Since Boulos et al. [16] evaluated this procedure in bacteria and demonstrated its ease 20 years ago, this method is widely used to assess bacterial viability. This method allows staining of viable and dead cells in a single step, and the acquisition of results is rapid. In addition to quantify *Leptospira* viability by fluorescent measurement, staining of *Leptospira* by Syto9 and PI can be measured by flow cytometry [17] (*see* Chapter 4, Fontana et al.) and fluorescence microscopy [18]. It should be pointed out however that several studies have raised some concerns about the Syto9/PI staining; in fact, under certain conditions, this method has led to under- or overestimation of the number of viable cells [19–21].

In conclusion, when *Leptospira* viability needs to be estimated, multiple approaches are available. The traditional gold standard colony-forming unit method is labor-intensive but the most sensitive and linear method. The Alamar Blue® and LIVE/DEAD

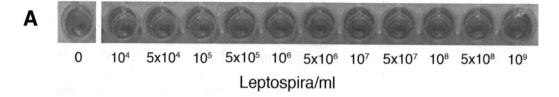

0   10⁴ 5x10⁴ 10⁵ 5x10⁵ 10⁶ 5x10⁶ 10⁷ 5x10⁷ 10⁸ 5x10⁸ 10⁹

Leptospira/ml

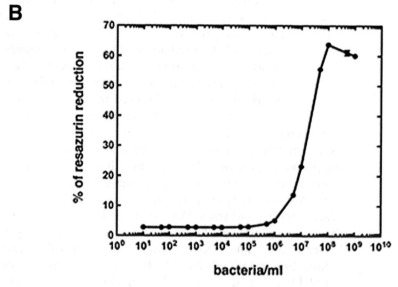

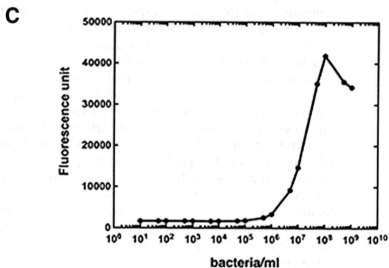

Fig. 2 Linear relationship between the number of *Leptospira* and resazurin reduction. Exponentially growing *Leptospira* were serially diluted to obtain the indicated concentration and 100 μl of *Leptospira* were transferred in a flat bottom 96-well plates (TPP® Tissue Culture Test Plate in (**a**) and (**b**); μClear Black 96-well plate in (**c**)). 80 μl of EMJH medium and 20 μl of 10× resazurin solution (provided by the Alamar Blue® Assay) were added. The plates were incubated 24 h at 30 °C. (**a**) The change of the blue resazurin solution into a pink resorufin solution upon reduction is shown. (**b**) The percentage of resazurin reduction was calculated as described in Fig. 3 and plotted in function of the bacterial concentration (**step 7** in Subheading 3.3). (**c**) The fluorescence was measured ($\lambda_{exc} = 560$ nm, $\lambda_{em} = 590$ nm) and plotted in function of the bacterial concentration (**step 7** in Subheading 3.3)

$$\% \ Reduction \ of \ resazurin = \frac{(E_{oxi}600 \times A_{570}) - (E_{oxi}570 \times A_{600})}{(E_{red}570 \times C_{600}) - (E_{red}600 \times C_{570})} \times 100$$

Fig. 3 Calculation of percentage of resazurin reduction. This formula is used to calculate the percentage of resazurin reduction with $E_{oxi}600$, the molar extinction coefficient of oxidized Alamar Blue at 600 nm (117216); $E_{oxi}570$, the molar extinction coefficient of oxidized Alamar Blue at 570 nm (80586); $E_{red}600$, the molar extinction coefficient of reduced Alamar Blue at 600 nm (14652); $E_{red}570$, the molar extinction coefficient of reduced Alamar Blue at 570 nm (155677); A_{570}, absorbance of the tested well at 570 nm; A_{600}, absorbance of the tested well at 600 nm; C_{570}, absorbance of the negative control well at 570 nm; and C_{600}, absorbance of the negative control well at 600 nm

*Bac*Light® assays are not laborious and well-adapted to high-throughput studies, but their sensitivity and linearity are limited to a short range of bacterial concentrations. We therefore would like to emphasize that accurate and meaningful bacterial viability assessment might require, under some circumstances, validation by two different approaches.

2 Materials

2.1 *Leptospira* Cultivation

1. Safety cabinet level 2.

2. EMJH albumin supplement: 10% (w/v) bovine serum albumin, 0.004% (w/v) zinc sulfate, 0.015% (w/v) magnesium chloride, 0.015% (w/v) calcium chloride, 0.1% (w/v) sodium pyruvate, 0.4% (w/v) glycerol, 1.25% (v/v) Tween 80, 0.0002% (w/v) vitamin B12, 0.05% (w/v) ferrous sulfate (added at the last moment) in sterile water for injection (WFI) (*see* **Note 1**).

3. EMJH base: dissolve 2.3 g of Difco *Leptospira* medium base EMJH (Becton Dickenson) in 900 ml sterile WFI. Autoclave the solution.

4. EMJH medium: add 100 ml of EMJH albumin supplement to 900 ml of EMJH base. Adjust the pH to 7.5 and filter sterilize the solution.

5. Spectrophotometer.

6. Disposable serological plastic pipettes.

7. Culture flasks and/or tubes.

8. Incubator with orbital shaker at 30 °C.

9. Dark-field microscope.

10. Petroff-Hausser counting chamber.

2.2 Determination of Colony-Forming Unit

1. Safety cabinet level 2.
2. P1000, P200, P20 micropipettes and sterile tips.
3. Sterile 1.5 ml polypropylene tubes.
4. EMJH medium (*see* Subheading 2.1).
5. Concentrated EMJH medium: dissolve 2.3 g of Difco *Leptospira* medium base EMJH (Becton Dickenson) in 560 ml sterile WFI. Add 100 ml of albumin supplement. Adjust the pH to 7.5 and filter sterilize the solution.
6. Agar solution: Dissolve 12 g of agar noble in 340 ml of sterile WFI. Autoclave the solution.
7. Solid EMJH agar medium: melt 340 ml of agar noble solution and keep it at 50–55 °C in a water bath. Pre-warm 660 ml of concentrated EMJH medium at 50–55 °C. Mix the agar noble solution with the concentrated EMJH medium. Pour 25–30 ml of the solution into petri plates (*see* **Note 2**).
8. 90 mm × 14.2 mm petri dishes.
9. Sterile inoculation loop or glass spreader or glass beads.
10. Microbiological incubator at 30 °C.
11. Aluminum foil.
12. Parafilm.

2.3 Measurement of Resazurin Reduction by the Alamar Blue® Assay

1. Safety cabinet level 2.
2. P1000, P200, P20 micropipettes and sterile tips.
3. 10× resazurin solution (Alamar Blue® Assay).
4. EMJH medium (*see* Subheading 2.1).
5. Sterile 96-well flat bottom microplates for absorbance or fluorescence reading.
6. Microplate reader for absorbance or fluorescence reading.
7. Plastic storage box (to accommodate 96-well microplates).
8. Microbiological incubator at 30 °C.

2.4 DNA Staining with Syto9 and PI

1. Safety cabinet level 2.
2. P1000, P200, P20 micropipettes and sterile tips.
3. Sterile 15 ml conical polypropylene tubes.
4. Sterile 1.5 ml polypropylene tubes.
5. 10 ml disposable serological sterile plastic pipettes.
6. Sterile ultrapure H_2O.
7. 0.5% NaCl solution: 0.5% NaCl in ultrapure sterile H_2O.
8. Water bath at 55 °C.
9. Centrifuge with rotor for 2600 × *g*.

10. Syto9 and PI nucleic acid dye mix: Prepare a mixture of 10 μM of Syto9 and 60 μM of PI in ultrapure sterile H_2O (*see* **Note 3**). Keep the stock and the mixed solutions at −20 °C and protected them from light.

11. 96-well flat bottom microplates for fluorescence reading.

12. Microplate reader for fluorescence reading.

3 Methods

3.1 Preparation and Treatment of a Bacterial Suspension

1. Inoculate EMJH medium-containing flasks or tubes with *Leptospira* cultures to achieve a final concentration of ≈10^7 bacteria/ml as assessed by absorbance measurement at 420 nm (OD_{420} ≈ 0.01–0.02) or enumerating the bacteria under a dark-field microscope with a Petroff-Hausser chamber.

2. Cultivate *Leptospira* at 30 °C with shaking at 100 rpm until they reach the suitable and desirable growth phase (*see* **Note 4**).

3. If comparing different *Leptospira* strains, adjust all samples to the same bacterial concentration with EMJH medium (*see* **Note 5**).

4. Divide the bacterial suspension in different samples according to the number of treatments you want to apply to the bacterial suspension.

5. Apply the treatment to the samples, e.g., temperature shift and addition of chemicals or antibiotics, for a given time. You should have the proper control of the treatment, i.e., a sample that is not subjected to the treatment (*see* **Note 6**).

3.2 Assessing Survival by Determination of Colony-Forming Unit (cfu)

The colony-forming unit method is considered as the most quantitative and linear method if the appropriate dilution is applied to the bacterial suspension.

1. Treated and non-treated *Leptospira* samples are tenfold diluted serially. For instance, 15 μl of bacterial suspension are added to 135 μl of EMJH medium in 1.5 ml polypropylene sterile tubes. A wide range of tenfold dilutions is necessary to determine the appropriate dilution allowing obtaining readily countable isolated colonies on agar plate (*see* **Note 7**).

2. Place 5–6 cleaned and sterile glass beads at the surface of the EMJH agar plates. Add 100 μl of the diluted treated or non-treated leptospires samples onto EMJH agar plates in duplicate, and spread the cells by agitating the plates (*see* **Note 8**).

3. Allow some time for the plated bacterial suspension to penetrate the agar and remove the glass beads (*see* **Note 9**). Wrap the plates in Parafilm to avoid drying off the medium and in

aluminum foil to protect them from light and lower risk of contamination.

4. Incubate the plates at 30 °C for 3–4 weeks (*see* **Note 10**).

5. Count the number of colonies on each plate and calculate the mean of the replicates (*see* **Note 11**). Multiplicate this number by the dilution factor and by 10 to obtain the number of cfu/ml. Calculate the ratio of cfu of the treated sample to the cfu of the non-treated sample. The percentage of survival after treatment is determined by multiplying this ratio by 100.

3.3 Assessing Survival by Measuring Cellular Reducing Environment

We describe the protocol to measure *Leptospira* spp. viability with the Alamar Blue® Assay (*see* **Note 12**). You can perform this assay with the bacterial suspension as prepared in Subheading 3.1 or directly perform the cell treatment in 96-well plates, depending on the nature of the treatment or whether cell viability will be simultaneously assessed by different techniques.

1. Transfer 100 μl of treated cells or non-treated control cells into each well of a 96-well plate (*see* **Note 13**), and add 80 μl of EMJH medium to each well. Proceed directly to **step 3**.

2. If the treatment is being performed directly in a 96-well plate, pipette 100 μl of *Leptospira* suspension obtained at **step 3** of Subheading 3.1 into each well of the plate. Add the desired amount of the chemical used in the survival test, and complete with EMJH medium to a final volume of 180 μl. Mix by pipetting and incubate the plate at 30 °C for a given time.

3. Add 20 μl of the 10× resazurin solution beforehand warmed at room temperature. Mix by pipetting (*see* **Note 14**).

4. Prepare a negative control sample that does not contain *Leptospira*: pipette 180 μl of EMJH medium and add 20 μl of the 10× resazurin solution. Mix by pipetting (*see* **Note 15**).

5. Place the 96-well plate into a plastic box and incubate the plate at 30 °C (*see* **Note 16**) until the reaction is completed, as seen by the appearance of a pink solution (*see* **Note 17**).

6. You can record the results simply by taking a picture (Fig. 2a).

7. The quantification can be done by absorbance measurement at 570 (absorbance maximum of resorufin) and 600 (absorbance maximum of resazurin) nm (Fig. 2b). The optical zero is made using the negative control well that does not contain any cell. Percentage of resazurin reduction is calculated using the equation indicated in Fig. 3 (as recommended by the manufacturer). The quantification can be also done by fluorescence measurement with an excitation and emission wavelengths of 560 nm and 590 nm, respectively (Fig. 2c) (*see* **Note 18**).

3.4 Quantitative Survival Test Based on Membrane Integrity

We describe the LIVE/DEAD *Bac*Light staining method to assess *Leptospira* spp. viability. In this method, the viability will be quantified by measurement of Syto9 and PI fluorescences. The ratio of green (Syto9) to red (PI) fluorescence is proportional to the relative number of live bacteria. A standard curve is necessary to verify the linearity of the assay and determine the proportionality factor between the fluorescence ratio and percentage of live bacteria.

3.4.1 Preparing a Standard Curve

1. Cultivate *Leptospira* in 20 ml of EMJH medium at 30 °C to reach the exponential phase and a concentration of at least 10^8 bacteria/ml.

2. Transfer 10 ml of the culture in two different sterile conical polypropylene tubes.

3. Obtain killed *Leptospira* by incubating one of the samples 2 h at 55 °C in a water bath, while the other sample (live bacteria) is maintained at 30 °C (*see* **Note 19**).

4. Harvest the bacteria by centrifugation for 15 min at $2600 \times g$ at room temperature (*see* **Note 20**).

5. Wash the bacteria twice with 10 ml of 0.5% NaCl (*see* **Note 21**).

6. Resuspend the bacteria in 10 ml of 0.5% NaCl.

7. Adjust the two samples (live and killed bacteria) to the same concentration (as assessed by absorbance measurement at 420 nm or enumeration with a Petroff-Hausser chamber under a dark-field microscope). We generally work with a bacterial suspension at about 10^8 *Leptospira*/ml.

8. Mix different proportions of killed and live *Leptospira* suspensions in sterile 1.5 ml polypropylene tubes in order to have, for instance, the following live/killed bacteria ratio: 0:100, 10:90, 25:75, 50:50, 75:25, 90:10, and 100:0.

9. In a 96-well flat bottom plate suitable for fluorescence measurement (*see* **Note 22**), mix 100 μl of bacterial suspension containing different proportions of killed and live *Leptospira* with 100 μl of Syto9/PI mixture. Mix thoroughly by pipetting up and down.

10. Incubate for 15 min at room temperature in the dark by enveloping the plate in aluminum foil.

11. Measure the fluorescence of each well with a plate reader (excitation at 485 nm and emission at 530 nm for the green fluorescent Syto9 (F_{530}); excitation at 485 nm and emission at 630 nm for the red fluorescent PI (F_{630})).

12. Calculate the ratio of Syto9 fluorescence (F_{530}) to that of the PI (F_{630}) for each sample.

13. The standard curve is obtained by plotting the F_{530}/F_{630} value for each well corresponding to the different live/killed bacteria

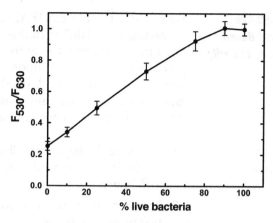

Fig. 4 Linear relationship between the percentage of live *Leptospira* and Syto9/PI fluorescence. Live and heat-killed *Leptospira* were obtained and treated with the Syto9/PI mixture in a flat bottom 96-well plates (μClear Black 96-well plate) as described in Subheading 3.4.1. Green Syto9 fluorescence (F_{530}; λ_{exc} = 485 nm, λ_{em} = 530 nm) and red PI fluorescence (F_{630}; λ_{exc} = 485 nm, λ_{em} = 630 nm) were measured. The F_{530}/F_{630} values (green Syto9-to-red PI fluorescence ratio) were plotted versus the percentage of live bacteria

ratio versus the percentage of live bacteria (Fig. 4). A fitted linear regression method will provide with an equation to deduce the percentage of live bacteria from the F_{530}/F_{630} value of any tested samples (*see* **Note 23**).

3.4.2 Survival Test of a Bacterial Suspension

1. Harvest the bacteria (treated and non-treated control samples obtained in Subheading 3.1) by centrifugation for 15 min at $2600 \times g$ at room temperature in conical polypropylene tubes (*see* **Note 20**).

2. Wash the cells twice with 0.5% NaCl and resuspend them in 0.5% NaCl (*see* **Note 24**).

3. Adjust all the samples to the same concentration (as assessed by absorbance measurement at 420 nm or enumeration with a Petroff-Hausser chamber under a dark-field microscope) (*see* **Note 25**).

4. In a 96-well flat bottom plate suitable for fluorescence measurement (*see* **Note 22**), mix 100 μl of bacterial suspension washed in 0.5% NaCl with 100 μl of Syto9/PI mixture. Mix thoroughly by pipetting up and down.

5. Incubate for 15 min at room temperature in the dark by enveloping the plate in aluminum foil.

6. Measure the fluorescence of each well with a plate reader as described in **step 11** in Subheading 3.4.1.

7. Calculate the ratio of Syto9 fluorescence (F_{530}) to that of the PI (F_{630}) for each sample.

8. You can deduce the percentage of live bacteria from the F_{530}/F_{630} value using the standard curve and its proportionality factor (Fig. 4; *see* **Note 23**). Alternatively, you can normalize the fluorescence ratio of the treated sample with that of the non-treated control one and calculate a percentage of live cells.

4 Notes

1. We use autoclaved glassware dedicated only to EMJH medium preparation. To avoid contaminating the glassware with components that could prevent growth of *Leptospira*, we rinse beforehand the glassware with sterile water for injection (WFI), and all the chemical stock solutions are prepared with sterile WFI.

2. We recommend pouring rather thick EMJH agar plates to avoid drying during the long storage at 30 °C.

3. This Syto9/PI combination has given satisfactory results; however it might be necessary to adapt dye concentrations to the condition the assay will be performed. We have used the LIVE/DEAD™ BacLight™ Bacterial Viability Kit from Thermo Fisher Scientific. This kit provides with 3.34 mM Syto9 and 20 mM PI solutions. Separated Syto9 and PI solutions allow finding the optimal dye combination for your specific assay condition.

4. We usually perform survival tests with exponentially growing *L. interrogans* serovar Manilae, which are at a concentration of about 10^8 bacteria/ml ($OD_{420} \approx 0.1$–0.3). These values might vary depending of laboratory conditions including the oxygenation level, the composition of EMJH medium, and the size of leptospiral strain used. We advise the experimenter to perform a growth curve with the strain used in survival test beforehand in order to have a knowledge of its growth feature in EMJH medium in his/her laboratory condition.

5. Rigorous viability comparison implies that all bacterial samples are taken at the same growth phase.

6. If a killing kinetic is performed, the non-treated control sample can be a bacterial suspension incubated in the same conditions and for the same duration without the killing agent. A sample of the bacterial suspension before the addition of the killing agent can also serve as non-treated control sample.

7. Finding the suitable dilution for plating is crucial for the accuracy of bacterial enumeration on agar plates. Indeed, to accurately count *Leptospira* on plates, it is important to obtain isolated colonies. Also, discrepancies in counting might occur when having too few (less than 10) or too many (more than

100) colonies on plate. Generally, sample dilution will also dilute the killing chemical agent used for the survival test to sublethal concentration. However, in some cases, it might be necessary to wash the bacteria with EMJH before plating them on solid media.

8. We generally use cleaned and autoclaved 4 mm glass beads from Sigma. You can also use a glass spreader or inoculation loops to spread bacteria, but this will make handling multiple plates difficult.

9. Glass beads are reusable. Wash them thoroughly with a decontaminating solution, rinse them extensively with distilled water, and autoclave them for further use.

10. It is important to ensure that the colonies will remain isolated. It might be necessary to check regularly the plates during their incubation at 30 °C as *Leptospira* spp. are motile and can produce spread large and confluent colonies when overgrown on agar plate.

11. The use of a black paper sheet facilitates the visualization of *Leptospira* colonies.

12. We generally use the Alamar Blue® Assay provided by Thermo Fisher Scientific Invitrogen but a resazurin solution from another manufacturer might give satisfactory results as well.

13. You have to anticipate which method will be used to quantify resazurin reduction. If fluorescence will be used, the reaction has to be performed in plates that are compatible with fluorescence measurement. We recommend using clear-bottom black-walled 96-well plates. We have obtained satisfactory results with the μClear Black 96-well plate provided by Greiner. But, these plates are not suitable to take a picture for recording the change in color upon resazurin reduction.

14. The optimum pH for resazurin reduction is around 7.0 [9]. In fact, resazurin and resorufin will be respectively pink and colorless under acidic pH. You have to ensure that the reaction mix for resazurin reduction is buffered.

15. It is crucial to verify that the chemical used as killing agent in the survival test does not interfere with resazurin reduction. This can be done by pipetting the chemical at the desired concentration into a well of the 96-well plate and completing with EMJH medium to have a final volume of 180 μl. Add 20 μl of the $10\times$ resazurin solution and mix by pipetting.

16. The optimal temperature for resazurin reduction is 37 °C, but we prefer to perform the resazurin reduction reaction at 30 °C to reduce evaporation. 200 μl of sterile deionized water can also be added to all outer-perimeter wells of the 96-well plate.

You can also seal the 96-well plate with Parafilm for a longer incubation or storage.

17. The time for the completion of the resazurin reduction reaction depends on the number of viable cells used in the assay. The change of blue to pink color can be seen after few hours, but it can also require longer incubation time (up to 3 days). Extended incubation times are not recommended to record the results. Indeed, the resazurin reduction into resorufin is irreversible but resorufin can be further reduced into hydroresorufin, a colorless and non-fluorescent chemical [8].

18. Resazurin reduction into resorufin is detectable only if *Leptospira* concentration is above 10^6/ml (Fig. 2). A good linearity is observed between resazurin reduction and *Leptospira* concentration only between 10^6 and 10^8 bacteria/ml, as assessed by absorbance (Fig. 2b) or by fluorescence (Fig. 2c).

19. We have verified that 2 h incubation at 55 °C results in more than 99.99% of *Leptospira* killing as measured by colony-forming unit on EMJH agar plate.

20. Do not centrifuge *Leptospira* at a speed above $2600 \times g$ as it is generally believed that this might result in damage to *Leptospira*.

21. Complete removal of EMJH medium is important as we have observed that this medium interferes with fluorescence measurement of Syto9 and PI. We have also verified that incubating *Leptospira* with 0.5% NaCl did not affect their viability. Noteworthily, *Leptospira* resuspended in 0.5% NaCl cannot be used in the Alamar Blue® Assay. Therefore, if different methods are simultaneously used to assess *Leptospira* viability, bacteria will have to be treated accordingly to each assay.

22. We generally use clear-bottom black-walled 96-well plates, and we have obtained satisfactory results with the µClear Black 96-well plate provided by Greiner.

23. The F_{530}/F_{630} value exhibits a linear relationship with the percentage of live bacteria. The linear regression by the least-squares method gave a R square coefficient of 0.9907, indicative of a very good linearity.

24. You should expect to lose some bacteria upon washing and a decreased bacteria concentration at the end of the washing steps.

25. We have observed that a bacterial concentration below 10^7 *Leptospira*/ml is not detected by this method. We generally perform this assay with a bacterial suspension at 10^8 *Leptospira*/ml.

Acknowledgments

We would like to thank Jan Bayram for his help with Fig. 1.

References

1. Goldman E, Green LH (2008) Practical handbook of microbiology, 2nd edn. CRC Press, Taylor and Francis Group, Boca Raton, Florida

2. Breed RS, Dotterrer WD (1916) The number of colonies allowable on satisfactory agar plates. J Bacteriol 1:321–331

3. Larson AD, Treick RW, Edwards CL et al (1959) Growth studies and plate counting of leptospires. J Bacteriol 77:361

4. Li S, Ojcius DM, Liao S et al (2010) Replication or death: distinct fates of pathogenic Leptospira strain Lai within macrophages of human or mouse origin. Innate Immun 16:80–92

5. Benaroudj N, Saul F, Bellalou J et al (2013) Structural and functional characterization of an orphan ATP-binding cassette ATPase involved in manganese utilization and tolerance in Leptospira spp. J Bacteriol 195:5583–5591

6. Kebouchi M, Saul F, Taher R et al (2018) Structure and function of the Leptospira interrogans peroxide stress regulator (PerR), an atypical PerR devoid of a structural metal-binding site. J Biol Chem 293:497–509

7. Rampersad SN (2012) Multiple applications of Alamar blue as an indicator of metabolic function and cellular health in cell viability bioassays. Sensors (Basel) 12:12347–12360

8. DeBaun RM, de Stevens G (1951) On the mechanism of enzyme action. XLIV. Codetermination of resazurin and resorufin in enzymatic dehydrogenation experiments. Arch Biochem Biophys 31:300–308

9. Barnes S, Spenney JG (1980) Improved enzymatic assays for bile acids using resazurin and NADH oxidoreductase from Clostridium kluyveri. Clin Chim Acta 102:241–245

10. Yajko DM, Madej JJ, Lancaster MV et al (1995) Colorimetric method for determining MICs of antimicrobial agents for Mycobacterium tuberculosis. J Clin Microbiol 33:2324

11. Murray CK, Hospenthal DR (2004) Broth microdilution susceptibility testing for Leptospira spp. Antimicrob Agents Chemother 48:1548–1552

12. Eshghi A, Lourdault K, Murray GL et al (2012) Leptospira interrogans catalase is required for resistance to H_2O_2 and for virulence. Infect Immun 80:3892

13. Pappas CJ, Benaroudj N, Picardeau M (2015) A replicative plasmid vector allows efficient complementation of pathogenic Leptospira strains. Appl Environ Microbiol 81:3176

14. Liegeon G, Delory T, Picardeau M (2018) Antibiotic susceptibilities of livestock isolates of Leptospira. Int J Antimicrob Agents 51:693–699

15. Stocks SM (2004) Mechanism and use of the commercially available viability stain, BacLight. Cytometry A 61A:189–195

16. Boulos L, Prevost M, Barbeau B et al (1999) LIVE/DEAD BacLight : application of a new rapid staining method for direct enumeration of viable and total bacteria in drinking water. J Microbiol Methods 37:77–86

17. Fontana C, Crussard S, Simon-Dufay N et al (2017) Use of flow cytometry for rapid and accurate enumeration of live pathogenic Leptospira strains. J Microbiol Methods 132:34–40

18. Chen X, Li S-J, Ojcius DM et al (2017) Mononuclear-macrophages but not neutrophils act as major infiltrating anti-leptospiral phagocytes during leptospirosis. PLoS One 12:e0181014

19. Stiefel P, Schmidt-Emrich S, Maniura-Weber K et al (2015) Critical aspects of using bacterial cell viability assays with the fluorophores SYTO9 and propidium iodide. BMC Microbiol 15:36–36

20. Kirchhoff C, Cypionka H (2017) Propidium ion enters viable cells with high membrane potential during live-dead staining. J Microbiol Methods 142:79–82

21. Rosenberg M, Azevedo NF, Ivask A (2019) Propidium iodide staining underestimates viability of adherent bacterial cells. Sci Rep 9:6483–6483

Chapter 21

Cultivation of *Leptospira interrogans* Within Rat Peritoneal Dialysis Membrane Chambers

Andre Alex Grassmann and Melissa J. Caimano

Abstract

In order to sustain its zoonotic lifecycle, leptospires must adapt to growth within the host milieu. Signals encountered within the mammal also trigger regulatory programs required by *Leptospira* for the expression of virulence-related gene products. The complex transcriptional, antigenic, and physiological changes leptospires undergo within the mammal are collectively referred to as "host adaptation." In this chapter, we describe the procedures for the generation of host-adapted *Leptospira* spp. by cultivation within dialysis membrane chambers (DMCs) implanted in rat peritoneal cavities. In this model, *Leptospira* spp. diluted in EMJH medium are sequestered within sterile dialysis membrane tubing closed at both ends. The chamber then is surgically implanted within the peritoneal cavity of a rat and incubated for 7–10 days. During this period, leptospires are exposed to many, if not all, of the physiological and nutritional cues required for host adaptation while at the same time protected from clearance by host innate and adaptive immune defenses.

Key words Host adaptation, Dialysis membrane chambers, In vivo model, Bacterial cultivation, Leptospirosis, Spirochetes

1   Introduction

L. interrogans is maintained in nature within a zoonotic cycle in which rats are the primary reservoir [1–3]. Infection begins when a reservoir or an incidental host, including humans, comes into contact with water or soil contaminated with urine from animal reservoir. Following acquisition, leptospires switch from a free-living to a parasitic lifestyle, a complex, multifactorial process often referred to as "host adaptation." Once in the bloodstream, leptospires rapidly disseminate to multiple tissues. The outcome of these early events differs dramatically depending on whether infection occurs in a reservoir-competent or reservoir-incompetent (i.e., incidental) host. In reservoir hosts, leptospires are cleared within

Electronic supplementary material: The online version of this chapter (https://doi.org/10.1007/978-1-0716-0459-5_21) contains supplementary material, which is available to authorized users.

Nobuo Koizumi and Mathieu Picardeau (eds.), *Leptospira spp.: Methods and Protocols*, Methods in Molecular Biology, vol. 2134, https://doi.org/10.1007/978-1-0716-0459-5_21, © Springer Science+Business Media, LLC, part of Springer Nature 2020

days from all sites except the kidney, where they colonize the proximal renal tubules [3–6]; within this immunoprivileged niche, leptospires are continuously bathed in nutrient-rich, glomerular ultrafiltrate, which closely resembles interstitial fluid [4, 5, 7]. Importantly, even at high inoculum (e.g., 10^7 organisms), rats are asymptomatic and, following renal colonization, continue to shed leptospires in urine for weeks [3–6]. In contrast, hamsters and guinea pigs, the species most commonly used to study acute leptospirosis [2, 3], are exquisitely sensitive to infection, succumbing to disease with inocula ≤ 10 organisms [8]. These observations imply that outcomes following infection are determined largely by the nature and intensity of the host inflammatory response [9–12]. These data argue that pathogenic *Leptospira* spp. employ similar, if not identical, genetic programs to establish themselves within asymptomatic (i.e., reservoir) and symptomatic (i.e., incidental) hosts despite the markedly divergent immunopathological events that follow infection. While little is known about how *Leptospira* spp. persist within mammals, numerous studies suggest that leptospires alter their transcriptomic and proteomic profiles in response to environmental signals encountered during infection [13–24] Although sufficient for growth in vitro, artificial medium almost certainly does not replicate the full range of environment signals encountered by leptospires either outside or inside the host. Analysis of host-adapted leptospires from naturally or experimentally infected hosts is limited by difficulties associated with recovering intact organisms free from host tissues or other contaminants. To circumvent these issues, we developed a facile model for generating large numbers of host-adapted *L. interrogans* within rat peritoneal cavities [13]. Leptospires diluted to low density (3000–10,000 per ml) within EMJH medium are placed within dialysis membrane tubing, tied off at both ends to form a chamber. The dialysis membrane chamber (DMC) is implanted in the peritoneal cavity of a rat and allowed to incubate for ~7–10 days, during which time leptospires reach densities (~10^8 per ml) comparable to those observed with in vitro cultures. The molecular weight cutoff of the dialysis tubing (8000 Da) allows for rapid exchange of nutrients and other small molecules within host interstitial fluid but protects the bacterium from the host immune system. Using this model, we routinely recover >10^9 viable host-adapted leptospires per DMC.

Because of the large numbers of host-adapted leptospires recovered, the DMC model provides a facile platform for genome-wide comparative transcriptomic and proteomic analyses. Using RNAseq, we identified >160 genes differentially expressed by *L. interrogans* serovar (sv.). Copenhageni strain (st.) Fiocruz L1-130 leptospires in DMCs were compared to their in vitro counterparts [13]. We subsequently reported that the proteome of DMC-cultivated leptospires undergoes posttranslational

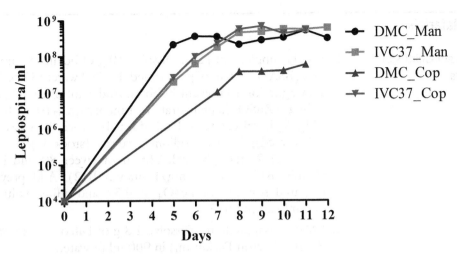

Fig. 1 *L. interrogans* serovar Copenhageni strain Fiocruz L1-130 and serovar Manilae strain L495 display similar growth rates in DMCs. Growth of Fiocruz L1-130 (Cop) and L495 (Man) strains was compared in DMCs and in vitro in EMJH at 37 °C (IVC37). Both DMCs and IVC37 cultures were started at a density of 10^4 leptospires per ml in 10 ml total volume. One animal per day for each strain was euthanized beginning either day 5 (Man) or day 7 (Cop). The content of each DMC was collected and counted in triplicate by dark-field microscopy using a Petroff-Hausser counting chamber. For IVC37, triplicate cultures of each strain were counted once a day starting on day 5 postinoculation until days 11 (Man) or 12 (Cop)

modifications that are not present in leptospires cultivated in vitro [21]. The DMC model also provides an appropriate biological framework for phenotypic characterization of a wide range of mutations. Because leptospires are separated from host tissues, the DMC model enables one to distinguish between gene products required for physiological adaptation and nutrient acquisition and those that serve virulence-related functions, such as adherence, immune evasion, and motility. Lastly, this model is also well suited to the investigation of environmental sensing, signal transduction, and the molecular pathways underlying differential gene expression by pathogenic *Leptospira* spp. within mammals.

The DMC model has been used successfully with several well-characterized virulent *L. interrogans* reference strains, including the highly virulent sv. Manilae st. L495, the sv. Copenhageni st. Fiocruz L1-130, and numerous transposon mutant strains. The Fiocruz L1-130 and L495 strains grow at similar rates within DMCs (Fig. 1) and yield comparable numbers of host-adapted organisms. However, saprophytic *L. biflexa* sv. Patoc st. Patoc1 and pathogenic *L. borgpetersenii* sv. Hardjo sts. HB203 and JB197 grew poorly or not at all within DMCs.

2   Materials

2.1   Leptospira spp. Culture

1. EMJH supplement [25]: Dissolve 10 g of bovine serum albumin (BSA) in 50 ml of prewarmed distilled water (37 °C), with slow agitation to prevent bubbles. Add 1 ml of $ZnSO_4$ (stock 0.04 g/l $ZnSO_4$ heptahydrated), 1 ml of $MgCl_2$ (stock 0.15 g/l $MgCl_2$ hexahydrated), 1 ml of $CaCl_2$ (stock 0.15 g/l $CaCl_2$ dehydrated), 1 ml of sodium pyruvate (stock 1 g/l sodium pyruvate), 2 ml of glycerol, 12.5 ml of Tween 80, and 1 ml of vitamin B12 (stock 2 mg/l vitamin B12). Add previously prepared solution of $FeSO_4$ at 0.5% and adjust volume to 100 ml and pH to 7.4.

2. EMJH basal medium: Dissolve 2.3 g of Difco *Leptospira* basal EMJH (Becton Dickinson) in 900 ml of water.

3. EMJH medium: 100 ml of EMJH supplement, 890 ml of EMJH basal medium, 10 ml normal rabbit serum, 1 g lactalbumin hydrolase, and 100 µl superoxide dismutase (stock 10 mg/ml in PBS). Adjust pH to 7.4 if necessary and filter sterilize (0.22 µm filter), aliquot in flask of smaller volume, and control sterility at 37 °C for 48 h and then store at 4 °C.

4. High-BSA EMJH: EMJH medium supplemented with an additional 10 mg/ml BSA (serum replacement grade)—final BSA concentration 20 mg/ml. Filter sterilize (0.22 µm filter) (*see* **Note 1**).

5. *Leptospira* spp. in vitro culture in mid- to late-logarithmic phase.

6. 15-ml polypropylene conical tubes, sterile.

7. Dark-field microscope.

8. Bacteriological incubators, at 30 °C and 37 °C.

9. Petroff-Hausser counting chamber.

10. BSC, biological safety cabinet (Biosafety Level 2; BSL-2).

2.2   Dialysis Membrane Chambers' Preparation and Filling

1. Ultrapure water.

2. EDTA: 1 mM solution in ultrapure water, pH 8.0.

3. Regenerated cellulose dialysis membrane tubing (6000 to 8000 MWCO, 32 mm width).

4. Filter units, 0.22 µm, 250 ml, sterile.

5. Three 2-L glass beakers, each containing a magnetic stir bar.

6. Hot plate with stirring option.

7. Extra-long blunt-end forceps, sterile.

8. Pipettor.

9. Scissors, iris, 4 in. (~10.2 cm).

10. Surgical gloves, individually wrapped, sterile.

11. Biological safety cabinet (Biosafety Level 2; BSL-2).

2.3 Surgical Implantation of DMCs

1. Adult Sprague-Dawley rats (175–200 g) (*see* **Note 2**).

2. Anesthetic cocktail: Ketamine 30–50 mg/kg, xylazine 3–5 mg/kg, and acepromazine 0.75–1 mg/kg.

3. Ophthalmic ointment.

4. Electric shaver.

5. Alcohol prep pads.

6. Carprofen: 5–10 mg per kg (*see* **Note 3**).

7. 70% ethanol.

8. Surgical gloves, individually wrapped, sterile.

9. Swab stick, 2% w/v chlorhexidine gluconate in 70% isopropyl alcohol (*see* **Note 4**).

10. Disposable serological pipets, 10 ml sterile, individually wrapped.

11. Surgical drape, cut into 12 in. (~46 cm) squares.

12. Gauze, 4 × 4 in., sterile (*see* **Note 5**).

13. Surgical instrument pack: one per animal, sterilized and kept within the same sterile package, containing scalpel blade holder, iris scissors (4 in./~10.2 cm), tissue forceps 1× 2 tooth dissecting or Adson-Brown (5½ in./~14 cm), tissue forceps blunt end (5½ in./~14 cm), and needle holder forceps with built-in scissors (e.g., Olsen-Hegar, 5½ in./~14 cm) (*see* **Note 6**).

14. 9-mm stainless steel wound closure clips and applicator.

15. Scalpel blades, No. 10.

16. Nonabsorbable suture with taper point needle attached (e.g., Ethicon 4-0, SH-1, 27 in. (~68.6 cm) coated VICRYL, violet-braided suture).

17. Biological safety cabinet (Biosafety Level 2; BSL-2).

18. Glass bead sterilizer, optional (*see* **Note 6**).

19. Circulating warm water blanket and pump.

20. 18-G, 1.5 in. needles attached to sterile 10 ml syringes.

3 Methods

3.1 Preparation of Sterile Dialysis Membrane Tubing

1. Fill three 2-L glass beakers as follows:

   Beaker 1: 1–1.5-L ultrapure water.

   Beaker 2: 1–1.5-L 1 mM EDTA (pH 8.0).

   Beaker 3: 1–1.5-L ultrapure water.

2. Add a magnetic stir bar to each beaker, cover with heavy-duty aluminum foil, and autoclave.

3. Place Beakers 1 and 2 on each of two hot plates and bring to a rolling boil with constant stirring.

4. Wearing sterile gloves, cut strips of dialysis membrane tubing, 7–9 in. (~18–23 cm) in length, using sterile scissors. Gently tie off one end of the tubing using a simple overhand knot by forming a loop and passing the free end of the tubing through the loop.

5. Place the tied tubing into Beaker 1 and replace the aluminum foil cover.

6. Repeat **steps 4** and **5** for each animal/DMC, plus 1–2 extra.

7. Boil tubing for 20 min with constant stirring. Be sure to keep the foil cover loose enough to allow steam to escape. As needed, use a sterile 10-ml serological pipette or extra-long forceps to push the tubing back into the boiling liquid.

8. After 20-min boiling, transfer tubing to Beaker 2 using sterile 10 ml pipette or extra-long forceps. Replace aluminum foil cover and boil tubing for 20 min. With constant stirring.

9. While strips are boiling in Beaker 2, place Beaker 3 on the hot plate and bring to a boil.

10. After 20 min boiling, transfer tubing to Beaker 3 as described in **step 8**.

11. After 20 min boiling, turn off heating/stirring for Beaker 3. Once cool to the touch, transfer Beaker 3 to a biological safety cabinet (BSC). To ensure sterility, the beaker containing the tubing should remain in the BSC until the tubing is transferred to a sterile container for storage.

12. Transfer dialysis membrane tubing to the bottom portion of a 0.22-μm filter unit using sterile blunt tip forceps. Replace the filter unit top and filter in ~200 ml of Ultrapure-Q water from Beaker 3. Seal the bottom portion of the filter unit using the sterile cap provided by the manufacturer (*see* **Note 7**). Going forward, the container should be opened only within the BSC.

3.2 *In Vitro* Cultivation *of* Leptospira *and Culture Dilution*

1. Transfer 10 ml of EMJH to a sterile 15-ml conical tube.

2. Inoculate with leptospires to give a final density of 10^3 to 10^5 leptospires/ml.

3. Incubate at 30 °C until the culture reaches mid- to late-logarithmic phase (usually 5–10 days).

4. On the day of DMC implantation, enumerate leptospires using Petroff-Hausser chamber or other suitable method.

5. Dilute the culture in 3–4 ml EMJH to 10^6 leptospires/ml.

6. Distribute 10 ml of high-BSA EMJH into sterile 50-ml conical tubes, one per DMC. Prewarm media to 37 °C in a bacteriological incubator.

7. Inoculate tubes with 10^5 leptospires/10 ml media (10^4 leptospires/ml) by adding 100 µl of 10^6 leptospires/ml dilution from **step 5**.

3.3 Preparation of DMCs and Surgical Implants

1. Prepare a workspace inside the BSC but first wiping down with 70% alcohol and then laying down a sterile 12-in. square drape to use as a workspace.

2. Place the following in the BSC: at least one disposable, individually wrapped serological pipet for each DMC; an automatic pipettor; one sterile 50-ml conical tube for each DMC; one sterile pack containing only scissors and blunt-end forceps for each strain being implanted; one sterile pack containing a complete set of surgical instruments for each strain; scalpel blades; sutures; applicator and clips; chlorhexidine/isopropanol swab sticks; gauze (*see* **Note 5**); and two pairs of gloves for each DMC to be implanted plus 1–2 extra pairs of gloves.

3. Expose the workspace to UV light for at least 15 min.

3.3.1 Preparation and Filling of DMC Tubing

1. Spray the outside of dilution tubes with 70% ethanol, wipe, and place inside BSC.

2. Put on a pair of sterile surgical gloves (*see* **Notes 8** and **9**) and loosen the caps of a sterile 50-ml conical tube and a *Leptospira* culture tube. Open a 10-ml serological pipette package and attach the pipette to automatic pipettor. Open the flask containing DMC tubing and place the bottle cap off to one side.

3. While still wearing gloves, carefully remove a strip of tubing from its storage container using sterile blunt-end forceps. Use your free hand to hold the tubing between index finger and thumb. With the other hand, remove the cap from the 50-ml conical tube containing diluted culture and transfer up to 10 ml to the tubing using a 10-ml disposable serological pipette (Fig. 2a). While continuing to hold the open end of the tubing, flatten the unfilled tubing between the fingers of your free hand to eliminate any air bubbles.

4. Gently twist the top end of the tubing to eliminate any remaining void volume and close off the top of the DMC (Fig. 2b). Use your thumb and index finger to hold on to the twisted area to prevent liquid from leaking out of the DMC while you tie off the top. Tie off the open end using a simple overhand knot (e.g., form a loop at the top of the DMC, passing the free end of the tubing through the loop, and then, without releasing your thumb and index finger, gently tighten the knot) (Fig. 2c). Bring the knot as close as possible to the liquid to ensure that the DMC is taut; this will help with maneuvering the chamber into the peritoneal cavity during surgery.

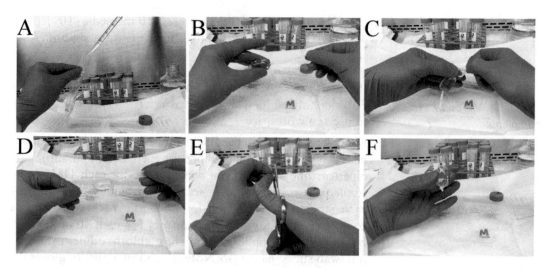

Fig. 2 Filling and preparation of dialysis membrane chambers. A dialysis tubing previously tied at one end and sterilized is filled with *Leptospira* culture using a 10-ml serological pipette (**a**); the air remaining inside the tubing is removed, the end is twisted (**b**) and tied using a simple overhand knot (**c**), and the remaining tubing (**d**) is trimmed away using sterile scissors (**e**). The DMC bag should be taut and around 1.5–2 in long (**f**)

5. Trim away excess tubing from the tied end using sterile scissors (Fig. 2d, e). The resulting filled DMC should be ~1.5–2 in. (Fig. 2f). Larger DMCs may interfere with normal intestinal or bladder functions.

6. Place filled DMC into a clean, sterile 50-ml conical tube.

7. Repeat **steps 2–6** until all bags are filled. Proceed as soon as possible to the surgical procedure (*see* **Note 10**).

3.3.2 Peritoneal Implantation Procedure

1. Turn on glass bead sterilizer and circulating water heating pad to warm up beforehand.

2. Open one pack of chlorhexidine/isopropanol swab stick (*see* **Note 4**), sterile surgical tools, scalpel blade, applicator and clips, suture, and surgical sterile gloves.

3. Anesthetize animal by intramuscular injection with the ketamine/xylazine/acepromazine cocktail.

4. Apply a small amount of ophthalmic ointment to each eye. Using forceps, gently place the animal's tongue to either side of the mouth; this is done to ensure that the animal's airway is not obstructed by the tongue or bedding material.

5. Administer preoperative analgesia (e.g., carprofen, subcutaneously) (*see* **Note 3**).

6. Gently shave the abdomen using clippers (*see* **Note 11**).

7. Remove away any loose trimmed fur using alcohol prep pads (*see* **Note 12**).

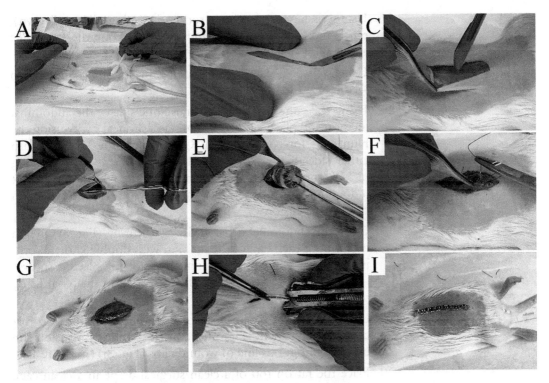

Fig. 3 Implantation procedure. After shaving, the surgical site is wiped using a chlorhexidine/isopropanol swab stick (**a**), and a small incision (~4–5 cm) is made in the skin using a scalpel blade (**b**). The fascia on either side is gently trimmed using the tip of the scalpel blade to release the muscle from the skin (**c**). A small opening in the abdominal muscle along the *linea alba* is first made using a scalpel and then extended using scissors (**d**). A DMC is inserted in the peritoneal cavity (**e**). The abdominal muscle is sutured, being careful to avoiding the DMC and internal organs (**f** and **g**). The edges of the skin incision site are gathered and held together using forceps (**h**) and then closed using wound clips (**i**)

8. Transfer the animal to the BSC. Place on a sterile drape in a supine position.

9. Prior to beginning the procedure, perform a "toe pinch" to ensure that the animal is properly anesthetized.

10. Use a chlorhexidine/isopropanol swab stick to aseptically clean the surgical site (Fig. 3a) (*see* **Note 13**).

11. Before proceeding, put on the pair of sterile surgical gloves (*see* **Note 14**).

12. Place a sterile scalpel blade in the scalpel blade holder (*see* **Note 6**). Carefully make a 4–5 cm incision through the skin only (Fig. 3b), starting ~2.5 cm below the ribcage, using the xiphoid process as a guide.

13. Using tissue forceps, pull up the skin on either side of the incision and gently trim the fascia connecting the skin to the abdominal wall using the same scalpel (Fig. 3c).

14. Repeat on the other side.

15. Using tissue forceps (two tooth dissecting/Adson-Brown), pinch and hold up the abdominal muscle along the *linea alba* and make a small incision.

16. Using scissors, extend the opening in abdominal muscle to ~4 cm (Fig. 3d).

17. Using the same tissue forceps, raise one side of the abdominal incision and place a DMC inside rat peritoneal cavity using the blunt-end forceps (Fig. 3e). Gently push the DMC toward either side of the peritoneal cavity to prevent it from being nicked/ruptured during suturing. Be sure to position it so that it does not become entangled in the intestines or interfere with bladder expansion.

18. Close the abdominal incision site by suturing (Fig. 3f). Begin by suturing each end of the incision with double knots. Working upward, close the remaining incision by placing sutures ~2–3 mm apart (Fig. 3g). Trim excess suture material using scissors.

19. Bring both sides of skin incision together using the two forceps, aligning them (Fig. 3h), and close the skin incision site using a contiguous line of wound clips (Fig. 3i) (*see* **Note 15**).

20. Place the rat on top of a clean surgical drape in a clean cage containing fresh bedding. Place the cage on top of the circulating water heating pad to maintain the appropriate core body temperature. Monitor the animal continuously until alert and responsive and then turn off heating pad.

21. Analgesia should be administered for at least 2 days postoperatively.

22. Animals should be monitored at least once daily for the first week postoperatively and then every other day thereafter. If animals show any sign of distress or discomfort, consult institutional veterinary staff immediately (*see* **Note 16**).

3.4 Recovery of Mammalian Host-Adapted Organisms from DMCs

1. At 7–12 days after implantation, euthanize animals by CO_2 asphyxiation or other approved method.

2. Place animal in a BSC in a supine position (Fig. 4a).

3. Using sterile surgical scissors, expose the abdominal wall.

4. Using sterile tissue forceps, lift one side of the sutured abdominal incision site and reopen using sterile scissors.

5. Locate and remove the DMC using blunt-end forceps.

6. Using sterile forceps to hold the DMC by one knot, remove contents using an 18-G, 1.5 in. needle attached to a sterile 10 ml syringe. Insert needle at the top of DMC to avoid leakage (Fig. 4b).

7. Transfer contents to a sterile 15 ml tube (*see* **Note 17**).

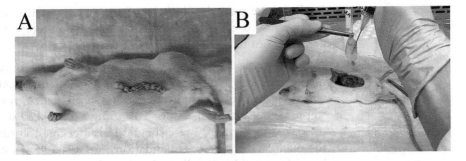

Fig. 4 Recovery of host-adapted leptospires. After 7–10 days postimplantation, the animal is euthanized and placed in supine position on a sterile workspace inside a biosafety cabinet (BSL-2). By this point, the incision site should be completely healed (**a**). The skin and abdominal wall are opened using scalpel blade and/or scissors, and the DMC is located. Using forceps to hold up the DMC, collect contents using an 18 G needle (1.5 in.) attached to 10 ml syringe (**b**) and transfer to a sterile tube for processing

8. Examine a small aliquot of the contents from each DMC for the presence of leptospires under dark-field microscopy.

9. Proceed to downstream applications (e.g., RNA and/or DNA isolation, analysis by one- and two-dimensional SDS-PAGE).

4   Notes

1. The quality of the BSA used for the cultivation of leptospires is critical. Please verify that the albumin used for DMC implantation supports the growth of virulent leptospires under standard in vitro growth conditions (e.g., when used as a component of EMJH medium).

2. This procedure is performed most often with Sprague-Dawley rats, but Wistar and Fisher rats also have been used successfully. While sex and age of rats do not appear to affect the experimental outcome (i.e., host adaptation), a minimal size/weight of 175–200 g is recommended. Smaller rats require a reduction in DMC volume. Larger DMCs are not recommended due to difficulty with handling when inserting into the peritoneal cavity.

3. Buprenorphine (0.05–0.1 mg per kg), administered subcutaneously every 6–12 h, also may be used for postoperative pain relief. However, animals may develop pica after receiving this drug. For this reason, carprofen is preferred.

4. Two percent chlorhexidine (w/v) in 70% isopropyl alcohol is preferred for preparation of the surgical site. Iodophor (e.g., Betadine) activity may be reduced in the presence of organic matter.

5. Sterile gauze pads should be on hand in case of the need to remove any excess blood around the incision site; they are generally not needed.

6. A clean sterile instrument pack for each animal is recommended. If this is not possible, then instruments should be wiped clean of blood and tissues with sterile gauze, rinsed in sterile saline, and sterilized for 15 s using a glass bead sterilizer or soaked in an appropriate chemical disinfectant according to the manufacturer's recommendations. If a glass bead sterilizer is used, remember to allow time for the instruments to cool before reusing them. If using a disinfectant, always rinse instruments with sterile saline before using on the next animal. When in use, the tips of instruments should be kept within the sterile field. Avoid touching unprepared areas. If this happens, replace with a new sterile instrument or re-sterilize using bead sterilizer or disinfectant as described above.

7. Dialysis membrane tubing may be prepared several days in advance and stored at 4 °C. Any unused tubing stored >2 months should be autoclaved to ensure sterility. For that, transfer to a 1-L glass beaker, cover with heavy-duty aluminum foil, and autoclave for a minimum of 15 min at 121 °C or 3 min at 131 °C. Once the beaker is cool to the touch, place in BSC and transfer to a clean sterile container as described in Subheading 3.1, **step 12**.

8. Be sure to select the appropriate size glove for each user. Gloves that are too large for the user's hands will make it more difficult to tie the dialysis tubing. Gloves that are too small may be challenging to put on or lead to cramping during the procedure.

9. The inner paper wrapping for sterile surgical gloves may be used as a clean sterile workspace.

10. DMCs should be filled and used the same day. A small amount (~10 ml) of high-BSA EMJH medium can be added to 50-ml conical tubes used to store filled DMCs to avoid drying.

11. Depilatory creams can irritate the skin and, therefore, are not recommended for hair removal. Preparation of the surgical site should take place at a location (bench or room) that is separate from where the surgical procedure will be performed.

12. Avoid wetting large areas of fur with chlorhexidine/isopropanol swab stick (or alcohol) because of the potential to induce hypothermia.

13. During the scrub, the process should begin along the incision line and extend outward and never from outward (dirty) toward the center (clean).

14. A fresh pair of gloves is used for each animal.

15. The skin excision site also may be closed using subcuticular stitches followed by liquid skin adhesive.

16. The principal investigator is responsible for ensuring that post-procedural care is provided as described in the approved animal use protocol. This plan should be developed in consultation with the veterinary staff.

17. Volume may be lost during implantation and incubation.

Acknowledgments

This work was supported by funding provided by the National Institutes of Health/National Institute of Allergy and Infectious Diseases (R01 AI029735, R21 AI39940, R21 AI126146, and R21 AI28379).

References

1. Ko AI, Goarant C, Picardeau M (2009) Leptospira: the dawn of the molecular genetics era for an emerging zoonotic pathogen. Nat Rev Microbiol 7(10):736–747. https://doi.org/10.1038/nrmicro2208

2. McBride AJ, Athanazio DA, Reis MG, Ko AI (2005) Leptospirosis. Curr Opin Infect Dis 18 (5):376–386

3. Adler B, de la Pena MA (2010) Leptospira and leptospirosis. Vet Microbiol 140 (3-4):287–296. https://doi.org/10.1016/j.vetmic.2009.03.012

4. Athanazio DA, Silva EF, Santos CS, Rocha GM, Vannier-Santos MA, McBride AJ, Ko AI, Reis MG (2008) *Rattus norvegicus* as a model for persistent renal colonization by pathogenic *Leptospira interrogans*. Acta Trop 105 (2):176–180

5. Bonilla-Santiago R, Nally JE (2011) Rat model of chronic leptospirosis. Curr Protoc Microbiol . Chapter 12:Unit 12E 13. https://doi.org/10.1002/9780471729259.mc12e03s20

6. Marshall RB (1976) The route of entry of leptospires into the kidney tubule. J Med Microbiol 9(2):149–152. https://doi.org/10.1099/00222615-9-2-149

7. Taal MW, Chertow GM, Marsden PA, Skorecki K, UYu ASL, Brenner BM (2012) Brenner and Rector's the kidney, 9th edn. Elsevier Mosby Saunders, Philadelphia

8. Villanueva SY, Saito M, Tsutsumi Y, Segawa T, Baterna RA, Chakraborty A, Asoh T, Miyahara S, Yanagihara Y, Cavinta LL, Gloriani NG, Yoshida S (2014) High virulence in hamsters of four dominant Leptospira serovars

isolated from rats in the Philippines. Microbiology 160(Pt 2):418–428. https://doi.org/10.1099/mic.0.072439-0

9. Monahan AM, Callanan JJ, Nally JE (2009) Review paper: host-pathogen interactions in the kidney during chronic leptospirosis. Vet Pathol 46(5):792–799. https://doi.org/10.1354/vp.08-VP-0265-N-REV

10. Haake DA, Levett PN (2015) Leptospirosis in humans. Curr Top Microbiol Immunol 387:65–97. https://doi.org/10.1007/978-3-662-45059-8_5

11. Fraga TR, Barbosa AS, Isaac L (2011) Leptospirosis: aspects of innate immunity, immunopathogenesis and immune evasion from the complement system. Scand J Immunol 73 (5):408–419. https://doi.org/10.1111/j.1365-3083.2010.02505.x

12. Zuerner RL (2015) Host response to Leptospira infection. Curr Top Microbiol Immunol 387:223–250. https://doi.org/10.1007/978-3-662-45059-8_9

13. Caimano MJ, Sivasankaran SK, Allard A, Hurley D, Hokamp K, Grassmann AA, Hinton JC, Nally JE (2014) A model system for studying the transcriptomic and physiological changes associated with mammalian host-adaptation by *Leptospira interrogans* serovar Copenhageni. PLoS Pathog 10(3):e1004004. https://doi.org/10.1371/journal.ppat.1004004

14. Artiushin S, Timoney JF, Nally J, Verma A (2004) Host-inducible immunogenic sphingomyelinase-like protein, Lk73.5, of

Leptospira interrogans. Infect Immun 72 (2):742–749

15. Dhandapani G, Sikha T, Pinto SM, Kiran Kumar M, Patel K, Kumar M, Kumar V, Tennyson J, Satheeshkumar PK, Gowda H, Keshava Prasad TS, Madanan MG (2018) Proteomic approach and expression analysis revealed the differential expression of predicted leptospiral proteases capable of ECM degradation. Biochim Biophys Acta Proteins Proteom 1866(5–6):712–721. https://doi.org/10.1016/j.bbapap.2018.04.006

16. Adhikarla H, Wunder EA Jr, Mechaly AE, Mehta S, Wang Z, Santos L, Bisht V, Diggle P, Murray G, Adler B, Lopez F, Townsend JP, Groisman E, Picardeau M, Buschiazzo A, Ko AI (2018) Lvr, a Signaling system that controls global gene regulation and virulence in pathogenic Leptospira. Front Cell Infect Microbiol 8:45. https://doi.org/10.3389/fcimb.2018.00045

17. Fraser T, Brown PD (2017) Temperature and oxidative stress as triggers for virulence gene expression in pathogenic Leptospira spp. Front Microbiol 8:783. https://doi.org/10.3389/fmicb.2017.00783

18. Lo M, Bulach DM, Powell DR, Haake DA, Matsunaga J, Paustian ML, Zuerner RL, Adler B (2006) Effects of temperature on gene expression patterns in *Leptospira interrogans* serovar Lai as assessed by whole-genome microarrays. Infect Immun 74(10):5848–5859

19. Patarakul K, Lo M, Adler B (2010) Global transcriptomic response of *Leptospira interrogans* serovar Copenhageni upon exposure to serum. BMC Microbiol 10:31. https://doi.org/10.1186/1471-2180-10-31

20. Nally JE, Chow E, Fishbein MC, Blanco DR, Lovett MA (2005) Changes in lipopolysaccharide O antigen distinguish acute versus chronic *Leptospira interrogans* infections. Infect Immun 73(6):3251–3260. https://doi.org/10.1128/IAI.73.6.3251-3260.2005

21. Nally JE, Grassmann AA, Planchon S, Sergeant K, Renaut J, Seshu J, McBride AJ, Caimano MJ (2017) Pathogenic Leptospires modulate protein expression and post-translational modifications in response to mammalian host signals. Front Cell Infect Microbiol 7:362. https://doi.org/10.3389/fcimb.2017.00362

22. Nally JE, Monahan AM, Miller IS, Bonilla-Santiago R, Souda P, Whitelegge JP (2011) Comparative proteomic analysis of differentially expressed proteins in the urine of reservoir hosts of leptospirosis. PLoS One 6(10): e26046. https://doi.org/10.1371/journal.pone.0026046

23. Nally JE, Timoney JF, Stevenson B (2001) Temperature-regulated protein synthesis by *Leptospira interrogans*. Infect Immun 69 (1):400–404. https://doi.org/10.1128/IAI.69.1.400-404.2001

24. Nally JE, Whitelegge JP, Bassilian S, Blanco DR, Lovett MA (2007) Characterization of the outer membrane proteome of *Leptospira interrogans* expressed during acute lethal infection. Infect Immun 75(2):766–773. https://doi.org/10.1128/IAI.00741-06

25. Johnson RC, Walby J, Henry RA, Auran NE (1973) Cultivation of parasitic leptospires: effect of pyruvate. Appl Microbiol 26 (1):118–119

Chapter 22

Use of Golden Syrian Hamster as an Animal Model to Study Leptospirosis-Associated Immune Responses

Julie Cagliero, Karl Huet, and Mariko Matsui

Abstract

Experimental infections greatly contribute to further deepen our knowledge of infectious diseases. In the case of leptospirosis, hamsters as well as gerbils and guinea pigs have been used as animal models of acute leptospirosis in studying the pathophysiology of the disease. Here we describe a typical *Leptospira* infection using golden Syrian hamsters. We will also present techniques we use to study the resulting bacterial burden and gene expression patterns in the host in order to decipher the innate immune response to leptospirosis.

Key words Leptospirosis, *Leptospira*, Syrian hamster, Experimental infection, Gene expression

1 Introduction

Leptospirosis is a re-emerging zoonotic disease caused by pathogenic spirochetal bacteria from the genus *Leptospira* and estimated to infect more than a million people with approximately 60,000 deaths annually [1]. Typically, asymptomatic reservoir animals, mostly rodents, carry the pathogen in their renal tubules and shed pathogenic spirochetes in their urine, contaminating the environment [2]. Accidental hosts of leptospires, like humans, can be infected through direct contact with urine from reservoir animals or contaminated soil or water, and most human infections are mild, with mainly flu-like symptoms, or even asymptomatic [3, 4]. However, roughly 10% of human leptospirosis cases develop into severe forms, among which are the Weil's disease and the severe pulmonary hemorrhagic syndrome, and lead to high leptospiremia, multi-organ failures, and a dramatically increased mortality rate.

This broad spectrum of symptoms renders the diagnosis difficult, as this disease can be mistaken with other acute febrile syndromes, and optimal treatment for leptospirosis remains a subject of debate [3, 5, 6]. Besides, to date, there is still no universal vaccine providing a long-lasting protection against leptospirosis

Nobuo Koizumi and Mathieu Picardeau (eds.), *Leptospira spp.: Methods and Protocols*, Methods in Molecular Biology, vol. 2134,
https://doi.org/10.1007/978-1-0716-0459-5_22, © Springer Science+Business Media, LLC, part of Springer Nature 2020

[7]. Moreover, leptospires are very successful pathogens that can infect all mammals, as well as birds, amphibians, reptiles, and possibly fish, and are able to escape from the host immune system. However, virulence factors associated with these pathogens and their interactions between leptospires and the host immune system remain largely unknown [8, 9].

Historically, animal models have been most helpful for studying leptospirosis, in particular to characterize the pathophysiology of the disease, the host immune response to infection, as well as pathogen-associated virulence factors. In 1915, Inada and Ido showed that the spirochetes they had found in the blood of a patient suffering from Weil's disease was the causative agent of the disease [10, 11]. To come to that conclusion, they performed experimental infections using monkeys, rabbits, rats, and guinea pigs and found that only guinea pigs developed clinical signs, including conjunctival congestion, jaundice, and hemorrhage. Moreover, the first described spirochete was found in the liver of a guinea pig injected with the blood of a patient suffering from Weil's disease [11]. Since then, susceptible animals, mainly hamsters, gerbils, and guinea pigs, have extensively been used to study acute lethal leptospirosis, whereas resistant mice and rats recapitulate sub-lethal leptospirosis and subsequent chronic renal colonization [4].

The hamster model has been described as a model of choice for accidental *Leptospira* infections, as are human infections, especially for the development of leptospiral vaccine and antibiotic treatment studies, for which data describing the host-*Leptospira* interactions are essential [12]. The latter is our topic of interest in the laboratory, and we aim for this chapter to describe a typical experimental infection procedure on golden Syrian hamsters we rely on for characterizing the host immune response to a *Leptospira* infection and investigating its role in the outcome of the disease [13]. Indeed, results from experiments on intraperitoneally infected hamsters notably suggest that a dramatic cytokine production imbalance might be involved in the development of severe forms of leptospirosis in susceptible hosts [14–16]. We will also describe the collection and conservation methods of blood and organs from infected and matching control animals destined to be submitted to total DNA or RNA extraction. Indeed, we will describe quantitative methods for assessing (1) bacterial burden, through the detection of leptospiral genes in total extracted DNA, and (2) host immune response by measuring the expression of immune mediators in total extracted RNA. In both cases, we will provide details on our use of quantitative PCR (qPCR), which has become the standard molecular tool for quantification purposes due to its high sensitivity. Moreover, this technique has proven particularly instrumental for animal models, like golden Syrian hamsters, lacking the immunological tools necessary for protein quantification. It should be noted that the relevance and availability of leptospirosis models is variable in many countries affected by leptospirosis, such as Australia. Importantly, all methods presented here on the hamster model can directly be transposed to the mouse model, using specific mouse qPCR primers.

2   Materials

2.1   Animals

1. Golden Syrian hamsters (*Mesocricetus auratus*): 6- to 8-week-old, male and female hamsters are used for infection experiments (*see* **Notes 1** and **2**). Body mass of laboratory hamsters between 6 and 8 weeks might be approximately 100 ± 10 g [17]. Hamsters are housed in cages with 1–5 animals per cage, in the animal house facility with controlled temperature and light conditions (ambient temperature, 20–24 °C; relative humidity, 55–65%; semi-artificial light-dark cycle of 12 L:12 D hours), except during experimental procedures. Food and water are provided ad libitum, and sunflower or millet seeds are distributed as part of their environmental enrichment, at the time of cage cleaning and/or daily check during an experiment.

2.2   Leptospira Culture

1. *Leptospira interrogans* serovar Icterohaemorrhagiae strain Verdun (LiVV; *see* **Note 3**): Leptospires are cultivated in glass tubes with screw cap filled with liquid Ellinghausen, McCullough, Johnson, and Harris (EMJH) medium at 30 °C under aerobic conditions (with loosely closed screw cap for tube aeration) and possibly with shaking to boost the bacterial growth if required. Cultures used for experimental infections should be in the course of their exponential phase of growth. Bacterial virulence is regularly maintained by intraperitoneally injecting hamsters and isolating leptospires from infected animals' blood.

2. EMJH basal salt solution: Dissolve 2.3 g of Difco™ Leptospira Medium Base EMJH (Becton Dickinson) in 900 mL of distilled or deionized water and autoclave.

3. Albumin supplement (for 1 L): Mix together 100 g of bovine serum albumin, 10 mL of $ZnSO_4$ (stock 0.4 g/100 mL of $ZnSO_4 \cdot 7H_2O$), 10 mL of $MgCl_2$ (stock 1.5 g/100 mL of $MgCl_2 \cdot 6H_2O$), 10 mL of $CaCl_2$ (stock 1.5 g/100 mL of $CaCl_2 \cdot 2H_2O$), 10 mL of sodium pyruvate (stock 10 g/100 mL of sodium pyruvate), 20 mL of bi-distilled glycerin (stock 20 g/100 mL of bi-distilled glycerin), 125 mL of Tween 80 (11.58 g of Tween 80 in sterile water, final volume of 125 mL), 1 mL of $CuSO_4$ (stock 0.3 g/100 mL of $CuSO_4 \cdot 5H_2O$), 100 mL of $FeSO_4$ (0.5 g of $FeSO_4 \cdot 7H_2O$ in 100 mL of sterile water), and 4 mL of 500 µg/mL vitamin

B12, and adjust pH to 7.4. Stock solutions except $FeSO_4$ and Tween 80 can be prepared in advance and stored at 4 °C for up to a year. $FeSO_4$ and Tween 80 solutions should be prepared immediately prior to albumin supplement preparation. Albumin supplement can be prepared in advance, aliquoted, and stored between −20 °C and + 4 °C for up to 18 months (*see* **Note 4**).

4. Complete EMJH medium: 900 mL of EMJH basal salt solution, 100 mL of albumin supplement, pH 7.4. Directly after preparation, required volumes of complete EMJH medium are distributed in glass tubes with screw cap that can be stored at 4 °C up to 6 months.

2.3 Experimental Infections

1. Hamsters in individual cages.

2. Petroff-Hausser counting chamber (0.02 mm depth) or similar counting chamber.

3. Infectious injectate: LiVV at the adequate concentration and volume (*see* Subheading 3.1).

4. Control injectate: Liquid EMJH alone (same volume with the infectious injectate).

5. 1 mL syringes equipped with 25G needle for injection.

6. 70% ethanol dispenser.

7. Gauze.

8. Sharps containers for biohazard material.

2.4 Blood and Tissue Collection

1. LiVV-infected and non-infected control hamsters.

2. Anesthesia/euthanasia station: CO_2 tank, with pressure-reducing valve, liter flow meter, and CO_2 chamber with lid.

3. Dissecting tools: scissors, tweezers, scalpel, table, pins.

4. Beaker with 70% ethanol for decontamination of dissecting tools.

5. 1 mL syringes and 25G needles (0.5 mm × 16 mm) for cardiac puncture.

6. Sharps containers for biohazard material.

7. 70% ethanol dispenser.

8. MagNA Lyser Green Beads tubes (Roche Life Science). For DNA extraction from organs, tubes should be filled with 360 µL of lysis buffer from QIAamp DNA Mini Kit (Qiagen; *see* Subheading 2.5) prior to tissue collection.

9. EDTA blood collector tubes.

10. RNAlater® Buffer (Qiagen).

11. PAXgene buffer from PAXgene Blood RNA tube (PreAnalytix, Qiagen).

2.5 Quantification of Bacterial Burden

1. Blood (stabilized in EDTA collector tubes) and organs (prepared in MagNA Lyser Green Beads tubes) collected from LiVV-infected and control hamsters.

2. MagNA Lyser Instrument.

3. QIAamp Blood Kit (Qiagen).

4. QIAamp DNA Mini Kit (Qiagen).

5. LightCycler FastStart DNA Master SYBR Green I (Roche Life Sciences).

6. Spectrophotometer for DNA and RNA quantification.

2.6 Quantification of Cytokine Gene Expression

1. Blood (stored in PAXgene RNA tubes) and organs (stored in RNAlater® reagent) from LiVV-infected and control hamsters.

2. High Pure tissue RNA Isolation Kit (Roche Life Science).

3. Transcriptor First Strand cDNA Synthesis Kit (Roche Life Science).

4. TURBO™ DNase (Thermo Fisher Scientific).

5. LightCycler FastStart DNA Master SYBR Green I (Roche Life Science).

6. Spectrophotometer for DNA and RNA quantification.

3 Methods

Carry out experiments at room temperature unless otherwise specified.

3.1 Experimental Infections

1. At 24 h prior to animal infections, house hamsters in individual cages for proper acclimatization (*see* **Note 5**). If animals were purchased from external suppliers, please consider longer acclimation period of time (>48 h) before experimental infection. All cages must be fitted with an individual ventilation system or kept in a ventilated storage unit, in a procedure room, separated from the animal maintenance facility. All animals should be carefully examined and weighted prior to *Leptospira* challenge.

2. Evaluate the density of the LiVV culture by dark-field microscopy and using a Petroff-Hausser or similar counting chamber (*see* **Note 6**).

3. Prepare the appropriate dose of leptospires that will be used for the animal infections. From previous published results, we know, for instance, that the LD50 is 10^8 leptospires per animal when using the hamster model and the LiVV strain (*see* **Note 7**). EMJH liquid medium is used to dilute the initial culture as well as control injectate for uninfected animals unless otherwise

needed (*see* **Note 8**). The volume that will be injected should not exceed 1 mL in total for each hamster.

4. Fill 1 mL syringes equipped with 25G needles (0.5 mm × 16 mm) with the volume that will be injected to animals, whether bacterial dose or control injectate. Do not lock the needles' caps back to avoid accidental infection.

5. Manually restrain the animals, exposing the abdomen and directing the head downward (*see* **Note 5** and [18]).

6. Spread the animals' abdomen with 70% ethanol for disinfection.

7. Inject the hamsters intraperitoneally with the appropriate dose of leptospires or the same volume of control injectate (*see* **Note 5**). Needle should be gently inserted into the left or right lower quadrant, avoiding the abdomen midline, and injections should preferably be done at once (*see* **Note 9**).

8. Following injections, hamsters should be weighted daily and examined twice per day for the onset of clinical signs and reaching of endpoint criteria (*see* **Note 10**). From that stage on, the entire content of infected animal cages, as well as their uninfected counterparts for safety, should be considered as contaminated since infected animals shed virulent bacteria in their urine.

3.2 Blood and Tissue Collection

1. LiVV-infected and uninfected control hamsters are subjected to a deep anesthesia through CO_2 inhalation. This procedure is considered as a deep and final anesthesia since animals will not recover from it. However blood collection is easier when the heart is still beating. Consequently, we consider that anesthesia is deep enough when breathing rate is lower than 3–5/min and animals no longer respond to pain stimuli (usually assessed through toe pinch).

2. Place deeply unconscious hamsters on their back, and pin them down on a dissecting table. Animals should be entirely disinfected with 70% ethanol and dissected in a laminar flow biological safety cabinet (*see* **Note 3**). Please note that the following next steps give storage conditions for tissue and blood samples according to the use they are destined to, especially for *Leptospira* burden quantification and cytokine gene expression, and are summed up in Table 1.

3. Collect blood through cardiac puncture [19] using 1 mL syringes with 25G needles (*see* **Note 9**), and immediately process and store:

   (a) In EDTA collector tubes that will be stored at 4 °C until DNA extraction.

Table 1
Blood and tissue sample storage conditions

		Total DNA extraction	Total RNA extraction
Blood	Storage material	EDTA collector tubes	PAXgene collector tubes
	Storage temperature	4 °C until DNA extraction, −20 °C for long-term storage	2 h at room temperature, followed by long-term storage at −20 °C until total RNA extraction
Tissue	Storage material	25 mg samples into MagNA Lyser green beads tubes containing 360 μL of QIAamp lysis buffer	Tissue pieces (~1 cm³) in 1.5 mL RNAlater reagent
	Storage temperature	4 °C until DNA extraction, −20 °C for long-term storage	2 h at room temperature, followed by long-term storage at −20 °C until total RNA extraction

Summary of stabilization reagents and storage conditions followed for organs and blood collected from LiVV-infected and uninfected control animals

(b) In PAXgene collector tubes for 2 h at room temperature to allow for RNA stabilization, and then store at −20 °C until total RNA extraction.

4. Cut the skin open with a set of scissors and tweezers from the throat to the rectum. Open the chest and peritoneal cavities with a scalpel.

5. As for animals' blood, any organ of interest can also be harvested, especially the liver and kidneys, and stored for further analysis:

   (a) For total DNA extraction, 25 mg samples should be placed into MagNA Lyser Green Beads tubes containing 360 μL of lysis buffer from DNA extraction kit and stored at 4 °C until DNA extraction.

   (b) For total RNA extraction, tissue samples (pieces of ~1 cm³) are placed in 1.5 mL RNAlater reagent for stabilization of total RNA at room temperature for 2 h before conservation at −20 °C until RNA extraction.

3.3 Leptospira Burden Quantification

3.3.1 Total DNA Extraction from Blood

1. Total DNA is extracted from blood samples kept in EDTA collector tubes, using the QIAamp Blood Kit according to manufacturer's instructions.

2. After washing steps, the total DNA is eluted in elution buffer or water.

3. Quantify total eluted DNA concentration by spectrophotometry.

3.3.2 Total DNA Extraction from Tissue Sample

1. Tissue samples in MagNA Lyser Green Beads tubes are first disrupted and homogenized using the MagNA Lyser Instrument during 50 s at $4000 \times g$.

2. DNA extraction is performed with QIAamp DNA Mini Kit according to manufacturer's instructions. After washing steps, the total DNA is eluted in elution buffer or water.

3. Quantify total eluted DNA concentration by spectrophotometry.

3.3.3 Bacterial Load Quantification by qPCR

1. Bacterial DNA in hamsters total eluted DNA from previous steps is analyzed by quantitative PCR (qPCR) assay (*see* **Note 11**). Primer sequences are provided in Table 2. Even though qPCR parameters should be adjusted depending on the instrument and reagents, they should include the following steps: 95 °C, 5–8 s (denaturation step); 58–62 °C, 5–12 s (annealing step); 72 °C, 10–12 s (elongation steps); 40–50 cycles.

2. Genomic DNA extracted from the *Leptospira* challenge strain (here LiVV) is used as a positive control. A standard curve obtained from serial tenfold dilutions of known numbers of leptospires is used for absolute quantification of *Leptospira* genome found in tissue samples from infected animals.

Table 2
Sequence of primers used for qPCR assays

Gene name	Primer sequence (5′ → 3′)	Amplicon size (bp)
Leptospira LipL32	(F) CGGACGGTTTAGTCGATG (R) GCATAATCGCCGACATTC	228
Hamster β-actin	(F) TCTACAACGAGCTGCG (R) CAATTTCCCTCTCGGC	357
Hamster GAPDH	(F) CCGAGTATGTTGTGGAGTCTA (R) GCTGACAATCTTGAGGGA	170
Hamster IL-1β	(F) ATCTTCTGTGACTCCTGG (R) GGTTTATGTTCTGTCCGT	156
Hamster TNF-α	(F) AACGGCATGTCTCTCAA (R) AGTCGGTCACCTTTCT	278
Hamster IL-6	(F) AGACAAAGCCAGAGTCATT (R) TCGGTATGCTAAGGCACAG	252
Hamster IL-10	(F) TGGACAACATACTACTCACTG (R) GATGTCAAATTCATTCATGGC	308
Hamster MIP-1α/CCL3	(F) CTCCTGCTGCTTCTTCTA (R) TGGGTTCCTCACTGACTC	210
Hamster IP-10/CXCL10	(F) CTCTACTAAGAGCTGGTCC (R) CTAACACACTTTAAGGTGGG	150

3. Results are then expressed as the number of *Leptospira* genome equivalents per μg of DNA extracted from tissue samples or blood (*see* **Note 12**).

3.4 Cytokine Gene Expression Analysis

All steps should be carried out using a refrigerate block or on ice to avoid RNA degradation, except otherwise cited.

3.4.1 Total RNA Extraction from Blood

1. Total RNA is purified from blood collected as described in Subheading 3.2 and stored in PAXgene collector tubes using the PAXgene blood RNA system according to manufacturer's instructions.

2. Following the washing steps, RNA is eluted in 80 μL elution buffer, and 40 μL of RNA eluate is treated for 30 min at 37 °C with TURBO DNase (2 units) for elimination of residual genomic DNA.

3. Incubate the DNase-treated RNA at 65 °C for 10 min for complete secondary and tertiary structure denaturation.

4. Before long-term storage at −80 °C, purified RNA is quantified by spectrophotometry through measurement of the optical density (OD) at 260 nm.

3.4.2 Total RNA Extraction from Organs

1. Prior to RNA extraction from organs collected as advised in Subheading 3.2, tissue samples kept in RNAlater reagent are first disrupted and homogenized using the MagNA Lyser Instrument during 50 s at 4000 × *g* and then collected after centrifugation for 2 min at 18,000 × *g*.

2. Total RNA is extracted from tissue lysates using the High Pure tissue RNA Isolation Kit.

3. Following the washing steps, RNA is eluted in 80 μL elution buffer, and 40 μL of RNA eluate is treated for 30 min at 37 °C with TURBO DNase (2 units) for elimination of residual genomic DNA.

4. Incubate the DNase-treated RNA at 65 °C for 10 min for complete secondary and tertiary structure denaturation.

5. Before long-term storage at −80 °C, purified RNA is quantified by spectrophotometry through measurement of the optical density (OD) at 260 nm.

3.4.3 RT-qPCR on Total RNA from Blood and Organs

1. Approximately 1 μg of total extracted RNA from blood and tissue from infected and control animals are retro-transcribed into cDNA using the Transcriptor First Strand cDNA synthesis kit. Program for cDNA synthesis is as follows: activation step, 25 °C, 10 min; RT step, 55 °C, 30 min; and inactivation of enzyme, 85 °C, 5 min. cDNA can be conserved between +4 °C and −20 °C until qPCR assays.

2. cDNA from previous steps are analyzed by qPCR assay (*see* Subheading 3.3.3 and **Note 11**). Primer sequences for the main immune mediators targeted in our experiments are provided in Table 2.

3. Gene expression results are expressed as relative normalized target gene expression (e.g., of immune mediators). Values can be calculated using the ratio of the level of expression in infected animals to the level of expression in uninfected animals used as calibrators (*see* **Notes 12 and 13**).

4   Notes

1. All animal experiments are prepared and conducted according to the guidelines of the Animal Care and Use Committees of the Institut Pasteur, follow ethical and regulatory standards of the European Union Legislation governing the care and use of laboratory animals (Directive 2010/63/EU), and are performed by an authorized trained person.

2. Although it was suggested that men are more prone to develop severe forms of leptospirosis and be hospitalized, studies comparing data from male and female patients with the same level of exposure show no difference in the severity of the illness [20–22]. However, the impact of sex on the severity of leptospirosis remains an open question, and, in two recent studies comparing experimentally infected male and female hamsters, the authors conclude that males are more susceptible to infection than females [23, 24]. We thus recommend matching of animals by gender.

3. Pathogenic *Leptospira* species are Biosafety Level 2 (BSL-2) pathogens, and *L. interrogans* is described as pathogenic for humans. Ensure that all experiments involving viable bacteria are completed in a containment level 2 laboratory and that appropriate risk assessments are in place. Cultures and infected materials must be handled in a laminar flow biological safety cabinet. Any material that has been in contact with infected animals is considered as contaminated and should be disinfected with bleach or autoclaved for appropriate waste disposal.

4. Alternately, a commercial enrichment additive, Difco™ Leptospira Enrichment EMJH (Becton Dickenson), can be purchased to supplement the EMJH medium.

5. Personnel involved in the procedure should be trained for proper hamsters handling and restraint, to avoid injury and/or contamination. Indeed, hamsters are generally calm and docile animals, but they may bite or scratch if they are stressed or surprised. For handling, always approach hamsters

from behind and place a hand palm down over the animal, preferably thumb over the neck. During leptospires injection procedures, make sure hamsters are secure while their abdomen exposed by grasping enough loose skin over the neck and back. Please *see* [25] for additional information on hamsters handling and restraint.

6. The chamber is used according to manufacturer's instructions. After completing the count, remove the reusable cover glass, and clean the counting chamber and the cover glass with ethanol. Dry the counting chamber with a soft cloth or wipe.

7. Previous results established that a dose of 10^8 leptospires from the LiVV strain per intraperitoneally infected hamster was a relevant and reproducible model of median lethal dose (LD50) [16]. However, infectious dose may vary a lot depending on the bacterial strain and the infection route that are being used for experimental procedures.

8. Alternatively, leptospires can be centrifuged at $3400 \times g$ for 10 min and resuspended in PBS. In that case, PBS alone will be injected in control non-infected animals.

9. Following use, syringes should be promptly disposed in the sharps container. Needles should not be bent, sheared, replaced in the needle sheath or guard, or removed from the syringe following use. Sharps containers must be disposed as hazardous waste and properly treated prior to final disposal.

10. Examination of challenged hamsters relies on the scoring of clinical signs of illness, mostly weight loss, low alertness, ruffled head and fur, hunched posture, prostration, and small eyes. In case of lethal challenges, death can closely follow the onset of clinical signs; thus endpoint criteria should be carefully defined and should include anorexia, dehydration, and weight loss of >10% of initial weight (*see* **Note 1** and [26]). However these endpoint criteria largely depend on the infecting *Leptospira* strain. In the case of an infection with the LiVV strain, lethally infected animals usually die at day 4 post-infection, without any clear clinical signs except for prostration that can be observed a few hours before death. In any case, moribund animals or survivors checked for sub-lethal infections are euthanized through CO_2 asphyxia.

11. qPCR assays are performed on a LightCycler 480 real-time PCR system using the LightCycler FastStart DNA Master SYBR Green I and the LightCycler 480 software (version 1.5.0) according to manufacturer's instructions and as previously described [14]. Acquisition of fluorescence for threshold cycle (C_T or Crossing Point C_p) calculation is processed during elongation step of the qPCR, each qPCR reaction being performed twice (for experimental replicates). The specificity of

amplification is confirmed by analysis of the melting curves of the PCR products. Results are validated only when C_T values are under the limit value of 40 cycles and with an acceptable reproducibility between qPCR replicates (<5% of variation).

12. Statistical analysis is performed using appropriate software (e.g., GraphPad Prism).

13. The level of expression of each target gene is normalized to the levels of glyceraldehyde-3-phosphate dehydrogenase (GAPDH) and β-actin as housekeeping (HKG) genes, using the qbase+ software and considering gene specific amplification efficiencies (AEs). AEs are determined by generating standard curves from serial dilutions of a known concentration of purified DNA specific for the target of each reference gene.

References

1. Costa F et al (2015) Global morbidity and mortality of leptospirosis: a systematic review. PLoS Negl Trop Dis 9(9):e0003898

2. Adler B, de la Pena Moctezuma A (2010) Leptospira and leptospirosis. Vet Microbiol 140 (3-4):287–296

3. Haake DA, Levett PN (2015) Leptospirosis in humans. Curr Top Microbiol Immunol 387:65–97

4. Gomes-Solecki M, Santecchia I, Werts C (2017) Animal models of leptospirosis: of mice and hamsters. Front Immunol 8:58

5. Levett PN (2001) Leptospirosis. Clin Microbiol Rev 14(2):296–326

6. Naing C, Reid SA, Aung K (2017) Comparing antibiotic treatment for leptospirosis using network meta-analysis: a tutorial. BMC Infect Dis 17(1):29

7. Vernel-Pauillac F, Werts C (2018) Recent findings related to immune responses against leptospirosis and novel strategies to prevent infection. Microbes Infect 20(9–10):578–588

8. Picardeau M (2017) Virulence of the zoonotic agent of leptospirosis: still terra incognita? Nat Rev Microbiol 15(5):297–307

9. Werts C (2018) Interaction of Leptospira with the innate immune system. Curr Top Microbiol Immunol 415:163–187

10. Kobayashi Y (2001) Discovery of the causative organism of Weil's disease: historical view. J Infect Chemother 7(1):10–15

11. Inada R et al (1916) The etiology, mode of infection, and specific therapy of weil's disease (spirochaetosis icterohaemorrhagica). J Exp Med 23(3):377–402

12. Haake DA (2006) Hamster model of leptospirosis. Curr Protoc Microbiol. Chapter 12: p. Unit 12E 2

13. Cagliero J, Villanueva SYAM, Matsui M (2018) Leptospirosis pathophysiology: into the storm of cytokines. Front Cell Infect Microbiol 8:204

14. Matsui M et al (2011) Gene expression profiles of immune mediators and histopathological findings in animal models of leptospirosis: comparison between susceptible hamsters and resistant mice. Infect Immun 79 (11):4480–4492

15. Matsui M et al (2017) High level of IL-10 expression in the blood of animal models possibly relates to resistance against leptospirosis. Cytokine 96:144–151

16. Vernel-Pauillac F, Goarant C (2010) Differential cytokine gene expression according to outcome in a hamster model of leptospirosis. PLoS Negl Trop Dis 4(1):e582

17. Gattermann R et al (2002) Comparative studies of body mass, body measurements and organ weights of wild-derived and laboratory golden hamsters (*Mesocricetus auratus*). Lab Anim 36(4):445–454

18. Donovan J, Brown P (2006) Handling and restraint. Curr Protoc Immunol. Chapter 1: p. Unit 1 3

19. Donovan J, Brown P (2006) Blood collection. Curr Protoc Immunol. Chapter 1: p. Unit 1 7

20. Morgan J et al (2002) Outbreak of leptospirosis among triathlon participants and community residents in Springfield, Illinois, 1998. Clin Infect Dis 34(12):1593–1599

21. Van CT et al (1998) Human leptospirosis in the Mekong delta, Viet Nam. Trans R Soc Trop Med Hyg 92(6):625–628

22. Leal-Castellanos CB et al (2003) Risk factors and the prevalence of leptospirosis infection in a rural community of Chiapas. Mexico Epidemiol Infect 131(3):1149–1156

23. Gomes CK et al (2018) Lethality of leptospirosis depends on sex: male hamsters succumb to infection with lower doses of pathogenic Leptospira. BioRxiv 322784, https://doi.org/10.1101/322784

24. Tomizawa R et al (2017) Male-specific pulmonary hemorrhage and cytokine gene expression in golden hamster in early-phase *Leptospira interrogans* serovar Hebdomadis infection. Microb Pathog 111:33–40

25. Donovan J, Brown P (2006) Handling and restraint. Curr Protoc Immunol;Chapter 1: Unit 1.3

26. Lourdault K et al (2014) Oral immunization with *Escherichia coli* expressing a lipidated form of LigA protects hamsters against challenge with *Leptospira interrogans* serovar Copenhageni. Infect Immun 82(2):893–902

Chapter 23

Evaluation of Vaccine Candidates against Leptospirosis using Golden Syrian Hamsters

Karen V. Evangelista and Kristel Lourdault

Abstract

Leptospirosis is a major public health problem, especially in developing countries. Current vaccine studies focus on identifying *Leptospira* proteins that elicit protective immunity. Here, we describe a method to assess recombinant proteins for their ability to protect hamsters from fatal infection against *Leptospira* and to provide sterilizing immunity.

Key words *Leptospira interrogans*, Leptospirosis, Vaccines, Hamsters, Recombinant protein, ELISA, Immunization

1 Introduction

Leptospirosis, caused by pathogenic strains of *Leptospira*, is the most widespread zoonosis in the world, affecting both humans and animals. Humans are infected either by direct contact with an infected animal's urine or indirectly through contaminated water. Leptospiral infection can lead to a broad range of symptoms from flu-like febrile illness to multi-organ complications characterized by jaundice, acute hepatic and renal dysfunction, and lung hemorrhage. There are more than one million severe human cases of leptospirosis every year, and up to 10% are lethal [1].

Effective methods to prevent leptospirosis infection include avoidance of high-risk activities such as wading in flood and contaminated waters, wearing protective equipment such as boots if exposure is unavoidable, and vaccination against the bacteria [2]. Currently, no leptospirosis vaccine is widely available for humans. Most vaccines on the market are for veterinary use and are either heat-killed or formalin-killed leptospires. They primarily target lipopolysaccharide, whose structure differs among the more than 300 serovars of *Leptospira* [3]. These vaccines provide

Nobuo Koizumi and Mathieu Picardeau (eds.), *Leptospira spp.: Methods and Protocols*, Methods in Molecular Biology, vol. 2134, https://doi.org/10.1007/978-1-0716-0459-5_23, © Springer Science+Business Media, LLC, part of Springer Nature 2020

short-term immunity and limited cross-protection against heterologous serovars of pathogenic leptospires [4, 5].

With the availability of whole-genome sequences from multiple strains of *Leptospira* [6–8], current vaccine research has focused on identifying conserved leptospiral proteins, which have the potential of eliciting cross-protective immunity. Surface-exposed outer membrane proteins that are recognized by the host immune response early during the infection and play a role in bacterial virulence are ideal targets for vaccine development.

Vaccine studies require the use of animals. Hamsters are one of the most common animal models of leptospirosis because they are highly susceptible to leptospiral infection and exhibit clinical symptoms that mimic severe human leptospirosis [9]. Animal experiments necessitate the use of low in vitro passage, highly virulent strains of *Leptospira* spp., and the preliminary determination of the median endpoint dose (ED_{50}), i.e. the number of bacteria with which 50% of animals meet the endpoint criteria. The ED_{50} depends on the serovar and strain used, the number of in vitro passages, and the animal model. A valid challenge requires that more than 80% of the unvaccinated control animals succumb to lethal infection [10], but the goal is that all animals in the control group reach the endpoint criteria.

Here we describe how a leptospiral protein is tested as vaccine candidate using golden Syrian hamsters. The coding sequences of the protein lacking the signal peptide are amplified from the *Leptospira* genome, cloned, and expressed as a recombinant protein, often with a small tag used for affinity purification of the protein. The purified protein (antigen) is mixed with an adjuvant and injected into hamsters. Animal serum is collected prior to and during the immunization protocol to measure the humoral immune response generated by the candidate protein using enzyme-linked immunosorbent assay (ELISA). Kidney colonization in survivors is assessed by in vitro culture in EMJH medium and quantitative PCR. An ideal subunit vaccine must have the ability to protect hamsters from lethal infection and provide sterilizing immunity.

Using these methods, we were able to show that hamsters immunized with a C-terminal fragment (domains 7–13) of leptospiral immunoglobulin (Ig)-like protein A (LigA), an outer membrane lipoprotein, survived intraperitoneal bacterial challenge with *L. interrogans* (Fig. 1) and generated a humoral immune response against LigA (Fig. 2). However, LigA failed to protect against kidney colonization by *L. interrogans* [11].

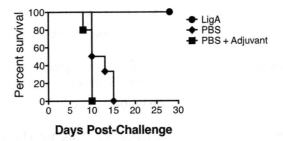

Fig. 1 Survival of LigA-immunized hamsters after lethal challenge. Groups of hamsters were immunized subcutaneously three times with LigA (100 μg), PBS, or PBS with adjuvant. Animals were challenged intraperitoneally with 1×10^4 *L. interrogans* serovar Copenhageni strain Fiocruz L1-130 at day 0 and were monitored daily for 28 days [12]

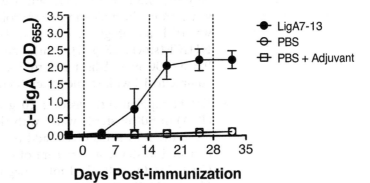

Fig. 2 IgG response to immunization with purified recombinant LigA protein. Hamster sera were collected weekly during the immunization protocol. Anti-LigA antibody levels (IgG) were measured in triplicate by ELISA. For each sample, the IgG level is read at OD_{655}, and the pre-immune read is subtracted to it. Each data point represents the mean of IgG response of groups of hamsters immunized with LigA, PBS, or PBS with adjuvant at different time points. Error bars indicate standard deviation. Vertical dotted lines show the immunization days

2   Materials

2.1 Vaccine Preparation

1. Phosphate-buffered saline (PBS): 1.5 mM KH_2PO_4, 155.2 mM NaCl, 2.7 mM $Na_2HPO_4 \cdot 7H_2O$, and pH 7.2. Add about 100 ml water to a glass beaker. Weigh 210 mg of KH_2PO_4, 9 g of NaCl, and 726 mg of $Na_2HPO_4 \cdot 7H_2O$ and transfer to the beaker. Add water to bring volume to 1 l. Mix and adjust pH to 7.2 with HCl. Autoclave and then store at room temperature (RT).

2. 0.1–1 mg/ml purified leptospiral protein in PBS (*see* **Note 1**).

3. Complete Freund's adjuvant (CFA) (*see* **Notes 2** and **3**).

4. Incomplete Freund's adjuvant (IFA).

5. 5 ml glass syringes.

6. Three-way plastic stopcock with Luer lock.

7. 19 G needles.

8. Pipettors and sterile barrier pipette tips.

2.2 Animals

1. Female golden Syrian hamsters (*Mesocricetus auratus*), 4 weeks old (*see* **Note 4**).

2. Precision balance.

2.3 Bacterial Strain

1. *Leptospira* culture for challenge: low in vitro passage pathogenic strain. For example, *L. interrogans* serovar Copenhageni strain Fiocruz L1-130 [12].

2. Ellinghausen-McCullough-Johnson-Harris (EMJH) basal medium [13, 14]: Use sterile double-deionized water in all media and media component preparation. Dissolve 1 g of Na_2PO_4, 0.3 g of KH_2PO_4, and 1 g of NaCl in 900 ml sterile water. Add the following stock solutions prepared in sterile water: 1 ml of glycerol (stock, 10 ml/100 ml), 1 ml of NH_4Cl (stock, 25 g/100 ml), and 1 ml of thiamine (stock, 0.5 g/100 ml). Adjust pH to 7.4 using NaOH then bring volume to 1 l with sterile water. Sterilize by autoclaving.

3. EMJH supplement: Dissolve 10 g of bovine serum albumin (BSA) in 50-ml sterile water with slow and constant stirring. Once the BSA is dissolved, add the following stock solutions prepared in sterile water: 1 ml of calcium chloride (stock, 1 g $CaCl_2 \cdot 2H_2O$/100 ml), 1 ml of magnesium chloride (stock, 1 g $MgCl_2 \cdot 2H_2O$/100 ml), 1 ml of zinc sulfate (stock, 0.4 g of $ZnSO_4 \cdot 7H_2O$/100 ml), 0.1 ml of copper sulfate (stock, 0.3 g $CuSO_4 \cdot 5H_2O$/100 ml), 10 ml of ferrous sulfate (stock, 0.5 g $FeSO_4 \cdot 7H_2O$/100 ml), 1 ml of vitamin B_{12} (stock, 0.02 g/100 ml), and 12.5 ml of Tween 80 (stock, 10 g/100 ml). Adjust pH to 7.4 using NaOH and then bring volume to 100 ml with sterile water. Sterilize by filtration through a 0.22 μm filter.

4. EMJH liquid medium: Add 100 ml of EMJH supplement to 890 ml of EMJH basal medium. Add 1 g of lactalbumin hydrolysate, 1 mg of superoxide dismutase (SOD), 40 mg of sodium pyruvate, and 10 ml of heat-inactivated (56 °C for 30 min) rabbit serum. Adjust pH to 7.4 if necessary and filter sterilize (0.22 μm filter). Aliquot into smaller volumes and store at 4 °C.

2.4 Immunization and Challenge

1. 1 ml syringes with 25 G needles.

2. 1 ml insulin syringes U-100 with 26 G x 1/2″ needles.

3. 1.5 ml microcentrifuge tubes.

4. Refrigerated centrifuge.

2.5 Blood and Tissue Collection

1. 100–200 μl capillary tubes.

2. Ophthalmic ointment and cotton balls for post-bleeding hemostasis.

3. Precision balance.

4. Isoflurane and isoflurane vaporizer.

5. Sterile scalpels, scissors, and forceps.

6. 5 ml syringes with 25 G × 5/8″ needle.

7. 5 ml EDTA tubes (BD Vacutainer).

2.6 ELISA

1. 96-well microtiter plates.

2. Purified protein (diluted to 1 μg/100 μl PBS).

3. Multichannel pipettors or 96 well automated washer.

4. Protein-free blocking buffer (PFB) (Pierce).

5. Horse radish peroxidase (HRP)-conjugated goat anti-Syrian hamster immunoglobulin G (IgG) diluted in PFB (1:5000).

6. 3,3′,5,5′-tetramethylbenzidine (TMB) substrate solution.

7. ELISA plate reader.

8. Pipettors and sterile barrier pipette tips.

2.7 Tissue Colonization

1. 15 ml tube with 10 ml semisolid EMJH supplemented with 100 μg/ml 5-fluorouracil (5-FU): Add 1.5 g of Noble agar to 880 ml EMJH basal medium and autoclave. Cool to ~50 °C and then add 100 ml of EMJH supplement, 1 g of lactalbumin hydrolysate, 1 mg of SOD, 40 mg of sodium pyruvate, 10 ml of heat-inactivated rabbit serum, and 10 ml of 0.01 g/ml 5-FU, sterilized through a 0.22 μm filter. Prepare 10 ml aliquots in 15 ml tubes and store at 4 °C.

2. Dark-field microscope.

3. Tissue homogenizer/disruptor.

4. DNeasy Blood and Tissue Kit (Qiagen).

5. Spectrophotometer.

6. qPCR master mix (Bio-Rad iTaq Universal Probes Supermix).

7. Primers at the concentration of 10 μM and probe at the concentration of 5 μM [15].

Name	Oligonucleotides (5′ → 3′)
LipL32-45F	AAG CAT TAC CGC TTG TGG TG
LipL32-268F	GAA CTC CCA TTT CAG CGA TT
LipL32-189P	FAM AA AGC CAG GAC AAG CGC CG BHQ1

8. Quantitative thermal cycler.

3   Methods

All steps involving *Leptospira* must be performed in a biosafety cabinet under BSL-2 containment. All animal experiments must be approved by the local Institutional Animal Care and Use Committee and performed under Animal Biosafety Level 2 (ABSL-2) conditions.

3.1   Vaccine Preparation

1. Draw up 2 ml of CFA (or IFA) into a 5 ml glass syringe with a 19 G needle (*see* **Note 5**).

2. In another 5 ml glass syringe, draw up 2 ml of purified protein (0.1–1 mg/ml). For the negative control group, use 2 ml of PBS.

3. Remove the 19 G needles and connect the two syringes to a plastic three-way Luer lock.

4. Mix the CFA (or IFA) and the purified protein by plunging the solutions into opposite syringe 15–20 times to create a white emulsion (*see* **Note 6**). For the control group, follow the same protocol with PBS.

5. Transfer all of the protein-adjuvant or PBS-adjuvant emulsion into one syringe and remove the stopcock. Remove the air bubbles and transfer the emulsion to a clean glass tube.

3.2   Hamster Immunization

1. Three days prior to the first immunization, collect the blood from hamsters by retro-orbital bleed using a capillary tube (*see* **Note 7**). Immediately transfer the blood to a microcentrifuge tube. Process the blood as described in Subheading 3.4.

2. On day 0, inject hamsters subcutaneously with 200 μl of antigen-CFA emulsion (10–100 μg protein) using a 1 ml syringe with a 25 G needle (*see* **Note 8**). For the negative control group, use 200 μl of PBS or PBS-CFA emulsion.

3. On day 14 and 28, inject hamsters subcutaneously with 200 μl of antigen-IFA emulsion (10–100 μg protein) using 1 ml syringe with a 25 G needle. For the negative control groups, inject 200 μl of PBS or PBS-IFA emulsion.

4. Starting on day 4 until day 32, collect the blood weekly from hamsters by retro-orbital bleed. Transfer the blood to a microcentrifuge tube. Process the blood as described in Subheading 3.4.

3.3   Challenge and Evaluation of the Protection

1. Determine the culture density of the *Leptospira* challenge strain (*see* **Note 9**) by dark-field microscopy using a Petroff-Hausser chamber (*see* **Note 10**).

2. Prepare appropriate dilution of the bacteria in liquid EMJH based on the ED_{50}. For *L. interrogans* serovar Copenhageni strain Fiocruz L1-130, prepare a suspension with a density of

1×10^3 to 10^4 bacteria/ml (50–500 times ED_{50}) (*see* **Note 11**).

3. On day 35 (1 week after the last immunization), challenge hamsters intraperitoneally with 1 ml of *Leptospira* suspension using an insulin syringe U-100 with a 26 G $\times$ ½" needle (*see* **Note 12**).

4. Weigh animals daily and monitor them twice a day after challenge for endpoint criteria, i.e., loss of appetite, gait or breathing difficulty, prostration, ruffled fur, or weight loss of $\geq$10% of the animal's maximum weight.

5. Animals that meet endpoint criteria are euthanized by isoflurane inhalation followed by immediate thoracotomy.

6. Twenty-eight days post-challenge, euthanize any hamsters that survived the infection using the same protocol.

7. Draw survival curves for each group (Fig. 1) and analyze statistical differences among groups using the Fisher test.

8. Collect the blood by postmortem cardiac puncture using 3 ml syringe with a 25 G $\times$ 5/8" needle. Transfer the blood into a 5 ml EDTA tube and mix well by inverting the tube five to six times. Immediately extract the DNA as described in Subheading 3.6, **step 2**.

9. Collect one kidney from each animal and inoculate in semisolid EMJH for culture as described in Subheading 3.5. Collect the other kidney and liver in individual dry tubes for quantification of tissue colonization by qPCR as described in Subheading 3.6.

3.4 Evaluation of the IgG Response by Enzyme-Linked Immunosorbent Assay (ELISA)

Allow all ELISA reagents to reach room temperature before use.

1. On the day of collection (i.e., before immunization and at day 4, 11, 18, 25, 32), prepare serum from the blood collected via retro-orbital bleed. Incubate tube containing the blood at room temperature for 30–45 min to allow clotting. Spin the blood at $2000 \times g$ for 15 min in a refrigerated centrifuge. Aspirate the supernatant (serum) and transfer it into a clean microcentrifuge tube. Keep serum at −80 °C until use (*see* **Note 13**).

2. Add 100 μl of PBS containing 1 μg purified protein into a microtiter plate well, in triplicate (3 wells for each sample to be tested). Cover the plate and incubate overnight at 4 °C.

3. Aspirate or decant solution from wells and discard the liquid. Wash wells three times with 200 μl of PBS using a multichannel pipettor or an automated 96-well plate washer (*see* **Note 14**).

4. Add 200 μl of protein-free blocking buffer (PFB) into the wells. Incubate for 1 h at RT.

5. Aspirate or decant solution from wells and discard the liquid. Wash wells three times with 200 μl of PBS.

6. Add 100 μl of hamster sera (pre-immune or immune sera, 1:64,000 dilution in PFB) and incubate for 1 h at 37 °C (*see* **Note 15**).

7. Aspirate or decant solution from wells and discard the liquid. Wash wells three times with 200 μl of PBS.

8. Add 100 μl of HRP-conjugated goat anti-Syrian hamster IgG (1:5000 dilution in PFB) and incubate for 30 min at RT.

9. Aspirate or decant solution from wells and discard the liquid. Wash wells three times with 200 μl of PBS.

10. Add 100 μl of 3,3′,5,5′-TMB substrate solution. Incubate plate for 15 min at RT in the dark (*see* **Note 16**).

11. Read the absorbance of each well at 655 nm with an ELISA plate reader.

12. The antibody response can be expressed as the relative anti-antigen level, determined by subtracting the absorbance of the pre-immune sera (baseline or background read) from the immune sera collected at different time points (Fig. 2).

3.5 Evaluation of Kidney Colonization by Culture

1. Pulverize the kidney by passing it through a 3 ml syringe directly into a 15 ml tube containing 10 ml semisolid EMJH supplemented with 100 μg/ml 5-FU (*see* **Note 17**).

2. Make a 1:100 dilution into a new 15 ml tube containing 10 ml semisolid EMJH supplemented with 5-FU.

3. Incubate tube at 30 °C for 4 weeks.

4. Check cultures weekly by dark-field microscopy for up to 28 days. A positive culture appears in 7–14 days. A Dinger's ring may become visible ~0.5 cm under the surface of the agar.

3.6 Evaluation of Blood and Tissue Colonization by qPCR

1. Extract DNA from kidney and liver tissues. Dice 50–80 mg of the kidney or liver. Place the diced tissue in a 1.5 ml microcentrifuge tube with 500 μl PBS (*see* **Note 18**). Homogenize tissue with tissue disrupter for 1 min at 5 movements per second. Transfer volume corresponding to 25 mg of tissue (*see* **Note 19**) into a microcentrifuge tube and perform the DNA extraction by using the DNeasy Blood and Tissue Kit (Qiagen) following the manufacturer's instructions. Elute DNA in 100 μl elution buffer.

2. Extract DNA from 100 μl blood collected on EDTA using the same DNA extraction kit (*see* **Note 20**). Elute the DNA in 100 μl elution buffer.

3. Extract DNA from 5 ml *L. interrogans* culture at a density of 4×10^5 spirochetes/ml using the same DNA extraction kit,

and elute the DNA in 100 μl elution buffer. Determine the DNA concentration using a spectrophotometer at 260 nm (*see* **Note 21**).

4. Dilute *Leptospira* genomic DNA to 10,080 pg/μl. Prepare tenfold serial dilutions of the DNA for the standard curve (1×10^7 to 1×10^0 genomic equivalents (Geq)/5 μl) (*see* **Note 22**).

5. Prepare the following PCR mix using Bio-Rad iTaq Universal Probes Supermix.

Reagent	1× reaction
Master mix 2×	10 μl
Primer 1, 250 nM (lipL32-45F)	0.5 μl
Primer 2, 250 nM (lipL32-268R)	0.5 μl
Probe, 150 nM (lipL32-189P)	0.6 μl
Water	3.4 μl
DNA	5 μl
Final volume	20 μl

6. Run the following program on a quantitative thermal cycler: 95 °C for 3 min followed by 40 cycles of amplification (95 °C for 15 s and 60 °C for 1 min). Process all samples and standards in duplicates.

7. Generate a standard curve, using the thermal cycler software. Plot Ct (cycle threshold) against the number of Geq/5 μl.

8. Determine the number of Geq/5 μl in the samples by using the standard curve generated by the thermal cycler software and threshold cycle obtained for each sample. A $C_t \geq 36$ is considered as no amplification.

9. Calculate the number of Geq per gram of tissue or μl of the blood using the formula Geq/g = Geq/5 μl × elution factor × volume factor (*see* **Note 23**).

---

4 Notes

1. Purified *Leptospira* protein for vaccine testing is prepared by recombinant DNA technology. The selected *Leptospira* gene is cloned and expressed as a recombinant protein in *E. coli*. Antigen preparations should be in PBS, sterile, and free of urea, acetic acid, endotoxin (LPS), and other toxic solvents.

2. Complete Freund's adjuvant (CFA) is extensively used and considered to be the most effective adjuvant for consistently

producing high-titer antibodies against a wide variety of antigens. CFA comprises of inactivated mycobacteria emulsified in mineral oil. CFA is administered for the initial immunization, while incomplete Freund's adjuvant (IFA), which lacks mycobacteria, is used for subsequent immunizations [16].

3. Although CFA is commonly used in animal experiments, the efficacy of leptospiral vaccines must also be determined using adjuvants approved for human use. Other adjuvants, such as aluminum hydroxide and aluminum phosphate, are less toxic and therefore more acceptable in human preparations [17].

4. Female hamsters are conventionally used in *Leptospira* animal studies because they are less aggressive than males. However, a recent study suggests that male hamsters are more susceptible to leptospirosis [18] compared to females infected with the same number of bacteria. With this result, along with the US National Institutes of Health's policy that recommends the use of animals of both sexes [19], performing vaccine studies with both male and female hamsters in one group should be considered.

5. CFA is a potent inflammatory agent. Accidental splashing in the eyes with CFA can cause profound sloughing of skin or loss of sight. Inadvertent needle punctures in humans can cause a positive TB test and can result in inflammatory lesions and abscess formation in tuberculin-positive individuals [20]. Wear safety glasses and gloves when handling CFA.

6. To prevent denaturation of conformational epitopes, antigen-CFA (or antigen-IFA) emulsions are prepared immediately before immunization. Because heat is generated during the emulsion preparation process, syringes should be laid on ice to keep the emulsion close to 4 °C. Some volume will be lost in the mixing process; the volumes described here produce sufficient amount to immunize up to 15 hamsters.

7. Perform retro-orbital bleed under general anesthesia. One hundred to 200 μl of blood is collected from the venous plexus using a capillary tube. The eye is made to bulge, and the capillary is inserted into the plexus with a gentle pressure and a small rotation of the capillary. The blood is allowed to flow into the capillary and is then transferred into a dry tube. After blood collection, place a dry cotton pad over the eye with gentle pressure to prevent retro-orbital hemorrhage and then apply an ophthalmic ointment. Perform only one bleed per week and alternate eyes to allow the tissue to heal. Monitor animals for signs of ocular damage including bulging, squinting, swelling, and eye discharge following the procedure. If an eye is damaged from the procedure, that eye may not be used again for blood collection.

8. Subcutaneous injection is administered on the scruff of the hamster's neck using a syringe with a 25 G needle. Use several injection sites over the hamster's back if larger volume of vaccine is administered.

9. We only use low passage *Leptospira* challenge strain, i.e., passaged fewer than five times in vitro after isolation from a mammalian host. Ensure the culture is in mid-log phase ($\sim 1 \times 10^8$ bacteria/ml) on the day of challenge.

10. Determine the density of *Leptospira* using a Petroff-Hausser chamber. Dilute the *Leptospira* culture 1:100 in EMJH. Place coverslip over the chamber in the center of the slide, and slowly dispense 10 μl of diluted bacterial suspension under the coverslip. Allow the liquid to fill the chamber grid by capillary action. Count bacterial cells within the center square (corresponding to 25 small squares) under a dark-field microscope. Calculate the density of *Leptospira* with the following equation: bacteria/ml = number of bacterial cells in all the 25 squares × dilution factor × 50,000. The value 50,000 is 50 (cell depth is 1/50 mm) × 1000 (1000 mm^3 = 1 ml).

11. The ED_{50} is the dose required for 50% of animals to meet the endpoint criteria. Endpoint criteria include loss of appetite, gait or breathing difficulty, prostration, ruffled fur, or weight loss of >10% or maximum weight. ED_{50} depends on the serovar and the strain used and the number of in vitro passages. *L. interrogans* serovar Copenhageni strain Fiocruz L1-130, commonly used for hamster immunoprotection studies, has an ED_{50} of ~20 leptospires [21]. For the challenge, the use of 1×10^3 to 1×10^4 (50–500 times the ED_{50}) has been previously described [11, 21, 22].

12. Intraperitoneal injection is the most common inoculation method for *Leptospira* studies [9]. However, IP inoculation does not reflect the natural transmission of *Leptospira*. Other routes of infection like through the skin (subcutaneous or intradermal injection) [23] or through mucous membranes (ocular inoculation) [24] that better simulate the natural entry of the bacteria can be considered.

13. Serum must be prepared from the blood immediately after it is collected by retro-orbital bleed. Because some antibodies lose activity on repeated freezing and thawing, aliquot serum in small volumes and store at −80 °C.

14. Manual wash consists of filling the wells with PBS and emptying them three separate times. Emptying the wells by slapping the plates against several layers of paper towels after each step can minimize bubbles in the wells. Do not allow the wells to dry. Avoid the formation of bubbles in wells because these may

affect the overall performance of the assay resulting in inaccurate readings.

15. The optimal antigen concentration, hamster sera dilution, and conjugated secondary antibody dilution that give the best signal with minimum background were determined experimentally by checkerboard analysis [25, 26].

16. The TMB substrate solution is colorless and should be stored at 4 °C. If the reagent turns blue, it may be contaminated and should not be used for the assay. If the negative control (PBS or no antigen) turns blue after development, the ELISA assay should be repeated with additional washing steps.

17. The kidney is cultured immediately after it is removed from the animal. Disperse the pulverized kidney on top of the semisolid EMJH. Use a sterile barrier pipette tip to swirl the kidney around to embed it in the gel.

18. Use screw-top microcentrifuge tubes to prevent tubes from opening and samples from spilling during the homogenization process.

19. Determine the volume (μl) of tissue to use for DNA extraction with the following formula: $x = 500 \, \mu l \times (25 \, mg)/mg$ of diced tissue. For example, if 50 mg diced tissue is suspended in 500 μl PBS, then 250 μl of homogenate will be used for DNA extraction.

20. DNA extraction from the blood collected on EDTA must be performed immediately after collection.

21. The standard curve is prepared with genomic DNA extracted from the same *Leptospira* strain used for the challenge.

22. A genome size of 4.6 Mb was used to determine the Geq concentration per microliter of purified DNA. A genomic DNA mass of 50,400 pg is equivalent to 1×10^7 *Leptospira* genome copies [6, 24]. To prepare 1×10^7 Geq/5 μl for the standard curve, purified leptospiral genomic DNA is diluted in distilled water to a concentration of 50,400 pg/5 μl. Tenfold serial dilutions are prepared to obtain 1×10^6 to 1×10^0 Geq/5 μl.

23. In this protocol, the elution factor is 20 because we used 5 μl of the 100 μl extracted DNA, and the volume factor is 40 because the DNA extraction was done with 25 mg (1 g/25 mg) of tissue. For the blood, the elution factor is 20 (5 μl of the 100 μl DNA), and volume factor is 1 (100 μl/100 μl).

References

1. Adler B, de la Pena MA (2010) *Leptospira* and leptospirosis. Vet Microbiol 140 (3-4):287–296. https://doi.org/10.1016/j.vetmic.2009.03.012

2. Haake DA, Levett PN (2015) Leptospirosis in humans. Curr Top Microbiol Immunol 387:65–97

3. Faine S, Adler B, Bolin C, Perolat P (1999) *Leptospira* and leptospirosis, 2nd edn. MediSci, Melbourne, VIC

4. Ellis WA (2015) Animal leptospirosis. Curr Top Microbiol Immunol 387:99–137

5. Adler B (2015) Vaccines against leptospirosis. Curr Top Microbiol Immunol 387:251–272

6. Nascimento AL, Ko AI, Martins EA, Monteiro-Vitorello CB, Ho PL, Haake DA, Verjovski-Almeida S, Hartskeerl RA, Marques MV, Oliveira MC, Menck CF, Leite LC, Carrer H, Coutinho LL, Degrave WM, Dellagostin OA, El-Dorry H, Ferro ES, Ferro MI, Furlan LR, Gamberini M, Giglioti EA, Goes-Neto A, Goldman GH, Goldman MH, Harakava R, Jeronimo SM, Junqueira-de-Azevedo IL, Kimura ET, Kuramae EE, Lemos EG, Lemos MV, Marino CL, Nunes LR, de Oliveira RC, Pereira GG, Reis MS, Schriefer A, Siqueira WJ, Sommer P, Tsai SM, Simpson AJ, Ferro JA, Camargo LE, Kitajima JP, Setubal JC, Van Sluys MA (2004) Comparative genomics of two *Leptospira interrogans* serovars reveals novel insights into physiology and pathogenesis. J Bacteriol 186(7):2164–2172

7. Ren SX, Fu G, Jiang XG, Zeng R, Miao YG, Xu H, Zhang YX, Xiong H, Lu G, Lu LF, Jiang HQ, Jia J, Tu YF, Jiang JX, Gu WY, Zhang YQ, Cai Z, Sheng HH, Yin HF, Zhang Y, Zhu GF, Wan M, Huang HL, Qian Z, Wang SY, Ma W, Yao ZJ, Shen Y, Qiang BQ, Xia QC, Guo XK, Danchin A, Saint Girons I, Somerville RL, Wen YM, Shi MH, Chen Z, Xu JG, Zhao GP (2003) Unique physiological and pathogenic features of *Leptospira interrogans* revealed by whole-genome sequencing. Nature 422 (6934):888–893

8. Fouts DE, Matthias MA, Adhikarla H, Adler B, Amorim-Santos L, Berg DE, Bulach D, Buschiazzo A, Chang YF, Galloway RL, Haake DA, Haft DH, Hartskeerl R, Ko AI, Levett PN, Matsunaga J, Mechaly AE, Monk JM, Nascimento AL, Nelson KE, Palsson B, Peacock SJ, Picardeau M, Ricaldi JN, Thaipandungpanit J, Wunder EA Jr, Yang XF, Zhang JJ, Vinetz JM (2016) What makes a bacterial species pathogenic?:comparative genomic analysis of the genus *Leptospira*. PLoS Negl Trop Dis 10(2):e0004403. https://doi.org/10.1371/journal.pntd.0004403

9. Haake DA (2006) Hamster model of leptospirosis. Curr Protoc Microbiol 12:2. https://doi.org/10.1002/9780471729259.mc12e02s02

10. Walker A, Olsen R, Toth M, Srinivas G (2018) Re-evaluating the LD50 requirements in the codified potency testing of veterinary vaccines containing *Leptospira* (L.) serogroup Icterohaemorrhagiae and L. serogroup Canicola in the United States. Biologicals 56:13–18

11. Evangelista KV, Lourdault K, Matsunaga J, Haake DA (2017) Immunoprotective properties of recombinant LigA and LigB in a hamster model of acute leptospirosis. PLoS One 12(7):e0180004. https://doi.org/10.1371/journal.pone.0180004

12. Ko AI, Galvao Reis M, Ribeiro Dourado CM, Johnson WD Jr, Riley LW (1999) Urban epidemic of severe leptospirosis in Brazil. Salvador leptospirosis study group. Lancet 354 (9181):820–825

13. Johnson RC, Harris VG (1967) Differentiation of pathogenic and saprophytic letospires. I. Growth at low temperatures. J Bacteriol 94(1):27–31

14. Ellinghausen HC Jr, McCullough WG (1965) Nutrition of *Leptospira pomona* and growth of 13 other serotypes: fractionation of oleic albumin complex and a medium of bovine albumin and polysorbate 80. Am J Vet Res 26:45–51

15. Stoddard RA (2013) Detection of pathogenic *Leptospira spp.* through real-time PCR (qPCR) targeting the *lipL32* gene. Methods Mol Biol 943:257–266. https://doi.org/10.1007/978-1-60327-353-4_17

16. Stills HF, Bailey MQ (1991) The use of Freund's complete adjuvant. Lab Animal Sci 20(4):25–31

17. Gupta RK (1998) Aluminum compounds as vaccine adjuvants. Adv Drug Deliv Rev 32 (3):155–172

18. Gomes CK, Guedes M, Potula HH, Dellagostin OA, Gomes-Solecki M (2018) Sex matters: male hamsters are more susceptible to lethal infection with lower doses of pathogenic *Leptospira* than female hamsters. Infect Immun 86 (10):e00369-18. https://doi.org/10.1128/IAI.00369-18

19. Clayton JA, Collins FS (2014) Policy: NIH to balance sex in cell and animal studies. Nature 509(7500):282–283

20. Hm C, August PJ (1976) Report of nine cases of accidental injury to man with Freund's

complete adjuvant. Clin Exp Immunol 24 (3):538–541

21. Lourdault K, Wang LC, Vieira A, Matsunaga J, Melo R, Lewis MS, Haake DA, Gomes-Solecki M (2014) Oral immunization with *Escherichia coli* expressing a lipidated form of LigA protects hamsters against challenge with *Leptospira interrogans* serovar Copenhageni. Infect Immun 82(2):893–902. https://doi.org/10.1128/IAI.01533-13

22. Coutinho ML, Choy HA, Kelley MM, Matsunaga J, Babbitt JT, Lewis MS, Aleixo JA, Haake DA (2011) A LigA three-domain region protects hamsters from lethal infection by *Leptospira interrogans*. PLoS Negl Trop Dis 5(12):e1422. https://doi.org/10.1371/journal.pntd.0001422

23. Coutinho ML, Matsunaga J, Wang LC, de la Pena MA, Lewis MS, Babbitt JT, Aleixo JA, Haake DA (2014) Kinetics of *Leptospira interrogans* infection in hamsters after intradermal and subcutaneous challenge. PLoS Negl Trop Dis 8(11):e3307. https://doi.org/10.1371/journal.pntd.0003307

24. Wunder EA Jr, Figueira CP, Santos GR, Lourdault K, Matthias MA, Vinetz JM, Ramos E, Haake DA, Picardeau M, Dos Reis MG, Ko AI (2016) Real-time PCR reveals rapid dissemination of *Leptospira interrogans* after intraperitoneal and conjunctival inoculation of hamsters. Infect Immun 84 (7):2105–2115. https://doi.org/10.1128/iai.00094-16

25. Engvall E, Jonsson K, Perlmann P (1971) Enzyme-linked immunosorbent assay. II. Quantitative assay of protein antigen, immunoglobulin G, by means of enzyme-labelled antigen and antibody-coated tubes. Biochem Biophys Acta 251(3):427–434

26. Crowther JR (1995) ELISA. Theory and practice. Methods Mol Biol 42:1–218

Chapter 24

Leptospira and Leptospirosis

Mathieu Picardeau

Abstract

Leptospira spp. are morphologically and phylogenetically unique bacteria that are ubiquitous in the environment. Pathogenic species are the agents of leptospirosis, an emerging zoonotic disease transmitted to both human and animals from the environment contaminated with the urine of reservoir animals. The taxonomy of *Leptospira* has recently undergone extensive revisions with the use of whole-genome sequences, thus expanding considerably the number of species.

Key words Taxonomy, Virulence factors, Pathogenesis, Genome

1 General Features of *Leptospira*: An Atypical Bacteria

Leptospira spp. are long, thin, flexible rods, 0.1 μm in diameter, and 6–12 μm in length, with a helical cell shape and periplasmic flagella. The envelope structure is similar to Gram-negative bacteria, and the cell surface is covered with lipopolysaccharides (LPS) which determine the antigenic diversity within the genus. Leptospires are not stained by gram staining, and dark-field microscopy is required for visualization of cells. *Leptospira* spp. are highly motile aerobic or microaerophilic bacteria. They use long-chain fatty acids as primary carbon and energy sources. In Ellinghausen-McCullough-Johnson-Harris (EMJH) culture medium, the optimum growth temperature is 28–30 °C, with a generation time of 6–20 h. Growth on solid agar medium results in the formation of subsurface colonies. *Leptospira* spp. comprise free-living saprophytes that are well adapted to soils and aquatic, but not marine, environments and pathogens that cause the disease leptospirosis in both human and animals. Animal reservoirs of pathogens consist of domesticated and wild mammalian hosts. Long-term survival of pathogens in water has also been described [1, 2].

Nobuo Koizumi and Mathieu Picardeau (eds.), *Leptospira spp.: Methods and Protocols*, Methods in Molecular Biology, vol. 2134, https://doi.org/10.1007/978-1-0716-0459-5_24, © Springer Science+Business Media, LLC, part of Springer Nature 2020

2 The Genus Leptospira: A Blooming Tree

The genus *Leptospira* forms a deep unique branch of spirochetes and is classified in the family *Leptospiraceae* which is currently divided into the genera *Leptospira*, *Turneriella*, and *Leptonema*. The genera *Turneriella* and *Leptonema* each contain a single species, and relative to *Leptospira*, very little information is available for these species [3]. The genus *Leptospira* is highly diverse and comprises 64 different species, which have been identified since the isolation of *L. interrogans* in 1915 [4, 5]. The availability of a new selective medium [6], the advent of relatively cheap whole-genome sequencing (WGS), and increased interest in metagenomic studies and soil microbial communities have resulted in expanding the number of *Leptospira* species from 22 in 2018 [7] to 64 in 2019 [4].

Phylogenetic analysis, initially based on 16S rRNA gene sequencing but now on whole-genome sequences, showed that the genus is separated in two clades: "saprophytes" containing species isolated in the natural environment and not responsible for infections and "pathogens" containing all the species responsible for infections in humans and/or animals, plus environmental species for which the pathogenicity remains unclear [4]. The two clades are further subdivided in four subclades called P1, P2, S1, and S2 [4]. The subclade P1 (formerly described as the pathogen group) comprises 17 species (*L. mayottensis*, *L. alexanderi*, *L. kirschneri*, *L. kmetyi*, *L. alstonii*, *L. adleri*, *L. barantonii*, *L. ellisii*, *L. dzianensis*, *L. gomenensis*, *L. putramalaysiae*, *L. tipperaryensis*, *L. borgpetersenii*, *L. interrogans*, *L. noguchii*, *L. santarosai*, *L. weilii*). Some species, such as *L. kmetyi*, *L. adleri*, *L. ellisii*, *L. gomenensis*, *L. barantonii*, *L. dzianensis*, and *L. putramalaysiae*, first identified in the environment, have never been isolated from infected animals or patients, suggesting that they are not true pathogens [4, 8]. The subclade S2, also called intermediate species, forms a group of 21 species distinct from the pathogens (*L. broomii*, *L. licerasiae*, *L. fainei*, *L. venezuelensis*, *L. wolffii*, *L. haakeii*, *L. hartskeerlii*, *L. saintgironsiae*, *L. neocaledonica*, *L. perolatii*, *L. dzoumogneensis*, *L. fletcheri*, *L. fluminis*, *L. johnsonii*, *L. koniamboensis*, *L. langatensis*, *L. sarikeiensis*, *L. selangorensis*, *L. semungkisensis*, *L. andrefontaineae*, *L. inadai*). Most of these species have been isolated from the environment, and their virulence status has not been proven in animal models. The saprophytes are then subdivided in subclades S1 (*L. terpstrae*, *L. vanthielii*, *L. yanagawae*, *L. brenneri*, *L. harrisiae*, *L. levettii*, *L. kemamanensis*, *L. bandrabouensis*, *L. bourretii*, *L. bouyouniensis*, *L. congkakensis*, *L. ellinghausenii*, *L. jelokensis*, *L. kanakyensis*, *L. montravelensis*, *L. mtsangambouensis*, *L. noumeaensis*, *L. perdikensis*, *L. biflexa*,

L. meyeri, L. wolbachii, L. idonii) and S2 (*L. ilyithenensis, L. kobayashii, L. ognonensis, L. ryugenii*).

3   Leptospirosis and Virulence Factors

Pathogenic *Leptospira* species such as *L. interrogans* are the agents of leptospirosis, which is a zoonotic disease responsible for more than one million cases and 60,000 deaths per year worldwide [9]. These are likely underestimates because of misdiagnosis and underreporting, particularly in regions where other diseases with similar nonspecific presentations, such as dengue and malaria, are prevalent. Leptospirosis is also emerging due to global climate changes resulting in more frequent and severe flooding events. Although leptospirosis has high burden and mortality and is treatable, it is not acknowledged as a neglected tropical disease; thus, efforts to raise awareness at the local and international levels are needed [10].

Transmission to humans usually occurs through contact with soils or surface waters contaminated with the urine of reservoir animals such as rats that are asymptomatic carriers of the pathogens. The severity of the disease appears to be mainly determined by the interaction of the virulence characteristics of the infecting strain, infecting inoculum size during environmental exposure, and host susceptibility factors. However, the basic biology and virulence factors of leptospires remain poorly characterized. This gap in our knowledge is largely due to the fact that this research area has been unattractive to both funders and researchers in the past decades [10]. As a consequence, for example, genetic manipulation of *Leptospira* remains relatively inefficient in comparison to other bacteria [11].

Pathogenic *Leptospira* spp. have developed different strategies such as rapid dissemination and ability to escape or hijack the host immune system to successfully establish and maintain an infection. The presence of endoflagella (or periplasmic flagella) enables the pathogens to rapidly cross the mammalian cell barriers, disseminate hematogenously, establish infection in target organs, and avoid flagellin recognition from the innate immune system, thus enabling the spirochetes to escape the immune attack during the infection [11, 12]. Leptospires produce several adhesins for binding to several components of host cells, including the extracellular matrix [13]. Under normal in vitro growth, the most abundant proteins are the lipoproteins LipL32, LipL41, LipL36, and Loa22 [14]. These lipoproteins could potentially interact with host cells and the immune system (complement, etc.) during the hematogenous dissemination of leptospires in the host. In addition, it has been shown that the leptospiral LPS, in contrast to other

Gram-negative bacteria, escapes human TLR4 recognition [15] and is recognized by TLR2 in human cells [16].

The first leptospiral genome sequence to be determined in 2003, that of *L. interrogans* serovar Lai, consists of a 4.33-Mb large chromosome and a 359-kb small chromosome [17]. Today, the genome sequences of hundreds of *Leptospira* strains have been determined, including representative of each of the 64 *Leptospira* species [4]. The genome of *Leptospira* is always composed of two chromosomes, and some plasmids have been recently identified [18–21]. Access to large sets of genome sequences has significantly improved our understanding of the emergence of virulence in *Leptospira*. Genomic comparisons between pathogenic and non-pathogenic species have revealed a number of important differences, suggesting that pathogens evolved from free-living ancestral species by successive gain and loss of genes/functions associated with the adaptation to new hosts [4, 8, 22]. For instance, it is possible to observe a gradient in the repertoire of genes encoding proteins (hemolysins, etc.) or protein domains (leucine-rich repeat, peptidases, etc.) known to be associated with virulence, with the most virulent species of subclade P1 having the most genes encoding these virulence factors [4].

4 Conclusion

Leptospira was identified as the causative agents of leptospirosis 100 years ago [5]. Despite some recent progress, although the burden of leptospirosis is comparable or even higher than diseases that are much better known, such as dengue or rabies [23], there is a considerable deficit in the understanding of basic aspects of the epidemiology of the disease and the biology of the bacterium responsible.

References

1. Trueba G, Zapata S, Madrid K, Cullen P, Haake D (2004) Cell aggregation: a mechanism of pathogenic Leptospira to survive in fresh water. Int Microbiol 7:35–40
2. Andre-Fontaine G, Aviat F, Thorin C (2015) Water borne leptospirosis: survival and preservation of the virulence of pathogenic Leptospira spp. in fresh water. Curr Microbiol 71:136–142
3. Picardeau M (2014) Family Leptospiraceae. In: Rosenberg E (ed) The prokaryotes. Springer-Verlag Berlin, Heidelberg, pp 711–729
4. Vincent AT, Schiettekatte S, Goarant C, Neela VK, Bernet E, Thibeaux R et al (2019) Revisiting the taxonomy and evolution of

pathogenicity of the genus Leptospira through the prism of genomics. PLoS Negl Trop Dis 13 (5):e0007270
5. Inada R, Ido Y, Hoki R, Kakeno R, Ito H (1916) The etiology, mode of infection and specific therapy of Weil's disease (Spirochaeta icterohaemorrhagiae). J Exp Med 23:377–403
6. Chakraborty A, Miyahara S, Villanueva SY, Saito M, Gloriani NG, Yoshida S (2011) A novel combination of selective agents for isolation of Leptospira species. Microbiol Immunol 55:494–501
7. Puche R, Ferrés I, Caraballo L, Rangel Y, Picardeau M, Takiff H et al (2018) Leptospira venezuelensis sp. nov., a new member of the

intermediate group isolated from rodents, cattle and humans. Int J Syst Evol Microbiol 68:513–517

8. Thibeaux R, Iraola G, Ferrés I, Bierque E, Girault D, Soupé-Gilbert ME et al (2018) Deciphering the unexplored Leptospira diversity from soils uncovers genomic evolution to virulence. Microb Genom 4

9. Costa F, Hagan JE, Calcagno J, Kane M, Torgerson P, Martinez-Silveira MS et al (2015) Global morbidity and mortality of leptospirosis: a systematic review. PLoS Negl Trop Dis 9:e0003898

10. Goarant C, Picardeau M, Morand S, McIntyre KM (2019) Leptospirosis under the bibliometrics radar: evidence for a vicious circle of neglect. J Glob Health 9:010302

11. Picardeau M (2017) Virulence of the zoonotic agent of leptospirosis: still terra incognita? Nat Rev Microbiol 15:297–307

12. Wunder EAJ, Figueira CP, Santos GR, Lourdault K, Matthias MA, Vinetz JM et al (2016) Real-time PCR reveals rapid dissemination of *Leptospira interrogans* after Intraperitoneal and Conjunctival inoculation of hamsters. Infect Immun 84:2105–2115

13. Murray GL (2015) The molecular basis of leptospiral pathogenesis. Curr Top Microbiol Immunol 387:139–185

14. Malmström J, Beck M, Schmidt A, Lange V, Deutsch EW, Aebersold R (2009) Proteomewide cellular protein concentrations of the human pathogen *Leptospira interrogans*. Nature 460:762–765

15. Nahori MA, Fournié-Amazouz E, Que-Gewirth NS, Balloy V, Chignard M, Raetz CR et al (2005) Differential TLR recognition of leptospiral lipid A and lipopolysaccharide in murine and human cells. J Immunol 175:6022–6031

16. Werts C, Tapping RI, Mathison JC, Chuang T-H, Kravchenko V, Saint GI et al (2001) Leptospiral lipopolysaccharide activates cells through a TLR2-dependent mechanism. Nat Immunol 2(4):346–352

17. Ren S, Fu G, Jiang X, Zeng R, Xiong H, Lu G et al (2003) Unique and physiological and pathogenic features of *Leptospira interrogans* revealed by whole genome sequencing. Nature 422:888–893

18. Zhu W, Wang J, Zhu Y, Tang B, Zhang Y, He P et al (2015) Identification of three extra-chromosomal replicons in Leptospira pathogenic strain and development of new shuttle vectors. BMC Genomics 16:90

19. Zhu WN, Huang LL, Zeng LB, Zhuang XR, Chen CY, Wang YZ et al (2014) Isolation and characterization of two novel plasmids from pathogenic *Leptospira interrogans* Serogroup Canicola Serovar Canicola strain Gui44. PLoS Negl Trop Dis 8:e3103

20. Schiettekatte O, Vincent AT, Malosse C, Lechat P, Chamot-Rooke J, Veyrier FJ et al (2018) Characterization of LE3 and LE4, the only lytic phages known to infect the spirochete Leptospira. Sci Rep 8:11781

21. Satou K, Shimoji M, Tamotsu H, Juan A, Ashimine N, Shinzato M et al (2015) Complete genome sequences of low-passage virulent and high-passage Avirulent variants of pathogenic *Leptospira interrogans* Serovar Manilae strain UP-MMC-NIID, originally isolated from a patient with severe leptospirosis, determined using PacBio single-molecule real-time technology. Genome Announc 3:e00882–e00815

22. Xu Y, Zhu Y, Wang Y, Chang YF, Zhang Y, Jiang X et al (2016) Whole genome sequencing revealed host adaptation-focused genomic plasticity of pathogenic Leptospira. Sci Rep 6:20020

23. Torgerson PR, Hagan JE, Costa F, Calcagno J, Kane M, Martinez-Silveira MS et al (2015) Global burden of leptospirosis: estimated in terms of disability adjusted life years. PLoS Negl Trop Dis 9:e0004122

Chapter 25

Laboratory Diagnosis of Leptospirosis

Nobuo Koizumi

Abstract

The diagnosis of leptospirosis depends on specific laboratory tests because nonspecific and diverse clinical manifestations make clinical diagnosis difficult and it is easily confused with other infectious diseases in the tropics. Suitable laboratory diagnostic tests vary depending on the stage of the disease, requiring the combination of diagnostic tests using appropriate specimens at each disease stage.

Key words Culture, Diagnosis, Microscopic agglutination test (MAT), Nucleic acid amplification test (NAAT), Serology

1 Introduction

Leptospirosis is an emerging zoonotic disease that is widespread globally. A World Health Organization (WHO) experts group estimated one million cases and 58,900 deaths due to leptospirosis worldwide each year, of which more than 70% occur in the tropical, poorest regions of the world [1]. Leptospirosis has been recognized as a major cause of acute undifferentiated fever in various parts of the world [2–6]. Nonspecific and diverse clinical manifestations make clinical diagnosis difficult, and it is easily confused with many other diseases in the tropics, such as dengue fever, malaria, and scrub typhus [7]. In addition, general clinical laboratory test findings are nonspecific. Therefore, the diagnosis of leptospirosis depends on specific laboratory tests. Table 1 describes the case definitions of leptospirosis, as recommended by a group of WHO experts [8].

Suitable laboratory diagnostic tests vary depending on the stage of the disease. During the acute phase, less than 7 days after onset, leptospires exist in the bloodstream (leptospiremia). Detection of *Leptospira* spp. in the blood by culturing and/or nucleic acid amplification tests (NAATs) is superior to antibody detection, but the timing of blood collection and the administration of antibiotics can affect their sensitivity and result in a false-negative result.

Nobuo Koizumi and Mathieu Picardeau (eds.), *Leptospira spp.: Methods and Protocols*, Methods in Molecular Biology, vol. 2134, https://doi.org/10.1007/978-1-0716-0459-5_25, © Springer Science+Business Media, LLC, part of Springer Nature 2020

Table 1
WHO case definitions of leptospirosis [8]

Confirmed cases. Clinical signs and symptoms consistent with leptospirosis and any one of the following:
- Fourfold increase in microscopic agglutination test (MAT) titer in acute and convalescent serum samples
- MAT titer ≥1:400 in single or paired serum samples
- Isolation of pathogenic species from normally sterile site
- Detection of pathogenic species in clinical samples by histological, histochemical, or immunostaining technique
- Detection of pathogenic species DNA by PCR

Probable cases. Clinical signs and symptoms consistent with leptospirosis and one of the following:
- Presence of IgM or a fourfold increase of antibody titer in immunofluorescence assay in acute and convalescent serum samples
- Presence of IgM antibodies by ELISA or dipstick
- MAT titer ≥1:100 in single acute-phase serum sample in non-endemic regions

Leptospires in the blood can be detected on the first few days after onset by dark-field microscopy, but the results need to be supported by other laboratory methods irrespective of positive results. On the second week, leptospires are cleared from the bloodstream as antibodies are produced and higher numbers of leptospires are excreted in the urine. During this stage, the sensitivity of antibody detection increases, whereas the sensitivity of DNA detection in the blood decreases and that in urine may increase.

2   Culture

Leptospires can be isolated from clinical materials such as blood, cerebrospinal fluid (CSF), urine, and postmortem samples of various tissues. Cultures are incubated at 28–30 °C and checked regularly by dark-field microscopy. On primary isolation, the growth of leptospires is slow, their generation times can be up to 20 h, and the cultures are recommended to be maintained for up to 13 weeks [9]. Therefore, culture is not useful for early diagnosis, although it constitutes the definitive diagnosis of leptospirosis. The isolation of leptospires requires specialized culture media such as Ellinghausen-McCullough-Johnson-Harris (EMJH) and Korthof's media [10], and its sensitivity is as low as less than 10% in NAAT-positive patients [11], which hamper the routine application of culture in many clinical settings.

Blood culture should be performed as soon as possible after the onset of the disease during the leptospiremia phase and before the administration of antibiotics. For the blood culture, a few drops of blood are inoculated into 5–10 ml liquid or semi-solid culture media and, if possible, separated into multiple tubes. In addition to whole blood, a culture using deposits from centrifuged plasma gave a high yield [12]. Leptospires can survive in heparinized blood stored at room temperature, from which they can be recovered at the longest of 109 days after blood collection [12]. It has been shown that some in vitro-cultured *Leptospira* strains can survive in commercially available blood culture media for up to 2 weeks [13]. Girault et al. successfully isolated *L. borgpetersenii and L. interrogans* from commercial aerobic blood culture bottles. They transferred culture bottle samples that were *Leptospira* real-time polymerase chain reaction (PCR)-positive to EMJH medium within 2.5 days after blood collection and obtained isolates in one-third of the patients investigated [14].

Leptospires may be isolated by inoculating 0.5 ml CSF into 5 ml semi-solid culture medium on the first week of illness [9]. Urine is the most suitable specimen for the isolation during the leptospiruria phase (on the second week after onset). As the viability of leptospires is limited in acidic urine, urine needs to be inoculated into the culture medium within 2 h after voiding. Alternatively, the urine sample is neutralized or centrifuged to obtain pellets, which are inoculated into the culture medium [9]. 5-Fluorouracil and/or some antimicrobial agents may be used to reduce the risk of contamination [9, 10].

3 NAATs

NAATs have demonstrated superior sensitivity in the acute phase of leptospirosis to culture and serological diagnostics. Leptospiral DNA has been amplified from blood, urine, CSF, aqueous humor, and tissues. A number of conventional PCRs including nested PCR and real-time PCR, as well as isothermal DNA amplification techniques, have been developed for the early diagnosis of leptospirosis in the last three decades [15–19]. NAAT using real-time PCR has been evaluated in large populations in clinical settings and becomes the standard for early diagnosis [20–22]. However, the instruments used in real-time PCR are expensive and may be unavailable in resource-limited settings. Isothermal DNA amplification techniques such as loop-mediated isothermal amplification and recombinase polymerase amplification can amplify the target DNA sequence without thermal cyclers and can be integrated into "lab-on-a-chip," which enables NAATs at the point of care [23]. Extensive evaluations are needed for the utility of isothermal DNA amplification techniques for leptospirosis diagnosis in clinical settings [11, 24].

As with culture, blood samples are most suitable in the acute phase for NAATs. However, in clinical settings, there are controversial reports: one study showed that serum is a better specimen than whole blood [21], whereas another study demonstrated that sensitivity is much higher for whole blood than for serum

[22]. Although the sensitivity is low, leptospiral DNA was detected in urine samples as early as in plasma collected from patients in the acute phase [11]. As mentioned above, blood culture bottles could be used for *Leptospira* detection, but special DNA extraction methods must be employed [25]. Leptospiral DNA detection in blood culture bottles was conducted in a clinical setting but turned out to be a limited utility [26]. DNA detection can be performed using filter paper-dried serum samples stored at room temperature for up to 30 days, the sensitivity of which was 78.6% compared with that using fresh serum samples [27].

Some NAAT target genes are specific for pathogenic *Leptospira* spp., such as *lipL32*. There is increasing evidence that the intermediate subclade of *Leptospira* spp. is responsible for human leptospirosis [28–32]. Therefore, pathogenic species-specific NAATs potentially fail to diagnose leptospirosis caused by intermediate *Leptospira* spp. In addition, new *Leptospira* spp. in the pathogenic subclade have been identified from the environment, and a new degenerate primer/probe set for real-time PCR detecting *lipL32* has been described [33]. In contrast, NAATs targeting *rrs* detect all subclades, pathogens, intermediates, and saprophytes, but it has been reported that these primer (and probe) sets could detect DNAs from other bacteria, especially when they are applied to urine samples [34–36].

To improve the sensitivity of real-time PCR, RNA-based [reverse transcription (RT)] PCR has been employed because RNA transcripts are present in clinical samples much more than DNA targets which usually present one copy in a bacterial cell [37, 38]. A new real-time RT-PCR developed by Backstedt et al. demonstrated significantly higher sensitivity than DNA-based real-time PCR using clinical samples [37].

Clinical manifestations of leptospirosis are similar to other infectious diseases such as dengue fever, malaria, and scrub typhus in the tropics, and rapid differential diagnosis could result in appropriate management and improve the outcome. Multiplex real-time PCR assays, one of which detects *Leptospira* spp., dengue virus, and *Plasmodium* spp., and the other detects *Leptospira* spp., chikungunya virus, and dengue virus, have been developed [39, 40]. Their diagnostic accuracy should be evaluated.

4   Serological Diagnosis

As defined by the WHO, microscopic agglutination test (MAT) using paired serum samples is the definitive serological diagnosis: seroconversion or at least fourfold increase in titers must be observed between acute and convalescent serum samples because antibodies detected by MAT are present for months to years after infection. WHO also defines ≥400 MAT titers in a single sample as a confirmed case, but titers as high as ≥1600 have been recommended [9]. MAT detects serogroup-specific antibodies IgM and IgG, and the panel of antigens and live cultures should include serovars representative of all serogroups and local strains. Alternatively, the WHO recommends 19 serovars of 16 serogroups [7]. Filter paper-dried serum samples can be used for MAT: their titers were almost equal to those of fresh samples when stored at room temperature for up to 7 days [41].

The principle of MAT is simple, but the procedure requires technical expertise and the maintenance of a panel of *Leptospira* cultures, and quality control must be employed [42]. Furthermore, the sensitivity of MAT in the acute phase is low, and paired sera are needed to confirm the diagnosis. Therefore, a number of rapid screening tests for antibody detection in acute infection, such as macroagglutination test, indirect hemagglutination assay, immuno-fluorescence assay, microcapsule agglutination test, enzyme-linked immunosorbent assay (ELISA), dot enzyme immunoassay (dip-stick), lateral flow assay (LFA), and latex agglutination test, have been developed [9, 17, 43], and some of the whole-cell-based assay kits are commercially available (Table 2). These whole-cell-based assays are believed to be genus-specific and detect IgM (and/or IgG) antibodies from patients, regardless of infective serovars or serogroups, with easy and rapid formats. These assays detected anti-leptospiral IgM in the earlier course of the disease than MAT but still have low sensitivity during the acute phase. In addition, several recent studies using currently available kits indicate that the diag-nostic accuracies of these techniques are poor in some areas where leptospirosis is endemic [44–47]. This may be attributed to the persistence of anti-leptospiral IgM after infection, recurrent infec-tion in endemic areas, or cross-reaction with other infectious microorganisms. A meta-analysis of the published literature on ELISA also indicated high heterogeneity in the overall sensitivity and specificity estimates, whereas that on LFA showed that the evidence base for the diagnostic accuracy of LFA is at risk of bias [48, 49]. Therefore, the confirmation of rapid serological test results by MAT, NAAT, and/or culture is strongly recommended [48, 50]. Furthermore, as pointed out previously [17, 51], training of laboratory personnel is important even in the simple-to-interpret platform assays due to the considerable inter-observer variability in reading the test results [47]. To improve their diagnostic accuracy, researchers suggested setting local cutoff values instead of suppli-ers' definitions [43, 44, 52]. Furthermore, a new assay using whole-cell antigens from the intermediate *L. fainei* showed good diagnostic performance compared with whole-cell-based commer-cial kits [53].

Recombinant protein or peptide-based IgM assays such as ELISA, latex agglutination test, dipstick assay, and Dual Path Plat-form have been developed [15, 54, 55], of which some assays

Table 2
Commercial kit for serological diagnostics for leptospirosis[a]

Kit	Manufacture Website[b]	Technology
AccuDiag™ Leptospira IgM ELISA kit AccuDiag™ Leptospira IgG ELISA kit AccuDiag™ Leptospira IgG&IgM ELISA kit	Diagnostic automation/Cortez diagnostics, Inc. http://www.rapidtest.com/index.php? product=Parasitology-ELISA-kits&cat=17	ELISA
Panbio® Leptospira IgM ELISA	Abbott https://www.alere.com/en/home/product-details/ panbio-leptospira-igm-elisa.html	ELISA
SD Leptospira IgM ELISA	Abbott https://www.alere.com/en/home/product-details/ sd-bioline-leptospira-igm-elisa.html	ELISA
SERION ELISA *classic* Leptospira IgM SERION ELISA *classic* Leptospira IgG	SERION diagnostics https://www.serion-diagnostics.de/en/products/ serion-elisa-classic-antigen/leptospira/	ELISA
ImmuneMed Leptospira rapid ImmuneMed Leptospira IgM duo rapid	ImmuneMed http://immunemed.co.kr/en/page_id5148/	LFA[c]
Leptocheck-WB	Zephyr biomedicals http://www.tulipgroup.com/Zephyr_New/html/ product_specs/1_leptocheck.htm	LFA
OnSite Leptospira IgG/IgM combo rapid test	CTK biotech https://ctkbiotech.com/product/leptospira-igg-igm- combo-rapid-test-ce/#tab-product-specifications| demoTab1	LFA
Test-it Leptospira test kit	VIA Global Health https://www.viaglobalhealth.com/product/test-it- leptospira-test/	LFA
ImmunoDOT *Leptospira biflexa*	Innominata dba GenBio https://www.genbio.com/ordering.php	Dot EIA
Leptorapide	Linnodee diagnostics http://www.cowslips.com/linnodee_leptorapide_ agglutination.html	Latex agglutination test

[a]This table was adopted, updated, modified from Table 1 in ref. 17
[b]Accessed 25 Feb
[c]*LFA* lateral flow assay

exhibit compatible or even better sensitivity than whole-cell-based ELISA [11, 54–56]. These assays were evaluated in limited settings, and their utility has not been generalized. The combination of multiple protein antigens demonstrated an increase in sensitivity [57], and novel protein antigens have recently been identified by expression screening of a genomic DNA library and protein micro-array analysis using patient sera [58–60].

5 Antigen Detection

The detection of leptospiral antigens in clinical specimens has not been applied widely for the diagnosis of leptospirosis [9]. Several leptospiral antigens have been detected in urine. Uncharacterized 35-kDa component and LigA were detected in boiled urine pellets by monoclonal antibody (mAb)-based dot ELISA [61, 62]. Lepto-spiral lipopolysaccharide was detected in concentrated boiled urine by mAb-based LFA [63]. LipL32, LipL41, Fla1, sphingomyelinases, and HbpA were detected in urine precipitates by saturated ammonium sulfate by ELISA [64]. Leptospiral 3-hydroxyacyl-CoA dehydrogenase was detected in concentrated urine samples by immunoblotting [65]. Some of the studies detected antigens in urine samples from anti-leptospiral antibody-negative patients.

6 Conclusion

The combination of diagnostic tests using appropriate specimens at each disease stage contributes to the accurate understanding of the burden of leptospirosis, but definitive diagnostic methods such as isolation and MAT are not easily performed in resource-limited settings.

The early diagnosis of leptospirosis is crucial for appropriate treatments without delay and the improvement of patient outcomes, but it remains a challenge, especially in resource-limited settings. The sensitivity of real-time PCR in the early stage of the disease is up to more than 80% [22], but it is unavailable in resource-limited peripheral health facilities. Current rapid serological tests have variable diagnostic accuracy and low sensitivity in the early phase, which are recommended to be confirmed by MAT. Therefore, rapid, accurate diagnostic tests for acute leptospirosis, which are applicable in resource-limited settings, remain to be developed.

Current reference tests, culture and MAT, are imperfect tests [66], and the combination of diagnostic tests using appropriate specimens at each stage should also be used for the estimation of the diagnostic accuracy of newly developed diagnostics.

Acknowledgments

The writing of this review was supported by the Research Program on Emerging and Re-emerging Infectious Diseases (19fk0108049j1903) from the Japan Agency for Medical Research and Development (AMED).

References

1. Costa F, Hagan JE, Calcagno J et al (2015) Global morbidity and mortality of leptospirosis: a systematic review. PLoS Negl Trop Dis 9: e0003898. https://doi.org/10.1371/journal.pntd.0003898

2. Crump JA, Morrissey AB, Nicholson WL et al (2013) Etiology of severe nonmalaria febrile illness in northern Tanzania: a prospective cohort study. PLoS Negl Trop Dis 7:e2324. https://doi.org/10.1371/journal.pntd.0002324

3. Mayxay M, Castonguay-vanier J, Chansamouth V et al (2013) Causes of nonmalarial fever in Laos: a prospective study. Lancet Glob Health 1:46–54

4. Mørch K, Manoharan A, Chandy S et al (2017) Acute undifferentiated fever in India: a multicentre study of aetiology and diagnostic accuracy. BMC Infect Dis 17:665. https://doi.org/10.1186/s12879-017-2764-3

5. Mueller TC, Siv S, Khim N et al (2014) Acute undifferentiated febrile illness in rural Cambodia: a 3-year prospective observational study. PLoS One 9:e95868. https://doi.org/10.1371/journal.pone.0095868

6. Wangrangsimakul T, Althaus T, Mukaka M et al (2018) Causes of acute undifferentiated fever and the utility of biomarkers in Chiangrai, northern Thailand. PLoS Negl Trop Dis 12: e0006477. https://doi.org/10.1371/journal.pntd.0006477

7. World Health Organization (2003) Human leptospirosis: guidance for diagnosis, surveillance, and control. World Health Organization, Geneva

8. World health Organization (2011) Report of the second meeting of the leptospirosis burden epidemiology reference group. https://apps.who.int/iris/bitstream/handle/10665/44588/9789241501521_eng.pdf?sequence=1. Accessed 19 Feb 2019

9. Levett PN (2001) Leptospirosis. Clin Microbiol Rev 14:296–326. https://doi.org/10.1128/CMR.14.2.296-326.2001

10. Faine S, Adler B, Bolin C et al (1999) Leptospira *and Leptospirosis*, 2nd edn. MediSci, Melbourne

11. Kitashoji E, Koizumi N, Lacuesta TLV et al (2015) Diagnostic accuracy of recombinant immunoglobulin-like protein A-based IgM ELISA for the early diagnosis of leptospirosis in the Philippines. PLoS Negl Trop Dis 9: e0003879. https://doi.org/10.1371/journal.pntd.0003879

12. Wuthiekanun V, Chierakul W, Limmathurotsakul D et al (2007) Optimization of culture of *Leptospira* from humans with leptospirosis. J Clin Microbiol 45:1363–1365. https://doi.org/10.1128/JCM.02430-06

13. Griffith ME, Horvath LL, Mika WV et al (2006) Viability of *Leptospira* in BacT/ALERT MB media. Diagn Microbiol Infect Dis 54:263–266. https://doi.org/10.1016/j.diagmicrobio.2005.11.001

14. Girault D, Soupé-Gilbert ME, Geroult S et al (2017) Isolation of *Leptospira* from blood culture bottles. Diagn Microbiol Infect Dis 88:17–19. https://doi.org/10.1016/j.diagmicrobio.2017.01.014

15. Toyokawa T, Ohnishi M, Koizumi N (2011) Diagnosis of acute leptospirosis. Expert Rev Anti-Infect Ther 9:111–121. https://doi.org/10.1586/eri.10.151

16. Ahmed A, Grobusch MP, Klatser PR et al (2012) Molecular approaches in the detection and characterization of *Leptospira*. J Bacteriol Parasitol 3:133. https://doi.org/10.4172/2155-9597.1000133

17. Picardeau M, Bertherat E, Jancloes M et al (2014) Rapid tests for diagnosis of leptospirosis: current tools and emerging technologies. Diagn Microbiol Infect Dis 78:1–8. https://doi.org/10.1016/j.diagmicrobio.2013.09.012

18. Ahmed A, van der Linden H, Hartskeerl RA (2014) Development of a recombinase polymerase amplification assay for the detection of pathogenic *Leptospira*. Int J Environ Res Public Health 11:4953–4964. https://doi.org/10.3390/ijerph110504953

19. Waggoner JJ, Pinsky BA (2016) Molecular diagnostics for human leptospirosis. Curr Opin Infect Dis 29:440–445. https://doi.org/10.1097/QCO.0000000000000295

20. Thaipadunpanit J, Chierakul W, Wuthiekanun V et al (2011) Diagnostic accuracy of real-time PCR assays targeting 16S rRNA and *lipl32* genes for human leptospirosis in Thailand: a case-control study. PLoS One 6:e16236. https://doi.org/10.1371/journal.pone.0016236

21. Agampodi S, Matthias M, Moreno A et al (2012) Utility of quantitative polymerase chain reaction in leptospirosis diagnosis: association of level of leptospiremia and clinical manifestations in Sri Lanka. Clin Infect Dis 54:1249–1255. https://doi.org/10.1093/cid/cis035

22. Riediger IN, Stoddard RA, Ribeiro GS et al (2017) Rapid, actionable diagnosis of urban epidemic leptospirosis using a pathogenic *Leptospira lipL32*-based real-time PCR assay. PLoS Negl Trop Dis 11:e0005940. https://doi.org/10.1371/journal.pntd.0005940

23. Mauk MG, Song J, Liu C et al (2018) Simple approaches to minimally-instrumented, microfluidic-based point-of-care nucleic acid amplification tests. Biosensors (Basel) 8:17. https://doi.org/10.3390/bios8010017

24. Sonthayanon P, Chierakul W, Wuthiekanun V et al (2011) Accuracy of loop-mediated isothermal amplification for diagnosis of human leptospirosis in Thailand. Am J Trop Med Hyg 84:614–620. https://doi.org/10.4269/ajtmh.2011.10-0473

25. Villumsen S, Pedersen R, Krogfelt KA et al (2010) Expanding the diagnostic use of PCR in leptospirosis: improved method for DNA extraction from blood cultures. PLoS One 5:e12095. https://doi.org/10.1371/journal.pone.0012095

26. Dittrich S, Rudgard WE, Woods KL et al (2016) The utility of blood culture fluid for the molecular diagnosis of *Leptospira*: a prospective evaluation. Am J Trop Med Hyg 94:736–740. https://doi.org/10.4269/ajtmh.15-0674

27. Nhan TX, Teissier A, Roche C et al (2014) Sensitivity of real-time PCR performed on dried sera spotted on filter paper for diagnosis of leptospirosis. J Clin Microbiol 52:3075–3077. https://doi.org/10.1128/JCM.00503-14

28. Matthias MA, Ricaldi JN, Cespedes M et al (2008) Human leptospirosis caused by a new, antigenically unique *Leptospira* associated with a *Rattus* species reservoir in the Peruvian Amazon. PLoS Negl Trop Dis 2:e213. https://doi.org/10.1371/journal.pntd.0000213

29. Slack AT, Kalambaheti T, Symonds ML et al (2008) *Leptospira wolffii* sp. nov., isolated from a human with suspected leptospirosis in Thailand. Int J Syst Evol Microbiol 58:2305–2308. https://doi.org/10.1099/ijs.0.64947-0

30. Chiriboga J, Barragan VA, Arroyo G et al (2015) High prevalence of intermediate *Leptospira* spp. DNA in febrile humans from urban and rural Ecuador. Emerg Infect Dis 21:2141–2147. https://doi.org/10.3201/eid2112.140659

31. Tsuboi M, Koizumi N, Hayakawa K et al (2017) Imported *Leptospira licerasiae* infection in traveler returning to Japan from Brazil. Emerg Infect Dis 23:548–549. https://doi.org/10.3201/eid2303.161262

32. Puche R, Ferrés I, Caraballo L et al (2018) *Leptospira venezuelensis* sp. nov., a new member of the intermediate group isolated from rodents, cattle and humans. Int J Syst Evol Microbiol 68:513–517. https://doi.org/10.1099/ijsem.0.002528

33. Thibeaux R, Girault D, Bierque E et al (2018) Biodiversity of environmental *Leptospira*: improving identification and revisiting the diagnosis. Front Microbiol 9:816. https://doi.org/10.3389/fmicb.2018.00816

34. Ganoza CA, Matthias MA, Saito M et al (2010) Asymptomatic renal colonization of humans in the peruvian Amazon by *Leptospira*. PLoS Negl Trop Dis 4:e612. https://doi.org/10.1371/journal.pntd.0000612

35. Villumsen S, Pedersen R, Borre MB et al (2012) Novel TaqMan® PCR for detection of *Leptospira* species in urine and blood: pit-falls of in silico validation. J Microbiol Methods 91:184–190. https://doi.org/10.1016/j.mimet.2012.06.009

36. Woods K, Nic-Fhogartaigh C, Arnold C et al (2018) A comparison of two molecular methods for diagnosing leptospirosis from three different sample types in patients presenting with fever in Laos. Clin Microbiol Infect 24:1017.e1–1017.e7. https://doi.org/10.1016/j.cmi.2017.10.017

37. Backstedt BT, Buyuktanir O, Lindow J et al (2015) Efficient detection of pathogenic leptospires using 16S ribosomal RNA. PLoS One 10:e0128913. https://doi.org/10.1371/journal.pone.0128913

38. Waggoner JJ, Balassiano I, Mohamed-Hadley A et al (2015) Reverse-transcriptase PCR detection of *Leptospira*: absence of agreement

with single-specimen microscopic agglutination testing. PLoS One 10:e0132988. https://doi.org/10.1371/journal.pone.0132988

39. Waggoner JJ, Abeynayake J, Balassiano I et al (2014) Multiplex nucleic acid amplification test for diagnosis of dengue fever, malaria, and leptospirosis. J Clin Microbiol 52:2011–2018. https://doi.org/10.1128/JCM.00341-14

40. Giry C, Roquebert B, Li-Pat-Yuen G et al (2017) Simultaneous detection of chikungunya virus, dengue virus and human pathogenic *Leptospira* genomes using a multiplex TaqMan® assay. BMC Microbiol 17:105. https://doi.org/10.1186/s12866-017-1019-1

41. Blanco RM, Romero EC (2012) Efficacy of serum samples stored on filter paper for the detection of antibody to *Leptospira* spp. by microagglutination test (MAT). J Immunol Methods 386:31–33. https://doi.org/10.1016/j.jim.2012.08.013

42. Goris MG, Hartskeerl RA (2014) Leptospirosis serodiagnosis by the microscopic agglutination test. Curr Protoc Microbiol 32:Unit 12E.5. https://doi.org/10.1002/9780471729259.mc12e05s32

43. McBride AJL, Athanazio DA, Reis MG et al (2005) Leptospirosis. Curr Opin Infect Dis 18:376–386

44. Desakorn V, Wuthiekanun V, Thanachartwet V et al (2012) Accuracy of a commercial IgM ELISA for the diagnosis of human leptospirosis in Thailand. Am J Trop Med Hyg 86:524–527. https://doi.org/10.4269/ajtmh.2012.11-0423

45. Tanganuchitcharnchai A, Smythe L, Dohnt M et al (2012) Evaluation of the standard diagnostics *Leptospira* IgM ELISA for diagnosis of acute leptospirosis in Lao PDR. Trans R Soc Trop Med Hyg 106:563–566. https://doi.org/10.1016/j.trstmh.2012.06.002

46. Chang CH, Riazi M, Yunus MH et al (2014) Limited diagnostic value of two commercial rapid tests for acute leptospirosis detection in Malaysia. Diagn Microbiol Infect Dis 80:278–281. https://doi.org/10.1016/j.diagmicrobio.2014.08.012

47. Dittrich S, Boutthasavong L, Keokhamhoung D et al (2018) A prospective hospital study to evaluate the diagnostic accuracy of rapid diagnostic tests for the early detection of leptospirosis in Laos. Am J Trop Med Hyg 98:1056–1060. https://doi.org/10.4269/ajtmh.17-0702

48. Signorini ML, Lottersberger J, Tarabla HD et al (2013) Enzyme-linked immunosorbent assay to diagnose human leptospirosis: a meta-analysis of the published literature. Epidemiol Infect 141:22–32. https://doi.org/10.1017/S0950268812001951

49. Maze MJ, Sharples KJ, Allan KJ et al (2018) Diagnostic accuracy of leptospirosis whole-cell lateral flow assays: a systematic review and meta-analysis. Clin Microbiol Infect. https://doi.org/10.1016/j.cmi.2018.11.014

50. Goris MGA, Leeflang MMG, Loden M et al (2013) Prospective evaluation of three rapid diagnostic tests for diagnosis of human leptospirosis. PLoS Negl Trop Dis 7:e2290. https://doi.org/10.1371/journal.pntd.0002290

51. Musso D, La Scola B (2013) Laboratory diagnosis of leptospirosis: a challenge. J Microbiol Immunol Infect 46(4):245–252. https://doi.org/10.1016/j.jmii.2013.03.001

52. Courdurie C, Le Govic Y, Bourhy P et al (2017) Evaluation of different serological assays for early diagnosis of leptospirosis in Martinique (French West Indies). PLoS Negl Trop Dis 11:e0005678. https://doi.org/10.1371/journal.pntd.0005678

53. Goarant C, Bourhy P, D'Ortenzio E et al (2013) Sensitivity and specificity of a new vertical flow rapid diagnostic test for the serodiagnosis of human leptospirosis. PLoS Negl Trop Dis 7:e2289. https://doi.org/10.1371/journal.pntd.0002289

54. Nabity SA, Ribeiro GS, Lessa Aquino C et al (2012) Accuracy of a dual path platform (DPP) assay for the rapid point-of-care diagnosis of human leptospirosis. PLoS Negl Trop Dis 6:e1878. https://doi.org/10.1371/journal.pntd.0001878

55. Nabity SA, Hagan JE, Araújo G et al (2018) Prospective evaluation of accuracy and clinical utility of the dual path platform (DPP) assay for the point-of-care diagnosis of leptospirosis in hospitalized patients. PLoS Negl Trop Dis 12:e0006285. https://doi.org/10.1371/journal.pntd.0006285

56. Kanagavel M, Shanmughapriya S, Anbarasu K et al (2014) B-cell-specific peptides of *Leptospira interrogans* LigA for diagnosis of patients with acute leptospirosis. Clin Vaccine Immunol 21:354–359. https://doi.org/10.1128/CVI.00456-13

57. Chen HW, Lukas H, Becker K et al (2018) An improved enzyme-linked immunoassay for the detection of *Leptospira*-specific antibodies. Am J Trop Med Hyg 99:266–274. https://doi.org/10.4269/ajtmh.17-0057

58. Raja V, Shanmughapriya S, Kanagavel M et al (2016) *In Vivo*-expressed proteins of virulent *Leptospira interrogans* serovar Autumnalis N2 elicit strong IgM responses of value in

conclusive diagnosis. Clin Vaccine Immunol 23:65–72. https://doi.org/10.1128/CVI.00509-15

59. Lessa-Aquino C, Borges Rodrigues C, Pablo J et al (2013) Identification of seroreactive proteins of *Leptospira interrogans* serovar Copenhageni using a high-density protein microarray approach. PLoS Negl Trop Dis 7:e2499. https://doi.org/10.1371/journal.pntd.0002499

60. Lessa-Aquino C, Lindow JC, Randall A et al (2017) Distinct antibody responses of patients with mild and severe leptospirosis determined by whole proteome microarray analysis. PLoS Negl Trop Dis 11:e0005349. https://doi.org/10.1371/journal.pntd.0005349

61. Saengjaruk P, Chaicumpa W, Watt G et al (2002) Diagnosis of human leptospirosis by monoclonal antibody-based antigen detection in urine. J Clin Microbiol 40:480–489

62. Kanagavel M, Shanmughapriya S, Aishwarya KVL et al (2017) Peptide specific monoclonal antibodies of Leptospiral LigA for acute diagnosis of leptospirosis. Sci Rep 7:3250. https://doi.org/10.1038/s41598-017-03658-0

63. Widiyanti D, Koizumi N, Fukui T et al (2013) Development of immunochromatography-based methods for detection of leptospiral lipopolysaccharide antigen in urine. Clin Vaccine Immunol 20:683–690. https://doi.org/10.1128/CVI.00756-12

64. Chaurasia R, Thresiamma KC, Eapen CK et al (2018) Pathogen-specific leptospiral proteins in urine of patients with febrile illness aids in differential diagnosis of leptospirosis from dengue. Eur J Clin Microbiol Infect Dis 37:423–433. https://doi.org/10.1007/s10096-018-3187-9

65. Toma C, Koizumi N, Kakita T et al (2018) Leptospiral 3-hydroxyacyl-CoA dehydrogenase as an early urinary biomarker of leptospirosis. Heliyon 4:e00616. https://doi.org/10.1016/j.heliyon.2018.e00616

66. Limmathurotsakul D, Turner EL, Wuthiekanun V et al (2012) Fool's gold: why imperfect reference tests are undermining the evaluation of novel diagnostics: a reevaluation of 5 diagnostic tests for leptospirosis. Clin Infect Dis 55:322–331. https://doi.org/10.1093/cid/cis403

INDEX

Nobuo Koizumi and Mathieu Picardeau (eds.), *Leptospira spp.: Methods and Protocols*, Methods in Molecular Biology, vol. 2134, https://doi.org/10.1007/978-1-0716-0459-5, © Springer Science+Business Media, LLC, part of Springer Nature 2020

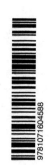